Efficient Cloud FinOps

A practical guide to cloud financial management and optimization with AWS, Azure, and GCP

Alfonso San Miguel Sánchez

Danny Obando García

‹packt›

Efficient Cloud FinOps

Group Product Manager: Preet Ahuja

Publishing Product Manager: Surbhi Suman

Senior Editor: Sujata Tripathi

Technical Editor: Yash Bhanushali

Copy Editor: Safis Editing

Project Coordinator: Ashwini Gowda

Proofreader: Safis Editing

Indexer: Hemangini Bari

Production Designer: Prashant Ghare

Marketing Coordinator: Rohan Dobhal

First published: February 2023

Production reference: 1310124

Published by Packt Publishing Ltd.

Grosvenor House

11 St Paul's Square

Birmingham

B3 1RB

ISBN 978-1-80512-257-9

www.packtpub.com

To my parents, who planted the seed of writing in me, and my brothers, for their continuous support. To Beatriz, my wife and life companion, whose support and love have been essential to completing this project. To all my colleagues and team at Santander and Avanade/Accenture, from whom I learned so much, without you this wouldn't be possible. To Danny, who has endured this adventure with me for a whole year without faltering. And last but not least, to Edgar Bahilo and my brother Ignacio San Miguel, who have kindly shared their expertise and wisdom in the last chapter of this book. And, of course, to the whole Packt team, for trusting us with this project and supporting us throughout its completion.

– Alfonso San Miguel Sánchez

To my sons, Gonzalo and Clara, whose innocence and eagerness to learn are my biggest inspirations. I want to also thank my parents, who have always supported me and shown me what respect and commitment are. To all my work colleagues who have shared this journey with me. Special mention to my friend Alfonso San Miguel, for letting me experience this adventure. And last but not least, to Eva, my wife and friend, who makes anything possible with her love and support.

– Danny Obando García

Contributors

About the authors

Alfonso San Miguel Sánchez is a multi-cloud architect, with a deep experience both on premises and in the cloud. He has always enjoyed being close to development teams, implementing coding, DevOps, and other methodologies into his way of working, with a strong focus on automation. Alfonso has a degree in computer science from Universidad Complutense de Madrid and an M.Sc. degree in machine learning. After his studies, he worked as a cloud architect for Tecnicas Reunidas, Avanade, and B2Impact, where he works now as a lead cloud architect. Though passionate about cloud governance, in the past few years, he has specialized in FinOps, aiming to develop an entire methodology around the practice.

Danny Obando García is a multi-cloud data architect, who has worked in various roles during his professional career, always aiming to create reliable and scalable data and infrastructure solutions by applying different frameworks and methodologies. Danny has a degree in computer science from **Universitat Oberta Catalunya** (**UOC**), which he complemented with an M.Sc. in artificial intelligence for financial markets. With a rich IT experience of about 15 years, he is currently leading data strategy for Holaluz, one of the biggest players in Spain's energy market. Before this, he had experience working and implementing FinOps for the biggest banking group in Spain.

About the reviewers

Israel Pérez Jiménez has more than 20 years of experience in IT. He has worked in multiple sectors such as transportation, banking, and the engineering industry, in varied roles such as project management, IT processes consultant, infrastructure, and digital transformation. He currently works for Tecnicas Reunidas as a lead systems architect, fully focused on cloud governance, modernization, and cost optimization, where he has been for more than 9 years. Architecting solutions is part of his DNA, with a strong focus on cost, security, automation, and reliability. From traditional on-premises environments to cloud solutions, his wide experience grants him a complete vision of IT challenges.

Ismael Doblas Bermudo began his journey as a cloud engineer. Through constant learning and training, he honed his skills as an architect to build more robust and scalable architectures. His focus on multi-cloud environments allows him a panoramic view, as well as a complete vision of the cloud's ever-evolving landscape, enabling him to navigate seamlessly across various public clouds. His knowledge of automation and IaC has also become a cornerstone of his approach, to ensure efficiency, consistency, and scalability. During the last years of his experience, during a pivotal juncture in his evolution, he chose to specialize in FinOps. He currently works as a global FinOps lead for a multinational company in the banking sector.

Eric Duquesnoy is a seasoned professional, currently the head of Cloud and DevOps consulting at ELCA Cloud Services. Based in Geneva, Eric leads strategic initiatives using his extensive expertise in cloud architecture (Azure, AWS) and FinOps. In addition to his professional activities, Eric is the founder of the *Silicon Chalet* meetup in Switzerland, a dynamic community shaping the future of technology. From 2019 to 2023, Eric held the position of head of Cloud at Eurovisions, where he contributed significantly to the advancement of the organization's cloud technology. Eric holds the FinOps Practitioner certification, which highlights his commitment to excellence in optimizing cloud spend and aligning financial strategies with business objectives.

Table of Contents

Part 2: Inform – How to Increase Cost Visibility

3

Designing and Executing the Tagging and Naming Convention Strategies 51

4

Estimating Cloud Solution Costs and Initiative Saving 83

5

Improving Cost Visibility with Dashboards and Reports   119

Part 3: Optimize – How to Get the Most out of Cloud Resources

6

Implementing IaaS Compute Optimization   145

7

Implementing PaaS and Other Compute Optimization Initiatives   179

8

Implementing Database Optimization   203

9

Implementing Storage Optimization   245

Part 4: Operate – How to Set Up a Governance Model around Cloud Costs

10

Designing and Implementing FinOps KPIs   295

11

Defining New FinOps Roles and Processes   321

Part 5: Hands-On Cost Optimization with Real-Life Use Cases and More

12

Case Studies for Cost Optimization   345

13

Wrapping up and Looking ahead   377

Preface

First and foremost, greetings and welcome to this book! Before we dive into it, we want to introduce you to the reasons why we wrote this book and set up the context.

The idea for this book was born after an intense experience of building up a FinOps practice together with a great team, which we worked on from scratch and created something that we felt was worth sharing with the FinOps community.

For almost two nonstop years, we were fully dedicated to FinOps, unlike other architects or engineers who divide their time between a lot of projects. We worked on FinOps governance and implementation in a really complex environment, where nothing was easy, but it was definitely satisfying to build it and see it grow.

It was two years full of research, learning every step of the way, thinking about what else to propose, coming out with new ideas and approaches, overcoming the different walls that were in front of us, solving problems, and adapting along the way.

Our goal is to share all of it with you, in the hope that it will aid you in your future experiences.

Who this book is for

This book is intended for cloud engineers, cloud and solutions architects, as well as DevOps and systems operations engineers interested in learning more about FinOps and cloud financial management for efficiently architecting, designing, and operating software solutions and infrastructure using public clouds. This book will also be useful for team leads, project managers, and financial teams interested in getting the most out of cloud resources.

Some prior knowledge of cloud computing and major public clouds will be needed to get the most out of this book, as in some sections, we will delve deeper into more technical work, terms, and examples.

What this book covers

Chapter 1, Introduction to FinOps Principles, provides an introduction to what FinOps is and why it is needed for organizations that are transitioning to or already in the cloud.

Chapter 2, Understanding How FinOps Fits into Cloud Governance, covers how FinOps interacts with different methodologies widely used in organizations, such as the **Well-Architected Framework**, **infrastructure as code**, **Agile project management**, and other key processes, such as **change management**. This chapter also covers how FinOps can adapt to organizations in different phases of their cloud journey, and the basic tools to perform cost analysis on Azure, AWS, and Google Cloud, as well as other market tools that are offered by other vendors outside of Microsoft, Amazon, and Google.

Chapter 3, Designing and Executing the Tagging and Naming Convention Strategies, provides a detailed explanation of why both tagging and naming convention strategies are essential for FinOps practices, as well as recommendations and tools that can be used to design, implement, and enforce your own strategies.

Chapter 4, Estimating Cloud Solutions Costs and Initiative Savings, provides a detailed description of all the migration models that can be used to migrate workloads to the cloud, as well as some key concepts about cloud costs that should be understood before going forward. It also covers how to leverage pricing calculators and REST APIs offered by cloud providers to create your own estimations, as well as how potential savings concepts can boost and drive your FinOps practices further.

Chapter 5, Improving Cost Visibility with Dashboards and Reports, provides an introduction to cloud billing data and the structure and fields of a cloud bill, as well as what dashboards and reports are and how they are different from each other. It also includes a lot of insights to improve the quality of your FinOps dashboards and reports using financial concepts and other key ideas, such as **unit economics**.

Chapter 6, Implementing IaaS Compute Optimization, provides an overview of FinOps initiatives that can be carried out on **infrastructure-as-a-service** compute services for cost optimization.

Chapter 7, Implementing PaaS and Other Compute Optimization Initiatives, provides an overview of FinOps initiatives that can be carried out in **platform-as-a-service** compute services for cost optimization, as well as other initiatives that are related to backup, licensing, and resource management best practices.

Chapter 8, Implementing Database Optimization, provides an overview of FinOps initiatives that can be carried out in database services for cost optimization. It also introduces a lot of key basic concepts around databases in general that are needed to fully understand the tools at our disposal for optimizing database services.

Chapter 9, Implementing Storage Optimization, provides an overview of FinOps initiatives that can be carried out in database services for cost optimization. It also explains in depth how the different storage paradigms work and some key concepts, such as redundancy, data temperature tiering, and the cost drivers of storage services.

Chapter 10, Designing and Implementing FinOps KPIs, covers what a KPI is and the different categories of KPIs that exist. Once the basic concepts have been introduced, it also provides a complete methodology to design and develop your own KPIs, with a lot of examples of FinOps KPIs that can be used as a starting point to create your own dashboards and reports.

Chapter 11, Defining New FinOps Roles and Processes, provides an overview of how to define and implement your own FinOps operating model, which includes the functions, capabilities, processes, and roles and responsibilities that enable FinOps practices to be part of the organization's DNA, as well as other key governance initiatives to enforce FinOps policies.

Chapter 12, Case Studies for Cost Optimization, presents two examples of real-life architectures to be optimized. In a step-by-step manner, we provide examples of different initiatives that we can use to optimize these solutions, analyzing throughout the process the impact on costs that these initiatives generate.

Chapter 13, Wrapping Up and Looking Ahead, provides a summary of sorts, where we reflect on what we've covered in this book and some challenges that FinOps practitioners may still be facing in the future. This chapter also covers two emergent fields of study that are on the rise, which are machine learning and sustainability, as well as the synergies to be found in each one with FinOps practices. To close the circle, this chapter also provides a self-assessment for you to evaluate what you have learned throughout this book.

To get the most out of this book

There are no specific requirements to follow along with this book. However, we have used certain conventions in the book, which we've explained as follows. Reviewing them will help you understand the content structure better.

Throughout this book, we will add some hints and important notes, for which we will use the following format:

> **Important note**
> This is a note, a comment, or an example.

When we dive deeper into the technical aspects of FinOps, we will include examples from all the major public clouds, which are currently the following:

- **Microsoft Azure**
- **Amazon Web Services (AWS)**
- **Google Cloud Platform (GCP)**

Note also that in the last chapter of the book, you will find a **self-assessment/knowledge check**, which you can use to evaluate what you have learned throughout each chapter of the book.

Across the book, in the more technical chapters, you will find references to production, preproduction, and development environments. They are defined as follows:

- **Production**: In this environment, our services are published to final users or used in business processes. The services of this environment are live, so everything should work perfectly. In this environment, changes that may impact users or business are done out of hours or in maintenance windows that are previously set and agreed upon.

- **Preproduction**: Also called staging or **User Acceptance Testing** (**UAT**), this environment should be as similar as possible to production. This environment is where key users can test applications before they are promoted to production, and also where contingency tests are carried out.

- **Development**: Development environments are where the development processes take place. In this environment, data is usually fictional or consists of dummy data, and the resources are downsized to optimize the costs, as their computing needs are way lower. These environments are where new code is thoroughly tested by developers, to add new features or solve bugs.

- **Sandbox**: A sandbox environment is an environment which is usually isolated from the rest and whose purpose is to freely experiment with cloud services and software development. Company and security policies are usually not that strict in sandbox environment, and it is often used to conduct **Proof of Concepts** (**PoCs**) in a controlled environment.

The currency for all the cost references, estimations, and calculations used in the book is the American dollar ($).

Some other conventions we have used are:

`Code in text`: Indicates code words in text, database table names, folder names, filenames, file extensions, pathnames, dummy URLs, user input, and Twitter handles. Here is an example: "From this point, let's say we want to check the current pricing. We should use `currentVersionUrl`."

When we show some examples of CLI commands to be used, the format used is as follows:

```
aws pricing describe-services --service-code AmazonEC2
```

A block of code is set as follows:

```
{
"companyname" : "imagineinc",
"businessunit" : "finance",
"city" : "madrid",
"region" : "spain"
}
```

Apart from these general notes, there are no requirements to navigate this book. A word of advice, though: *Chapters 6* to *9* do get really technical, which may be challenging for readers coming from other non-cloud backgrounds.

Get in touch

Feedback from our readers is always welcome.

General feedback: If you have questions about any aspect of this book, email us at `customercare@packtpub.com` and mention the book title in the subject of your message.

Errata: Although we have taken every care to ensure the accuracy of our content, mistakes do happen. If you have found a mistake in this book, we would be grateful if you would report this to us. Please visit `www.packtpub.com/support/errata` and fill in the form.

Piracy: If you come across any illegal copies of our works in any form on the internet, we would be grateful if you would provide us with the location address or website name. Please contact us at `copyright@packtpub.com` with a link to the material.

If you are interested in becoming an author: If there is a topic that you have expertise in and you are interested in either writing or contributing to a book, please visit `authors.packtpub.com`.

Share Your Thoughts

Once you've read *Efficient Cloud FinOps*, we'd love to hear your thoughts! Scan the QR code below to go straight to the Amazon review page for this book and share your feedback.

`https://packt.link/r/1805122576`

Your review is important to us and the tech community and will help us make sure we're delivering excellent quality content.

Download a free PDF copy of this book

Thanks for purchasing this book!

Do you like to read on the go but are unable to carry your print books everywhere?

Is your eBook purchase not compatible with the device of your choice?

Don't worry, now with every Packt book you get a DRM-free PDF version of that book at no cost.

Read anywhere, any place, on any device. Search, copy, and paste code from your favorite technical books directly into your application.

The perks don't stop there, you can get exclusive access to discounts, newsletters, and great free content in your inbox daily

Follow these simple steps to get the benefits:

1. Scan the QR code or visit the link below

https://packt.link/free-ebook/978-1-80512-257-9

2. Submit your proof of purchase
3. That's it! We'll send your free PDF and other benefits to your email directly

Part 1:
Get Started with FinOps

In this part, we will go through all the ideas behind FinOps practices, and we will also explain in depth why this buzzword is becoming so popular.

Once through the basics, we will deep dive into how FinOps practices can fit and benefit multiple organizations in different stages of their cloud journey, as well as the synergies to be found with widely used methodologies such as DevOps and Agile.

This part has the following chapters:

- *Chapter 1, Introduction to FinOps Principles*
- *Chapter 2, Understanding How FinOps Fits in Cloud Governance*

1

Introduction to FinOps Principles

This chapter aims to introduce the topic we are going to cover throughout the book, which is FinOps, a methodology relating to cloud financial optimization. We will explain why it is so important, and why it has become one of the biggest current IT trends.

In this chapter, we are going to cover the following main topics:

- What is FinOps, and why do we need another buzzword?
- Why FinOps?
- The three pillars of FinOps

What is FinOps, and why do we need another buzzword?

Let's start by asking the question, what is this new methodology that everyone is talking about and, suddenly, all companies are in need of? It's a new groundbreaking practice, or way of doing things, that everyone claims will allow you to take control of your cloud costs and revolutionize how we work with cloud resources!

We are, of course, referring to FinOps. Let's take a look at how it is defined by the FinOps Foundation:

"FinOps is an evolving cloud financial management discipline and cultural practice that enables organizations to get maximum business value by helping engineering, finance, technology, and business teams to collaborate on data-driven spending decisions."

This is an amazing definition, but we will also provide our own interpretation of the topic. For us, FinOps is about the following things:

- First, very much like DevOps, it is about teams working together to remove the barriers between them. Once these barriers are down, companies can benefit from this teamwork greatly by people sharing knowledge and working together to reach common objectives.

- It is also about creating a specialized team that oversees cloud projects and operations from a new and different perspective that did not exist before, which can act as a common bond between financial, architecture/engineering, business, and technical teams. This new team is especially important in big companies with different business units across the globe, where keeping standardized best practices on cost optimization and strong governance, while maintaining the independence of each business unit, can be a demanding and daunting task.

- This specialized FinOps team can also become technical and help other technical teams define, design, and implement plans to understand, control, and reduce costs. Often, this exercise leads to FinOps-specific solutions being developed to improve automation and control costs. FinOps training is also one of the biggest focuses of FinOps teams, which should educate all interested parties on cost optimization and FinOps concepts.

- FinOps practices also aim to change organizations and improve the governance processes themselves, by adapting them to the new reality that is the cloud, where the established rules of on-premises computing don't apply anymore and never will do.

- Finally, FinOps is a mechanism to pave the way for technical teams to be more accountable for cloud costs. If FinOps practice and the FinOps team help to shed some light on what the cost drivers are and how to control costs while workloads are optimized, it can be very beneficial for technical teams, which will take responsibility for costs after the FinOps ideas and initiatives are understood and incorporated.

All in all, it brings a new vision to the table that everyone will benefit from.

We will describe in more detail what the FinOps Foundation is and its purpose later in this chapter.

Side note

It is important to understand that *FinOps is not about generating savings.*

In a non-optimized environment, a lot of savings can be obtained by implementing FinOps initiatives and governance. However, the savings can be generated only to an extent, as eventually, when FinOps practices mature, there are no more savings to be made.

Due to this, the performance FinOps practices should not be entirely measured in savings KPIs, as these have a limited lifespan.

FinOps is not about saving money; it is about creating value. It is also about making the most out of a budget and organizations being able to invest even more money in the cloud, being confident that it will be done in an efficient and optimized way that leads to assured investment returns in the future.

Why FinOps?

FinOps is a cloud cost optimization/cloud financial management methodology, aimed at companies that already work in the cloud or are transitioning to it.

The need for FinOps becomes apparent when companies embark on their cloud journeys or strive for excellence once they are already settled in using it:

- For newcomers, when the journey to the cloud begins, it becomes apparent that the well-known on-premises rules regarding cost management or capacity management don't apply anymore. This requires a difficult adaptation as well as training. FinOps can help in this complex process, both optimizing and enabling cloud governance in many ways.

- If an organization has used the cloud for a few years already, its cloud spend will not be aligned with its initial budget for a number of reasons, such as non-optimized workloads that are migrated through lift-and-shift, constant struggles with inefficiencies, and difficulties in applying a standardized cloud governance and best practice baseline. FinOps can help mitigate the impact on costs by working to implement cost reduction plans, as well as help build tighter cloud governance to avoid past mistakes in the future.

The public cloud changed everything, and the result often was that you either spent too much or you spent too little on things that were not relevant to the cloud or beneficial to a company, offering a bad impression of cloud computing to the business and its stakeholders. FinOps aims to bring visibility, optimization, and governance to improve the cloud's image from a business perspective.

To fully understand why we have ended up where we are, and why this new methodology is needed, let's go back a few years, around five or six. The public cloud, after years of development and improvements, has become a more mature product, ready to change how IT services are purchased and delivered. The public cloud, therefore, made a big splash in IT market, and many companies were interested in this new paradigm.

Before the cloud

From a financial and capacity management perspective, IT equipment traditionally was acquired with the expectation that it would last a few years. It should not only be operative for at least five years but also be able to accommodate the estimated expected growth in capacity (mainly storage but also computing capacity) before reaching its end-of-life date.

This led to the underutilization of virtual machines when growth did not meet expectations and non-optimized workloads were not as impactful financially. Unneeded virtual machines were often left unused on servers for several months, and badly optimized workloads or underused servers did not hurt that much if there was spare capacity left.

Being on-premises made it really hard to scale down, as due to this infrastructure purchasing model, once you committed to acquiring new assets, you were expected to keep growing instead of downsizing.

The cloud comes into play

When the public cloud made its entrance, every company adapted as much as possible. Depending on each company's situation, the process was as follows:

- For newly established companies, it was way easier to create everything from ground zero without an on-premises legacy, following somewhat of a greenfield approach. However, cloud knowledge was not that mature and quick decisions were needed, which often led to non-ideal architectures or non-optimized solutions.

- Smaller companies that were already established, due to budget and resource limits, were not able to jump into the cloud as eagerly, so they mostly continued following the traditional approach, postponing their journey to the cloud.

- For big companies with big budgets and business cases to migrate to the cloud in place, this process was much harder. There were contracts with infrastructure vendors in place, clusters of recently renewed servers, and money spent on disaster recovery data centers, following the traditional infrastructure approach. The cloud then became something like a commodity for most organizations, where they put their non-critical services or development environments, but they didn't fully commit to a cloud-only operation. Cloud vendors often pressured their clients into monetary commitments that were often partially wasted, as migrations were hard to implement due to high price tags, limited budgets, and lots of technological debt from on-premises vendors and software.

A paradigm shift

Let's explain point by point how the cloud revolutionized everything to fully understand how the traditional on-premises approach was moved away from:

- With the *acquisition of assets and capacity management*, a major change occurs at a company, as the financial teams lose control of infrastructure spend and the technical teams still don't feel accountable for cloud costs. **Pay as you go** is the billing model the cloud uses, which is a huge benefit when you need to downsize, for example, but it is a heavy burden if your workloads are not optimized. This approach is totally opposed to traditional on-premises approach, where assets were bought all the time, and technical architecture was set up around these assets, adapting to the platform or products each company made use of.

- *Business and financial teams* transition from a standard model where most of the investments were capital costs (hardware, for example) that followed a certain life cycle, as well as a few operational costs (infrastructure operations support, for example) on top to keep the business running. In the cloud, we shift to a totally different model, where there are no big capital costs and almost all the expenditure is operational (we will cover these ideas later on in this chapter)

- *Access* to the company assets also evolves. In a traditional data center, the assets were often hosted locally, and technical teams set up all the technical architecture and security needed for them (e.g., internal networks, firewalls, physical security, remote access, and VPNs to connect from multiple offices), building walls and secured doors to protect the organization assets. Resources and workloads in the cloud are not hosted locally anymore, so access is managed in more advanced ways, such as point-to-site VPNs, allowing users to work remotely anywhere (the COVID pandemic changed everything in this regard, forcing companies to establish such mechanisms). Traffic travels through the internet, so security teams put their efforts into building endpoints, data, and identity bastions instead of physical and perimeter security.

- Regarding the *technical distribution* of assets, there is often only one tenant or enterprise domain that can span multiple offices. Identity management is often unified under one unique domain (AD or Samba, for example), and there are minimal trust relationships or connections with other tenants or environments. Once assets such as virtual machines are deployed, they usually keep the same configuration during their lifespan, with minimal configuration changes. When the cloud comes into play, resources are distributed across multiple tenants (for example, per business unit), with federation and B2B/B2C with other tenants. The workloads adapt to scale on demand, following elasticity and loosely coupled architecture principles, rapidly changing configuration and resources in a much more agile manner.

- The *delivery* of projects and IT services transitions from locally hosted delivery, where all responsibilities and management are in the remit of each organization, to delivery through cloud services, in which responsibility is shared due to IaaS, PaaS, and SaaS models. In the case of a traditional IT approach, when a company grew, new land, bigger offices, and data centers were acquired. Now, we can save up some money and just focus on technology, without the need to invest in real estate to host infrastructure. This makes the **Return of Investment** (**ROI**) for a business much faster, as you can spin up some cloud services in minutes and build a proof of concept, without any added complexity. Also, if you use fewer managed cloud services, such as PaaS or SaaS, you can even make some savings on architectural or technical resources, for example.

- *Security* also undergoes a major change. There are transitions from perimeter (e.g., firewalls and demilitarized or DMZ networks) and physical (e.g., protecting offices with alarms, security personnel, and cameras) security to identification (e.g., multifactor and conditional access), endpoint, and other methods to prevent unwanted access to company data or assets. Methodologies such as Zero Trust also make an appearance to better adapt security measures to this new situation.

- From a business *continuity* perspective, disaster recovery and contingency plans were built in-house, which took a lot of effort from a technical architecture perspective. Backup was also part of this exercise, where backup software from different vendors was used and stored in different ways. In the cloud, when using less managed services such as PaaS or SaaS, services are highly available by default, as information is kept in multiple data centers across the same region and SLA is granted from the cloud vendor. Backup functionality is built in to most services, making this process much easier, and high availability can be achieved by using multi-regional services and multiple redundancy options for storing our data accross multiple cloud regions. As a result of all this, having a disaster recovery data center does not make sense anymore for most companies, as you can attain highly reliable and fault-tolerant workloads and applications in different ways. Also, some savings can be generated by getting rid of the housing, power, electricity, and maintenance costs for secondary data centers.

All this information can be summarized using the following table:

Model	Traditional computing	Cloud computing
Acquisition	Buy assets Build technical architecture	Buy cloud PAYG services which require continuous cost optimization Build minimal technical architecture
Business	Pay for assets Administrative overhead CAPEX/capital expenditure Technical teams not accountable for infrastructure costs	PAYG Reduce admin functions by using automation and shifting to fully managed cloud services OPEX/operational expenditure Technical teams fully accountable for infrastructure costs
Access	Hosted locally Internal networks Corporate desktop PCs and laptops Local logon for access control	Hosted by third-party service and cloud providers Over the internet Any device Unified access control
Technical	Single tenant and non-shared Static and hard to transform and scale up or down Tightly coupled workloads	Multi-tenant with B2B/B2C Scalable, dynamic, adaptable, and elastic Loosely coupled workloads

Delivery	Delivery through local data centers Costly and lengthy deployments Land and staff expansion	Delivery through cloud services Much faster deployment and new services testing Faster ROI
Security	Physical and perimeter security to protect company assets	Identity, data, device, network, and infrastructure security to protect assets anywhere
Business continuity	Use of secondary data centers Backup and disaster recovery strategies built by technical teams	Highly available cloud services with SLA availability granted from the vendors Built-in backup and redundancy in less managed services such as PaaS and SaaS

Apart from these groundbreaking changes, there are also other factors to consider when setting up cloud operations. They are as follows:

- Lack of knowledge about how the cloud works and cloud offerings and technologies often create different degrees of change resistance in technical teams, and the cloud world can be perceived somewhat as an enemy. To overcome this, the best tool is to invest in technical training and work on a cloud enablement pitch, as well as having as much sponsorship from the C-level executives as possible.

- Hiring people with high cloud knowledge also becomes a challenge, as talent is scarce and there is a high demand for technical roles such as cloud engineers, architects, software developers, and DevOps experts. In the cloud, a new service or offering that has new features is released almost every day, so technical teams need to be up to date on everything and invest time and resources in constant training. It gets harder and harder to keep up with technology and adapt to it.

To avoid these issues, it is usually wise to avoid fully committing to one platform or product, easing the process of adapting to changes that may come in the future. Also, it is wise to wait for a new trend to mature a bit before jumping in so that you can benefit from community experience and global knowledge.

Now we have covered the differences between cloud and on-premises, let's address another key question to evaluate when considering a move to the cloud – how much does the on-premises setup cost?

Hidden on-premises costs

When companies begin their own **Journey to the Cloud** (**J2C**), a complete business case is always needed. A business case should always include at least the following:

- The actual cost of running on-premises infrastructure
- The projected costs of moving to the cloud
- The features of the chosen cloud and how cloud operations will change how operations are run
- The benefits of the cloud in terms of security, reliability, and performance
- The value from a business perspective due to the migration

Part of this business case elaboration exercise is to evaluate the current cost versus the estimated costs in the cloud, as well as all the benefits, improvements, and drawbacks that the cloud brings.

At some point, the following question will be raised – how much does it cost to run infrastructure on-premises?

> **Important note**
>
> Given our experience, we have often found out that no one really knows the exact costs of keeping on-premises running because it was never needed.
>
> Part of the FinOps team's responsibility is to aid in improving the visibility of these costs for a business.

Before getting deeper into the topic, let's introduce three key IT concepts, which are **Capital Expenditure (CAPEX)** costs, **Operational Expenditure (OPEX)** costs, and **Total Cost of Ownership (TCO)**. These important key concepts are defined as follows:

- **CAPEX:** This refers to the aforementioned capital or direct costs, which are fixed, one-time expenses incurred on the purchase of land, buildings, construction, and equipment used in the production of goods or the rendering of services. In the IT realm, it's the cost of the infrastructure hardware (e.g., servers, networking, and all IT equipment) and the software itself.
- **OPEX:** These costs include the operating expenses needed for the infrastructure to function, such as electricity, housing, and hosting.

- **TCO**: As stated in the Gartner Glossary (https://www.gartner.com/en/information-technology/glossary/total-cost-of-ownership-tco), TCO *"is a comprehensive assessment of information technology (IT) or other costs across enterprise boundaries over time. For IT, TCO includes hardware and software acquisition, management and support, communications, end-user expenses, and the opportunity cost of downtime, training, and other productivity losses."* It should include all CAPEX and OPEX costs.

To begin the exercise of analyzing on-premises costs, let's consider the different OPEX costs that come to mind when thinking about on-premises infrastructure:

- **Power, utilities, and generators**: These ensure that an infrastructure has enough electricity to function. For bigger companies, this often includes even a second circuit for equipment to function if there is an outage. These secondary circuits are rarely used, but they need to be in place to ensure business continuity.

- **Cooling**: Keeping servers at the right temperature for optimal performance is never cheap, and there are always maintenance costs involved.

- **Real estate**: Owning or renting a property where the infrastructure is hosted costs money and requires some degree of planning.

- **Physical security**: This point needs to always be considered, as no company can afford their servers to be tampered with or accessed by malicious third parties (to steal data, for example). These costs cover security guards, cameras, access control systems, and other security measures.

These indirect costs are inherent to IT infrastructure and are paid even in private cloud and hosting services models, even though the vendors or service providers that offer these services may be third parties external to an organization.

Now, let's move on to CAPEX. These costs are not easy to measure either, as some of them are paid in full before equipment is purchased, and since the life of a CAPEX generally extends beyond a fiscal year, amortization should be used to redistribute costs while keeping in mind the depreciation of assets over time.

In terms of CAPEX costs, we have the following:

1. **Hardware**: This is the cost of physical servers, storage or disk cabinets, and **Network Attached Storage (NAS)** devices. Usually, these assets are acquired for a price that includes a warranty for a limited time, around five years. When the warranty expires, it is possible to extend it further by paying additional fees to the vendor or other service providers. If your server or device is no longer supported, you won't have any help from the vendor, and the replacement parts can be impossible to attain.

- **Support**: When purchasing either hardware or software to be used in enterprise applications or environments, it is often necessary to have some degree of support from the vendor or a partner. These support fees are paid using different models.

- **Software and licenses**: Software and license purchases have varied purchasing models. It is possible for some software to acquire perpetual licenses, which avoids the management overhead that subscriptions pose. The fact that each piece of software has a different licensing model makes the process of evaluating costs more complex.

- **Backup and disaster recovery**: In order to run successful IT operations, it is important to ensure that, in the event of an issue or a hardware or software failure, there are mechanisms to keep workloads functioning and running. Designing solutions that include these mechanisms always poses additional costs, such as having a secondary copy of the infrastructure or storing backups in additional storage.

- **Networking**: This is the purchase of network devices such as firewalls and routers. These devices also require periodic maintenance.

- **IT maintenance**: Having a team that operates the infrastructure is not free.

- **Training**: To train the IT team to perform their management of the infrastructure, some degree of training is needed, which also comes with some associated costs, even if it is just the time spent by the technical teams to prepare documentation and train newcomers.

Example

To exemplify these concepts we have just described, let's use as an example an out-of-support old server.

This server, which has already passed its end-of-life date is still used to host some workloads. We have experienced from our discussions with clients and organizations that this kind of situation happens more often than we think.

The cost of this old server for financial teams is really good, as almost all the costs for this asset are OPEX now.

However, there are things that need to be taken into account:

The TCO for this asset is really low compared to a new server. This creates the false perception that savings are generated by having this outdated asset in operation.

Having an end-of-life server involves a high risk from any perspective, even if it is used as a passive node or as part of a secondary data center. The hardware can malfunction anytime, and spare parts may be impossible to get a hold of.

Old hardware is usually tied to old software as well. Old software tends to have security vulnerabilities that newer versions often address, which poses a potential risk that could impact business operations in many ways.

Having old hardware and software in place also means that our personnel needs to know how to properly manage this far from state-of-the-art systems, equipment and products, which can be no easy feat.

There is a visual concept that is often used to reflect these fundamental differences between on-premises and cloud costs, which is the iceberg of on-premises costs, as shown in the following figure:

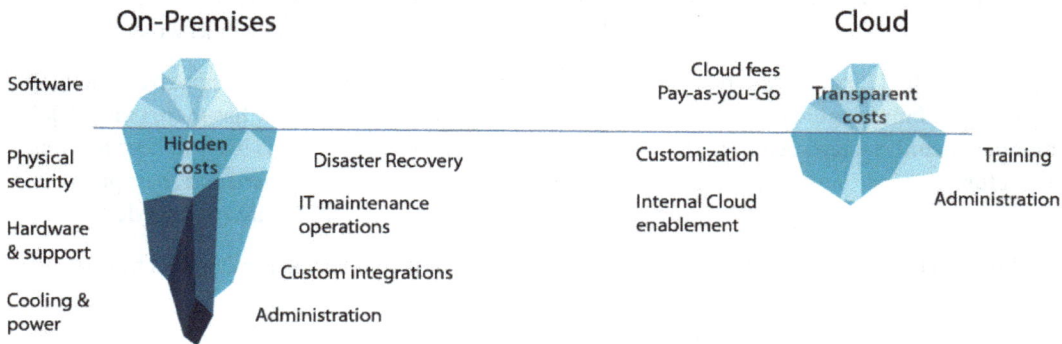

Figure 1.1 – On-premises versus cloud costs

Now that the fundamental differences in both approaches have been covered, let's go back in our time machine from past to present, to close the circle and fully understand the need for FinOps practices.

Back to the present

Let's travel back to present times with our FinOps-powered time machine. A few years have passed since the cloud's first appearance, and cloud platforms have evolved in a big way, increasing the portfolio and features of services offered, which enables you to use the cloud for enterprise-grade solutions with full confidence.

From an organizational point of view though, the financial result of transitioning to the cloud is that the initial expected budget for the cloud is nothing like the estimated costs. Most companies are still not fully committed to the cloud, facing the same challenges, and people with knowledge of the cloud are even harder to come by.

There is also one thing that is vital to understand when migrating to the cloud, which a lot of organizations fail to comprehend. If a workload is not optimized when hosted on-premises, it won't be optimized if it's migrated to the cloud either. Subsequently, if something does not work well before migration, all those issues and technical limitations due to badly architected solutions will also be transferred to the cloud.

In summary, ideally, all efforts should be put, in our opinion, into optimizing workloads *before* migration, ensuring that all the benefits of the cloud can be leveraged to its full extent once the applications and workloads are moved.

After some time has passed, non-optimized workloads will have become more common, both on-premises and in the cloud. It's time to face the truth – someone will need to get their hands dirty and tackle all these issues and limitations once and for all. This is where FinOps practitioners come into play.

The FinOps Foundation

The FinOps Foundation is a program of **the Linux Foundation** (https://www.linuxfoundation.org/) and is dedicated to advancing people who practice the discipline of cloud financial management.

It was founded by J.R. Storment (https://www.linkedin.com/in/jrstorment/) in 2019 and joined the Linux Foundation in 2020. According to its web page, it currently has more than 12,000 community members, representing more than 3,500 companies, and this figure is likely to keep growing in the future. It is a strong community where FinOps practitioners can share experiences, best practices, tools, and documents and get access to a lot of other assets to increase their FinOps knowledge.

In addition to this, it offers multiple certification programs and courses for FinOps practitioners to complete and tailor their FinOps education.

The FinOps Foundation created a complete framework (https://www.finops.org/framework) that describes how FinOps practices can take place. Essentially, it breaks down the practice into three different phases – **inform, operate, and optimize**. Be mindful that, in this book, we offer our own take on FinOps practices that, although similar, differ slightly from the FinOps Foundation approach. Instead of phases that must be followed one after another, we follow different **streams or pillars** that group together all the different FinOps initiatives we will work on. With this pillar-based approach, we can adapt to the specific needs of each company from a FinOps perspective, tackling the most pressing matters first and then working on different domains simultaneously, instead of following sequential phases one after another.

Let's move on to describe in detail what initiatives can be included in the Inform, Operate, and Optimize pillars.

The three pillars of FinOps

In this section, we will explain our view of the three fundamental pillars upon which a successful FinOps practice should be built.

Let's describe one by one the three pillars, which are Inform, Optimize, and Operate, and the key points to work on and develop in each one, alongside some examples to illustrate how to get started.

Inform

One of the most important things every FinOps practice must do is to improve the visibility of cloud costs for all the technical, financial, and project management teams. Everyone needs to be initially aware of how costs are distributed between, for example, business units, projects, offices or locations, and cloud regions. This process is often called **cost allocation**, as it allows you to divide costs across different units.

In many organizations, this is key to enabling two important cloud cost visibility exercises, which are showback and chargeback:

- **Showback** refers to the operation of dividing the costs of IT services across different IT departments and providing the obtained information to management
- **Chargeback**, conversely, goes one step beyond and bills the different departments or business units for IT assets, licenses, training, and any kind of relevant work or service

This may seem like a simple thing to do, but in reality, it is a complicated matter, to say the least, as it often implies working on naming conventions, tagging, and other Operate pillar initiatives (which we will describe in detail later on this section), which are not present or not consistently applied in a lot of organizations. There is also another issue that arises here, which is how complexity rises when applying chargeback and showback on a shared infrastructure or other shared service costs. We will cover this in later chapters.

Another part of the Inform pillar, and one of the first starting points of any FinOps implementation, is to undertake a FinOps maturity review, assessing the status of an organization in terms of cloud cost optimization and cloud financial management. This review will serve as a starting and reference point for the future, assessing and determining how far a company has gone in this cost optimization process.

Also, in our experience, once the FinOps practice begins and there is a team working on visualizing cloud costs, everyone working on the cloud suddenly begins to be a little more aware of how everything they do may have an impact on costs, and they start to think twice before upscaling a database for a particular service or requesting a bigger virtual machine for a small application.

Awareness across cloud technical teams is a really good thing when we deal with cloud costs, and it is often one of the quick wins that every FinOps practice can achieve in almost no time.

The result of the Inform exercise can be delivered by creating FinOps dashboards and reports, which can include the degree of application of the different FinOps initiatives that are proposed by the organization.

When we improve visibility on costs and track the application of different initiatives, there is no better driver for these initiatives than to calculate the projected savings of these initiatives when applied to their full extent.

In our view, FinOps work in the Inform pillar never ends, as we can continue to improve visibility and adapt it to each team's needs.

This diagram summarizes the key points to work on in the Inform pillar:

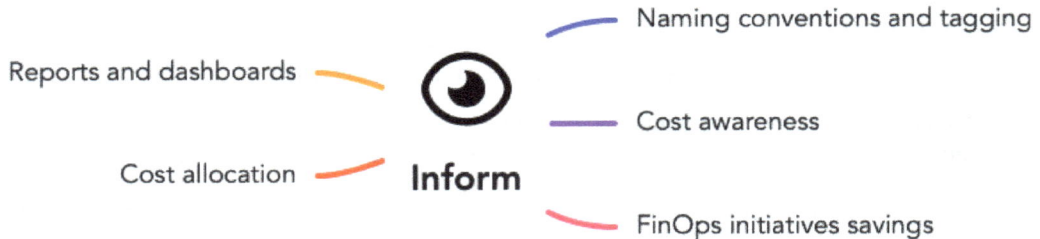

Figure 1.2 – The Inform pillar key points

Let's try to summarize all this work with a small example to better understand the key questions to ask and answer during this process.

Example (the Inform pillar)

Do you know how costs are distributed per environment in your organization? How much does each environment cost?

The first thought that comes to mind is that costs for production and preproduction should be similar, with development environment costs way behind, assuming stable workloads that have been deployed for a long time. For new projects, development can sometimes have higher costs if a project has not gone live yet.

Each organization and workload should have its own cost variables, so the first step is to make this cost distribution visible for everyone to see. Once it's visible, it's our job to make sense of it (working together with technical teams) and decide whether it's aligned with the use case of each environment and the status of each workload.

If it's not, it's time to dig deeper and check for inefficiencies.

This is the train of thought we should follow when working in the Inform pillar, which is raising questions and searching for answers that make sense, while we keep investigating and searching for inefficiencies along the way.

This example only covers a really small part of the Inform pillar but does illustrate the methodology to be followed when working to improve cost visibility.

Optimize

The Optimize pillar is about making the most effective use of cloud resources. This pillar requires a huge amount of research and technical work on all the cost drivers of cloud services and offerings in the cloud.

Apart from understanding how costs work and what drives them up or down, it is important to have deep architectural knowledge, to discern what is needed and when, as well as to be able to reduce unnecessary compute, features, or configurations that raise costs without justification.

In this pillar, we learn how to make use of new purchasing models such as Reserved Instances and Saving Plans, Spot Virtual Machines, and all the possible ways to improve governance based on costs in the cloud.

We also learn other cost optimization strategies, such as licensing, compute, storage, and database optimization, as well as other general concepts such as autoscaling and power scheduling.

Part of this Optimize path involves training and working together with technical teams, as well as helping them apply the initiatives that are proposed by FinOps teams and providing them aid in the form of assets, such as scripts, automation, and dashboards, to help them streamline and standardize the implementation of different initiatives that everyone will benefit from.

This diagram summarizes the key points to work on in the Optimize pillar:

New purchasing models —   Cost drivers

Cost optimization training — **Optimize**   Cost optimization initiatives

Cost optimization assets

Figure 1.3 – The Optimize pillar key points

Example (the Optimize pillar)

The best way to begin the Optimize pillar is to propose a series of quick wins, which are initiatives that provide cost savings with little to no impact on running workloads, such as deleting orphaned resources or decommissioning unused services. We will delve deeper into this topic in upcoming chapters.

After the implementation of some quick wins, we can begin to implement cost optimization initiatives that are easier to implement in new projects, while we remediate non-optimized workloads, using potential cost savings as a way to prioritize one initiative over another.

Once the practice is established, we can tackle the most difficult initiatives that require additional planning and careful analysis, such as making use of Reserved Instances, Saving Plans, and Spot Virtual Machine purchase models.

While we optimize, we cannot forget to make the advances visible for all stakeholders and interested parties to see. By doing so, we effectively promote FinOps practices and create confidence in this methodology.

Operate

In our experience talking and working with different organizations and clients, we have seen a lot of one-time cost optimization exercises that reduce costs. These exercises require a lot of effort and resources, but they provide great value.

The problem with this approach is that, if the roots of a practice are not deep enough, the same errors will eventually resurface again, which takes organizations back to square one, rendering all the effort and resources that were invested useless in the long run.

In this pillar we will also work in designing and implementing new KPIs and measures, which will show objectively how well we are doing in objective terms to a business, financial teams, and stakeholders, and bringing to the table a fresh point of view on optimization, modernization and any other topic that can be valuable.

The Operate pillar is about governance. With this pillar, our proposal is to explore the roots that are needed to build a strong practice. You can begin this pillar by working on naming conventions or tagging, including defining new FinOps roles and processes, enabling organizations to stay one step ahead and not keep making the same errors. This makes FinOps an iterative exercise that improves on each iteration as the teams involved mature and acquire cost optimization expertise.

This diagram summarizes the key points to work on in the Operate pillar:

FinOps KPIs

FinOps roles

Cultural change

Operate

FinOps processes

Figure 1.4 – The Operate key points

Let's illustrate with a simple example how we can get started in the Operate pillar.

Example (the Operate pillar)

One of the first steps in the Operate pillar can be simply establishing a quarterly review for cloud cost optimization and the cloud costs themselves. Technical and architectural teams can liaise with financial and FinOps teams, discuss the cloud costs, and prepare a report for the meeting, such as a manually created cost report in the initial phases of FinOps adoption, which can be replaced later on with a report that is generated automatically through an automated process.

This recurring meeting is not that hard at all to set up, and it can be a great starting point to make cost reviews an iterative exercise while fostering collaboration between teams.

Summary

As we are sure you have already determined by what we have covered so far, FinOps is definitely not a one-size-fits-all methodology.

The practice needs to be tailored and adapted to each company's needs; it's about doing things in a different way, which is always a challenge in organizations.

FinOps practitioners need to develop a strong knowledge and understanding of what a company needs and how to create value in small, easily obtainable steps from day one, without overreaching and trying to change everything in a month.

It is never easy to change your mindset, so taking *one step at a time* is often the best way forward.

In the next chapter, we will elaborate and thoroughly explore how FinOps practices can fit with other cloud governance methodologies and best practices.

2

Understanding How FinOps Fits into Cloud Governance

In this chapter, we will delve deeper into the FinOps domain. We will understand how FinOps practices fit in organizations, starting from its basic principles.

From its interaction with other methodologies and frameworks to the different approaches that we can use, depending on our organization's cloud maturity, we will understand how cost optimization has a place in every situation.

To close the chapter, we will also review different tools that can be leveraged to accompany our FinOps practices – beginning with tools that are included in each public cloud and then covering commercial all-in-one products and software that will help establish a practice with deep roots.

In this chapter, we will cover the following topics:

- The Well-Architected Framework – an introduction
- FinOps as part of bigger governance
- Tailoring a FinOps approach for every organization
- Selecting the right tools for the job

The Well-Architected Framework – an introduction

Before going forward, we want to introduce a key framework for optimization, one that is not only focused on cost management and optimization but also optimization in general.

The **Well-Architected Framework** is a framework that outlines the use of best practices when architecting and operating cloud solutions. It basically consists of the following series of pillars:

- **Cost**: This pillar is as straightforward as it seems. The idea is to iteratively optimize the cost of our solutions. Keep in mind that to optimize costs is not to reduce them but to make the most out of our cloud budget. *An example of a cost initiative is the shutdown of virtual machines during off-hours.*

- **Security**: This pillar describes how to make more secure solutions by protecting data, identity, and infrastructure in various ways. *An example of a security initiative is enforcing multi-factor authentication and conditional access for administrators.*

- **Operational excellence**: This pillar covers the improvement of processes and monitoring of workloads on daily operations, striving to be more effective and avoid putting effort into the wrong tasks. *An example of operational excellence is the implementation of a CI/CD and Infrastructure as Code for infrastructure management.*

- **Performance efficiency**: This pillar consists of designing solutions that work in an optimal way, taking up just enough capacity and resources and adapting to demand. *An example of performance efficiency is an application of horizontal scaling in a containerized web application to accommodate high-user-demand hours.*

- **Reliability**: The reliability pillar basically describes ways to make our workloads more fault-tolerant, with the ability to recover quickly and adapt in the event of contingencies. *An example of reliability is using a load balancer in an IaaS solution to distribute load between virtual machines in different availability zones.*

- **Sustainability**: This one is only present in the AWS Well-Architected version, and it is not currently part of Azure and GCP documentation. Regardless of this, we think it is an important pillar to cover as well. The sustainability pillar strives to design solutions that don't compromise the environment we live in, trying to reduce, for example, the carbon emissions of our workloads. We will cover this topic in *Chapter 13*, the last one of this book. *An example of a sustainability initiative is keeping track of our virtual machines' carbon emissions in operational dashboards and lowering them as much as possible.*

The Well-Architected Framework proposes an iterative approach to optimization based on these pillars. The idea is to constantly analyze our cloud solutions for possible improvement opportunities and implement them, improving overall quality and designing more and more optimal solutions that provide great value for business.

Each cloud provider offers its own take on this framework, with slight differences between them (apart from slight changes in the different pillar names), such as the following:

- In AWS, there are six pillars due to the inclusion of the sustainability pillar. This pillar tries to make sustainable solutions in the cloud, aiming to reduce carbon emissions as much as possible for more environment-friendly cloud environments.

- In Google Cloud, there is a sixth pillar that is the foundation for the rest of the pillars, called system design. System design dictates principles such as using fewer managed services, using lightly coupled architecture, and having proper documentation in place. Additionally, the pillar that covers security also includes privacy and compliance.

Each cloud provider provides recommendations and best practices to build cloud solutions around these pillars, including a lot of assets (such as code examples or architectural blueprints) and documentation to help cloud practitioners improve their designs. You can find the official web pages for each cloud provider at the following links. In our opinion, reading these resources is one of the first key steps on the road to becoming a FinOps practitioner:

- AWS Well-Architected Framework: `https://docs.aws.amazon.com/wellarchitected/latest/framework/welcome.html)`

- The Google Cloud Architecture Framework: `-https://cloud.google.com/architecture/framework`

For us, FinOps is one of the gears that need to be put in motion to create optimized cloud operations. Before we get deep into the specifics, we must understand the global picture.

With this said, let's move on to our next section, where we will analyze the integrations between FinOps and other methodologies and frameworks that are widely used.

FinOps as part of bigger governance

One of the things that FinOps practitioners need to grasp before delving deeper into the practice is how it integrates with other common methodologies and frameworks that are used in most organizations. By analyzing these synergies, we can understand how each framework can benefit from each other and how to get the most out of them.

Let's begin with Agile methodologies, one of the staples in modern-day project management.

FinOps + Agile methodologies

As we already explained, for us (the authors), FinOps is an iterative exercise. We must begin by applying *lowest hanging fruit* strategies to generate value as soon as possible, and then iterate to get to more complex work, with higher maturity requirements and initiatives.

This idea is referred to by the FinOps Foundation as the **crawl, walk, and run** approach. It suggests that, when applying FinOps in an iterative manner, the practice will improve and get more mature as we repeat and iterate. We must begin simple and then get complex as teams feel more comfortable and knowledgeable about everything FinOps.

Let's provide an example of Crawl, Walk, and Run:

FinOps pillar	Crawl	Walk	Run
Inform	Monthly reports with cost evolution per business unit/project.	Weekly reports with cost evolution per Business unit/project/environment.	Automated reports with a complete cost evolution. FinOps initiative tracking.
Optimize	50% virtual machine off-hours shutdown in development environments.	50% virtual machine off-hours shutdown in development and preproduction environments.	80% virtual machines off-hours shutdown in development and preproduction environments. Reserved capacity in 80% production.
Operate	Monthly meetings to review cloud costs.	Monthly meetings to review cloud costs. Weekly meetings to review and track FinOps initiative implementation status across business units.	Monthly meetings to review cloud costs. Weekly meetings to review and track FinOps initiative implementation across business units. FinOps cells in each business unit.

Table 2.1 – Crawl, walk, and run examples in the inform, optimize, and operate pillars

We have referred continuously to iterative exercises, and what's more iterative than **Agile methodologies**? FinOps and Agile methodologies are the perfect partners in crime, as they are based on the same concepts.

Agile methodologies are one of the most common project management processes, and they are based on constant iteration through *sprints* and collaboration, responding easily and faster to our clients' needs. Agile is a process that specifically tries to set itself apart from old **Waterfall** project management methodologies that were based on plan, design, build, test, and deliver project phases.

Agile project management, in our view, also tries to avoid having a really complex process that is difficult to follow in reality, so it proposes a really simple approach based on minimal phases that are easier to understand and follow.

Let's use a diagram to illustrate how a sprint can look in Agile project management (a project can have a lot of sprints before its completion):

Figure 2.1 – An example of an Agile sprint

Conversely, this is how a traditional Waterfall project management looks, covering the entire project life cycle:

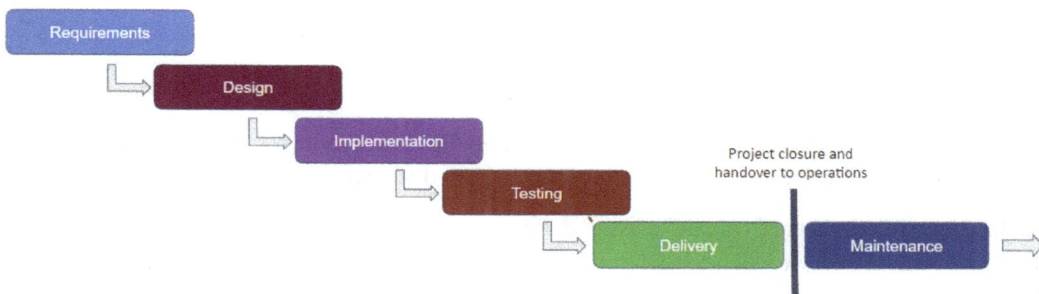

Figure 2.2 – An example of Waterfall project management

Using Agile methodologies, as opposed to Waterfall, can yield the following results for our projects:

- More customer satisfaction and alignment, as they are given results at the end of each sprint instead of waiting for the end of the project

- Better-quality products with fewer bugs, as testing is part of each sprint

- Better control and adaptability due to sprint lifespan and not rigid project planning

- Reduced risk of delivering a product that won't fulfill project requirements

In Agile project management, the bugs and features (bugs are usually found and pointed out as a result of testing processes and new features are proposed by end users) in our applications are part of the sprint planning, which means that we work simultaneously on solving current issues and providing new functionalities. These bugs and features are tracked as backlog tasks throughout an entire project.

As an example, the following diagram depicts how backlog items can be included in a sprint that will result in completed products:

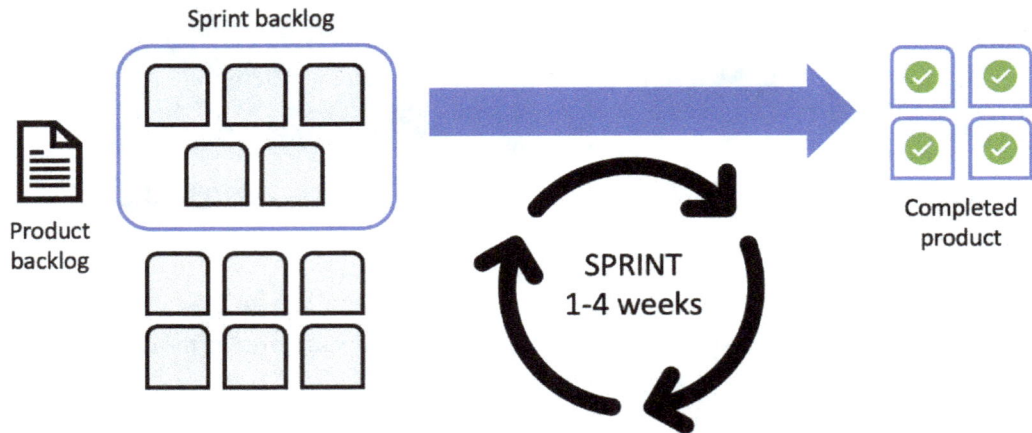

Figure 2.3 – Backlog sprint planning

Agile methodologies are the perfect way to implement FinOps practices in each of the pillars. FinOps initiatives can be divided into really small steps that we can distribute in sprints.

By doing this, we can create frequent "release" deliveries that will create value from the beginning of the project. In doing so, we will get feedback from key stakeholders, operations, and financial teams to focus on what's important, while creating value and making all interested parties aware of the work that is done in the FinOps domain, every step of the way.

We can create a backlog of FinOps initiatives to be applied in each of the inform, operate, and optimize pillars, which can be distributed in sprints based on value, complexity, and cost reduction.

We can even take *bug and feature tracking* ideas from Agile development, reporting non-optimized workloads as bugs or initiatives that we propose to apply as features, which we will then prioritize.

Bugs will then coexist with new features to be implemented, and we will decide what to fit in each sprint based on priority.

Be mindful that we are not dismissing Waterfall or traditional project management in any way but highlighting the synergies between Agile and FinOps methodologies. In our experience, in real-world scenarios, a lot of organizations actually opt for **hybrid project management approaches**, where they take the best out of Waterfall and Agile methodologies, seeking to get the best advantages of each one. In this hybrid approach, we use the project phases of Waterfall projects to structure the project work, combined with sprints and Agile planning for the work that needs to be done in each project phase.

With the interaction between these two key methodologies for modern organizations covered, let's now move to describe how Infrastructure as Code, CI/CD, and DevOps can synergize with FinOps.

FinOps, Infrastructure as Code, CI/CD, and DevOps

Infrastructure as Code (**IaC**) is a methodology that uses declarative programming languages and automation to automatically deploy and manage infrastructure. The code for this includes the resources and their complete configuration in an easily readable format.

CI/CD stands for **Continuous Integration/Continuous Deployment**, which are two key best practices that are essential in DevOps methodologies.

We can achieve IaC by using multiple programming languages, such as AWS CloudFormation or Terraform. However, having a programming language is not enough; we need some additional key ingredients to achieve end-to-end infrastructure management through IaC:

- A *code repository* where our IaC code will be stored, such as **GitHub** or **Azure DevOps repos**, where we can use code versioning and code repositories to manage our code, documentation, branches, and so on
- *Automation pipelines* that will use the code to implement IaC processes to review and deploy our infrastructure, such as **Azure DevOps pipelines**, **GitHub Actions**, and **Bitbucket**

Among the benefits of IaC are the following:

- The standardization of best practices, policies, and configurations across environments, business units, regions, or any other constraint
- A reduction in the risk of human error through automation
- It fosters collaboration and allows for version control
- It ensures the consistency and scalability of our infrastructure resources
- It eases security and compliance processes with clear configurations that are present in code and are human-readable

One of the best traits of IaC is how it eases standardization across organizations, and FinOps as well is based on setting good practices and cost optimization standards. There are many ways that IaC can be used to implement FinOps practices and initiatives:

- Implementing **policy as code** that adheres to FinOps policies, such as disallowing the use of high-performance (and costly) virtual machines in development and sandbox environments.
- It may seem unrelated, but having IaC in place simplifies a great deal the process of implementing a proper *naming convention* and *tagging policies* for our resources. We will cover this in depth in *Chapter 3*, but these two key governance initiatives can unlock incredible value for cost allocation purposes.

- Implementing FinOps initiatives and configurations as standards in code. We can do so by creating modules with best practices and configurations already built in, which can be used for other teams. We often refer to these modules as *golden images* that we can define and prepare for any resource.

- We can even go the extra mile and define/implement key assets for FinOps, such as cost alerts, budgets, and other FinOps-specific resources.

- We can even use some tools that we can incorporate into our IaC processes to estimate the impact on the cost of our deployments before their execution, such as *Infracost*. We will cover this in our *Other interesting tools* section later in this chapter. As a small example, let's see what IaC implementing policy as code looks like in Azure using Terraform. We can see in the following diagram how the policies are also declared in code, preventing non-compliant resources from being deployed:

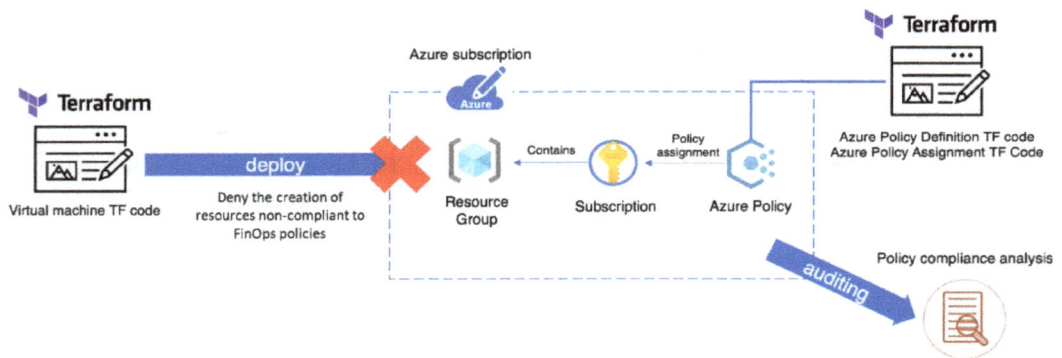

Figure 2.4 – An example of policy as code in Terraform and Azure

By avoiding the manual deployment of resources, we also ensure that all the changes in cloud resources, such as a new virtual machine or configuration change, always come from the same source. This allows FinOps practitioners to have a common place where they can review planned changes, checking whether they are compliant with FinOps policies and initiatives on cost optimization, tagging, or other relevant aspects of the practice.

Let's analyze this idea further in the next section.

FinOps and change management

Change management is one of the key processes for organizations under the IT service management discipline, and it dictates and describes the standardized process that should be followed to ensure that changes are properly analyzed, prepared, and implemented. Additionally, it describes how changes should be informed to relevant stakeholders and interested parties.

This idea is aligned with FinOps initiatives, as FinOps teams need to oversee the resources that are created, checking whether they are fully compliant with FinOps policies or not.

As a small example to illustrate how FinOps and change management synergize, let's say we have CI/CD implemented on Azure DevOps using Terraform to deploy cloud resources.

Let's also assume that have in place a FinOps policy to disallow the creation of Azure Premium SSD Managed Disks in non-productive environments. This policy is not properly implemented in Azure Policy, but it is a rule that we apply in our cloud environments, tracked over time using FinOps dashboards.

As part of a *build and release* process and pipelines, we could add an additional gate before the Terraform `apply` command, allowing the FinOps team to approve whether the planned changes in infrastructure are compliant with our policy and other FinOps initiatives. Having this gate will improve the traceability of decisions taken before any change.

We can also use this unified change management to increase the visibility of the implementation of FinOps initiatives. To do so, we can assign pull requests to FinOps features registered in our sprint planning, which we can also correlate with cost reduction after implementing FinOps initiatives.

Part of the **IT Service Management** (**ITSM**) paradigm is to use ITSM-oriented tools such as ticketing products, such as *Jira* or *ServiceNow*, where we can trace open incidents and requests or track changes across our environments. Change management can, and should, also be implemented using such tools, ensuring that we have all the information about changes and justification of all decisions taken related to them.

Also, if we register the FinOps initiatives, once we have agreed on their implementation, as changes in our *ITSM tools* as tickets, we can assign them to specific teams and track their implementation, giving us full traceability that will simplify correlating savings with FinOps initiatives.

Enabling traceability of our FinOps actions from the beginning of a project is always good in that it allows more visibility. It is often complex to calculate savings after the fact when we are asked to justify our work or show its impact.

With all these interesting integrations and synergies between different methodologies and frameworks with FinOps, let's now move on to the next section, in which we will try to adapt FinOps practices to different profiles of organizations.

Tailoring a FinOps approach for each organization

The path to implement FinOps in each organization is definitely not the same. In some organizations, the focus of these practices should have a technical focus, increasing an organization's overall knowledge on how to get the most out of cloud resources by educating it on key initiatives such as Reserved Instances, rightsizing, and other relevant cost optimization initiatives.

Conversely, in some scenarios, FinOps practice can shift value to the inform or operate pillar instead of the optimize pillar, allowing you to begin creating some visibility on how costs are distributed and getting a general understanding of cloud maturity before going forward, while defining FinOps processes to be followed to ensure continuous cost optimization.

In this section, we will try to describe, by using some scenarios, how to tailor a FinOps practice based on each organization's cloud maturity and other relevant factors.

Scenario 1 – companies not yet in the cloud or beginning their journey to it

When companies are taking their first steps on their journey to the cloud, there are many initial challenges to face, such as cloud training, the creation of a business case, and careful planning for cloud migration.

For these organizations, we consider that FinOps is applied in a **greenfield** environment.

A greenfield environment in the cloud is one that is created almost from scratch, without restrictions or constraints imposed by legacy applications or design and architectural decisions that were made long ago.

The upside of this is that there are no dependencies forcing us to make decisions, so we can implement state-of-the-art architectural best practices.

In an organization in this state, our recommendation is to take baby steps in all three FinOps pillars, so we begin creating trust around the practice while making advances on every front. Having a strong foundation to build upon is a great way forward.

As there should be only a few resources in our cloud environments, we should focus on optimization before migration, avoiding having non-optimized resources as much as possible.

Scenario 2 – companies already in the cloud but not mature enough or that have non-optimized workloads

For these organizations, we consider that FinOps is applied in a **brownfield** environment.

A brownfield environment in the cloud is an already established cloud environment in which we assume there may be non-optimized solutions, workloads that are difficult to modernize, and even legacy applications that we won't be able to change as easily. In a brownfield environment, we can expect a lot of lift-and-shift migrations, where non-optimized workloads are migrated as is without refactoring or modernization.

In this situation, the usual first steps are to try to generate some savings through quick wins while we focus on non-optimized workloads, the costs of which we can reduce easily. Quick wins are often needed, as there are probably cloud operational teams already in place that may not take the presence of FinOps teams, who could be seen as auditors of their work. Creating trust and confidence between teams is one of the first steps in these situations.

An additional step that is essential in brownfield environments is to implement a tagging policy as soon as possible, as the one in place is often incomplete and lacking. Without proper tagging policies, we will be unable to work on a lot of FinOps initiatives and processes.

While we work on the technical side challenges, we will also focus on creating thorough cost reports that we will digest so that we can fully understand where unwanted or excessive costs come from. A complete assessment will be needed so that we can identify the pain points, and we can work on it while we begin the implementation of some of the initiatives for cost optimization that have been proposed.

To finish the process, we will set up basic governance, maybe setting up some periodic meetings that can be used to review cloud costs and track basic FinOps initiatives.

Scenario 3 – big companies with strong cloud maturity

Along with scenarios 1 and 2, this is another common one, especially for bigger organizations.

We must take into account that not all organizations are able to justify, from a personnel cost perspective, having a FinOps division among their ranks, or even any FinOps-dedicated role at all. Additionally, cloud roles in general are harder and harder to come by, and FinOps practitioners can be hard to hire due to high demand.

More often than not, the organizations that have a full-blown team are the biggest ones, with more resources and budget available for the cloud. These organizations are also the ones that are most advanced and mature in their cloud operations.

For these organizations, which usually have strong technical knowledge, the value of FinOps often comes from the following:

- Having a team that oversees business units, projects, and teams across an organization, ensuring that best practices are followed and implemented. This team can set up standards that will be part of the company policies on cloud optimization, and they can also help in cost optimization tasks by training and creating valuable assets such as documentation and blueprints.

- Cost visibility improvement by creating standardized FinOps reports and dashboards, allowing you to track the practice across projects and business units. If a project or business unit is in dire need of optimization, having such visibility will help you to prioritize as well.

- Governance with FinOps functions and processes is another key point to work on. These big organizations are often complex, with a lot of business units and separate projects managed by different providers, which makes setting up centralized governance hard, to say the least. If FinOps practices are able to break through all this complexity and standardize the practice of cost optimization in an organization heavily invested in the cloud, long-term cost avoidance can be substantial.

These organizations are already multi-cloud or considering such an approach, benefitting from each public cloud's strengths and weaknesses, as they are able to distribute their workloads between multiple public clouds based on cost, simplicity, or any other valid criterion that is part of their cloud strategy. Multi-cloud environments add complexity and help justify FinOps practices, as our initiatives and ideas are concepts that can be implemented regardless of which public cloud you use. For example, in a multi-cloud organization, having a cost dashboard where we can aggregate the costs of multiple clouds can be invaluable.

Scenario 4 – companies focused on generating cloud cost savings

Finally, let's analyze our last scenario, which is organizations that want to only focus on reducing costs as much as possible.

We don't support this idea, as we don't think FinOps is about reducing costs at all, but we have seen during our experiences with multiple clients that this is what some companies want.

They don't want to hear about dashboards and visibility or governance; they just want their cloud costs reduced. This situation is not ideal as, even if the costs are lowered, without governance and visibility, we are bound to encounter the same pitfalls.

Regardless of this fact, we want to prepare you for this situation as well. With these companies, the best approach is to work along these lines:

- We can begin with a complete assessment of how optimized all the cloud workloads are, working from all angles, which we will go through in the later stages of this book.

- Quick wins will be a key driver to generate confidence in FinOps practices, especially at the beginning when the practice takes its first steps.

- Establishing mid-term and long-term cost reduction plans. After our analysis, we can propose mid- and long-term initiatives, such as modernization, by replacing old legacy solutions with state-of-the-art, cloud-native solutions, resulting in cost optimization as well as modernization. Keep in mind that, in order to make these initiatives attractive, we should calculate the potential savings; otherwise, without any incentive, the technical teams will not invest all the time required.

One key point to keep in mind about cloud savings is that we will only be able to generate savings up to a point. After reaching some maturity in FinOps, generating savings will be greatly reduced and much harder, so our recommendation is to avoid approaches to finance FinOps practices using the savings generated.

Remember that FinOps is not only about savings but also optimization.

Selecting the right tools for the job

Some of the key questions that come to mind when an organization's FinOps journey begins are as follows:

- Do I need specific tooling for FinOps practices?
- Should I invest in developing in-house assets and tools, or should I rely on assets that are already available?
- Should I invest in a market product such as CloudHealth?
- Will I have enough ROI to justify this investment?

It is never an easy question to answer. However, we think that building a roadmap and a clear path to follow is better than choosing a product and creating the practice around it.

The perfect product to accompany FinOps practices, in our view, does not exist, at least not yet. Good and complete solutions are not cheap, and more often than not, we need to combine different tools for different purposes, and even consider small development projects if we need to cover gaps that are not offered by market products. Keep in mind that a FinOps journey is different for every organization, as they don't all share the same needs and purposes.

In this section, we will explore some base tools that are offered in Azure, AWS, and Google Cloud natively, understanding what they offer and how to get the most out of them. Apart from these base tools, we will also review some market products available. Keep in mind that new FinOps tools are being released every day, and it is impossible for us to cover them all. Consider this subset of tools as a starting point for your FinOps tooling research.

Let's start with the base tools that are included in all three major public clouds, understanding what we have out of the box before considering other options.

Base tools

We should always look first at the set of tools that each cloud provider offers before considering paying top dollar for a market product, as there are a lot of things that are offered by these tools out of the box.

The richness that these tools offer and their compatibility with FinOps vary from one cloud provider to another. In general, our opinion is that AWS tools are by far the most advanced, providing all the information that is essential for FinOps practices at the deepest level. This does not mean that Azure and GCP don't offer interesting tools and services to analyze and optimize our costs; it is just that they are not as rich and complete as their AWS counterparts. For example, in Azure, calculating global coverage across subscriptions, a key KPI for Reserved Instances, is a complex challenge, while in AWS, we get this KPI from **Cost and Usage Dashboards Operation Solution** (**CUDOS**) reports out of the box. We will start with Azure, and then cover the AWS and Google Cloud tools.

Azure Advisor

Azure Advisor is a tool offered by Microsoft for free.

It scans your Azure subscriptions and provides recommendations based on all the Well-Architected Framework pillars, one of them being cost.

It is a good starting point for optimization, as it is able to detect orphaned resources and underused virtual machines, among other things.

The following is how Azure Advisor looks in Azure:

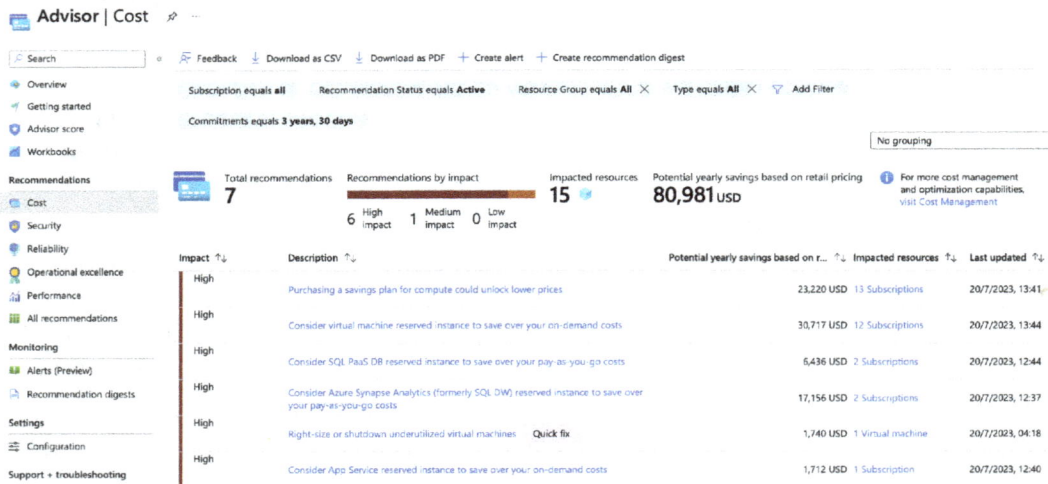

Figure 2.5 – Azure Advisor

As you can see in the preceding screenshot, there is a section with recommendations for each one of Azure Well-Architected pillars. Each recommendation for the cost pillar includes how much we can save by implementing it, as well as the considerations needed to do so.

However, we have one word of advice at the time of writing regarding Advisor recommendations:

- When considering rightsizing virtual machines using Advisor recommendations, keep in mind that the decision engine for which virtual machine to use is really simple and does not consider all variables (multiple NICs and other specific features of virtual machines are not checked by the tool). Also, it allows you to set a CPU threshold for recommendations, and usually only recommends rightsizing when usage is really low. For more information on how rightsizing works with virtual machines, refer to this link: https://learn.microsoft.com/en-us/azure/advisor/advisor-cost-recommendations#optimize-virtual-machine-vm-or-virtual-machine-scale-set-vmss-spend-by-resizing-or-shutting-down-underutilized-instances.

- Advisor will recommend using saving plans and reserved capacity to save money on virtual machines and all Azure services that support this purchase model. We will cover this topic later in the book in *Chapter 6*. Make sure to read it thoroughly, and understand the implications of Reserved Instances clearly before going forward with the recommendations.

- Keep in mind that rightsizing virtual machines could have an impact on Reserved Instances that are already in place. Analyze this before considering going forward with size changes in virtual machines.

Overall, despite being limited to some initiatives, Azure Advisor provides valuable information.

We don't think that following Azure Advisor recommendations to the letter is the best way to plan for Reserved Instances, but it can definitely help with rightsizing exercises on virtual machines. At the time of writing, you can only include and exclude subscriptions as well as resource groups using the Advisor's analysis. We would have liked to use tags to filter or see recommendations only scoped to a specific region, business unit, or project, but this is currently not possible.

Over time, the tool will improve, for sure, as well as the engine that does the processing, and it will provide even more valuable insights.

Outside of the FinOps domain, in our opinion, it is a great tool to accompany the Well-Architected Framework optimization in all the pillars.

Azure Cost Management

Azure Cost Management is the tool that Microsoft offers to review the costs of our Azure resources.

It allows you to filter costs per subscription or service, and it even provides tags to deep-dive into all you need on your cloud cost, offering a better understanding of how they evolve and why. Billing data for cost analysis is only available for the last 13 months, so keep in mind that if longer retention is needed, you should store the data elsewhere and use another product to visualize and analyze your cloud costs.

With Azure Cost Management, we can also get a centralized view of all of our reservations and saving plans to track usage and reservation terms.

To use this tool, we just need to type `Azure Cost Management` in the search bar in the Azure portal. This is how the tool looks in the portal:

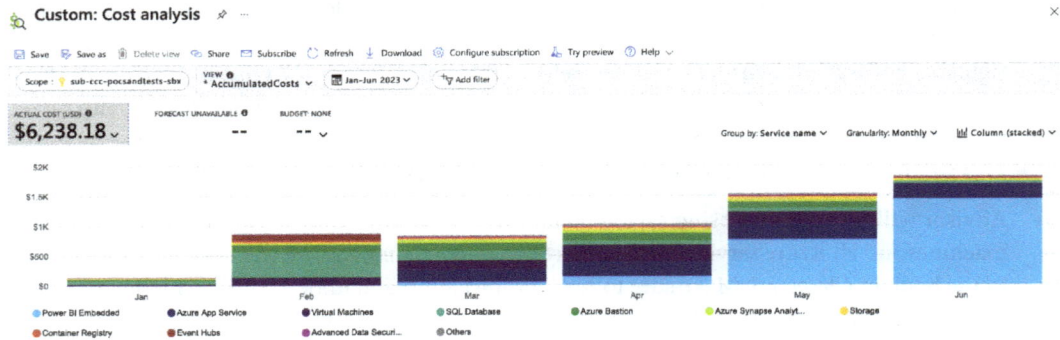

Figure 2.6 – Azure Cost Management

Apart from these cost analysis functionalities, Azure Cost Management offers an amazing additional feature (in our view), which is exporting all billing information across our subscriptions to an Azure storage account in the **Comma-Separated Values** (**CSV**) format. These files can be ingested in an ETL process and provide the basis for FinOps dashboards. Having this option allows you to use cost data in custom dashboards and reports.

However, there is one major drawback regarding Azure's billing structure, and that is that it is currently not possible to have an aggregated view (all costs of all subscriptions) under a management group. It is also not possible to see the aggregated costs of an organization using a **Cloud Service Provider** (**CSP**) (a common contract type with Microsoft to use their cloud services) model with this product, as Cost Management is limited to the subscription level. As you can imagine, if we have multiple subscriptions, this tool won't be enough for cloud cost controllers that need a complete view. One of the ways to cover this gap is to extract data programmatically or by using the **Cost Export** feature, aggregating it using Power BI or any other business intelligence product to create cost reports. Nonetheless, having this aggregated view of costs not available out of the box is definitely inconvenient.

Apart from the service that can be accessed through the portal, Microsoft has made programmatic access to all this costs related information available through the Azure CLI, PowerShell, and REST APIs. This gives you a lot of flexibility to work around cloud costs and create reports or dashboards, including all the information that is required from a cloud cost perspective.

All in all, we think that this product, despite its limitations, is a great tool for performing cost analysis swiftly and efficiently.

AWS Cost Explorer

AWS Cost Explorer is the built-in tool offered by AWS for cost analysis. It is similar to Azure Cost Management, as it lets you filter data in any way possible for deep analysis.

This is how the tool looks if we open it from the AWS Management Console:

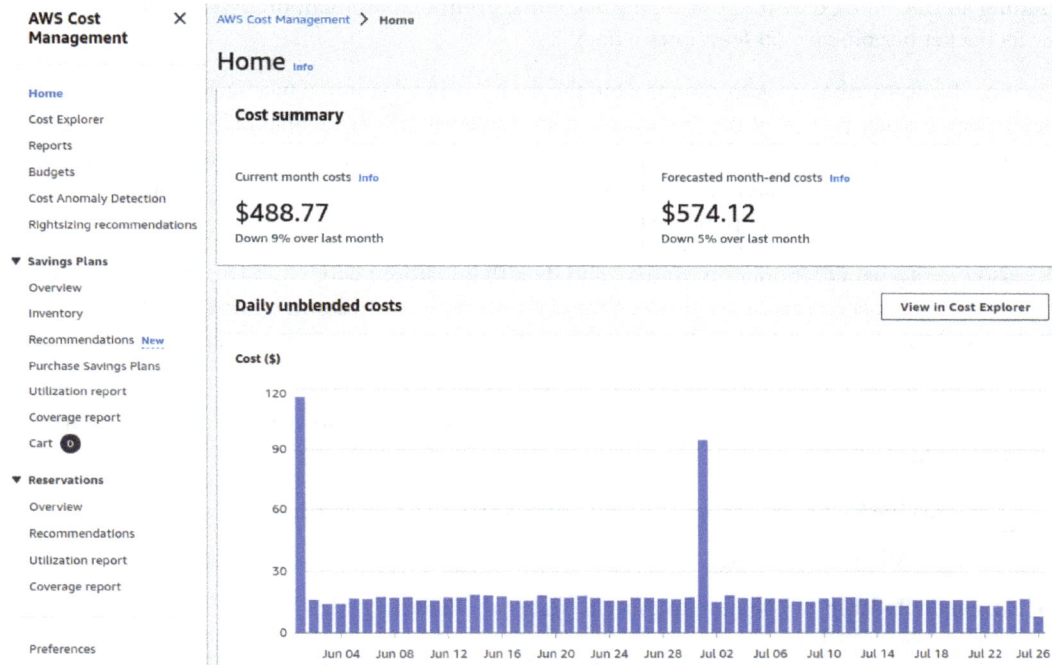

Figure 2.7 – AWS Cost Explorer

AWS Cost Explorer allows filtering using dimensions such as account, service, and other data grouping constraints. From AWS Cost Explorer, we can also check whether our instances are correctly sized, as well as get recommendations of instances that can be well suited for current usage, in addition to cost savings should we move to a proposed instance.

Cost analysis also includes the forecasting of cloud costs, based on last month's spend, and the latest changes to our service configuration.

Similar to the tools from Azure and GCP, AWS Cost Explorer offers a place to centrally manage our Savings Plans and Reserved Instances, where we can have out-of-the-box key KPIs such as coverage and utilization. From the section of the tool dedicated to Reserved Instances and Saving Plans, we will also get recommendations on what Savings Plans and Reserved Instances to purchase to generate the greatest savings.

AWS Cost and Usage Reports and CUDOS dashboards

AWS Cost and Usage Reports (**CUR**) provides a set of reports, including all relevant data for cost analysis and cost optimization.

When AWS CUR is created, it automatically sends a report to an AWS S3 bucket of your choice, including all the billing data for your billing accounts. From that moment on, AWS updates data in your S3 bucket periodically (at least once a day).

Activating this feature has no activation fee, but the data generated is stored in a bucket and, therefore, generates some costs. The rest of the resources that are deployed with the template also entail additional costs, so make sure to choose the right settings (for example, using Parquet as a compressed file format) to avoid surcharges after activating this feature. CUR is often the foundation for many tools that perform cost analysis and optimization.

AWS also offers a data dictionary to understand data that is stored with CUR (`https://docs.aws.amazon.com/cur/latest/userguide/data-dictionary.html`). We need to keep in mind that some columns will only appear when we use specific services and when they are part of our bill, such as fields dedicated to Reserved Instances.

Once the data for the reports is in place, we can use multiple products to visualize and analyze data, such as the following:

- Amazon QuickSight can be used to set up complete reports, showing data that is included in CUR
- Amazon Athena can be used to query data using SQL queries
- Amazon Redshift can be used to move S3 data into Redshift tables

Conversely, the CUDOS is one of the six dashboards that are included in AWS Cloud Intelligence Dashboards, which is a set of QuickSight dashboards offered by AWS out of the box and part of AWS Well-Architected Labs (`https://wellarchitectedlabs.com/cost/200_labs/200_cloud_intelligence/`).

We can deploy this dashboard by using the AWS CloudFormation CUDOS template in the S3 bucket configured with CUR, which will create all the required resources for CUDOS.

As a small example, this is one of the many sections that are included in the CUDOS dashboard once deployed:

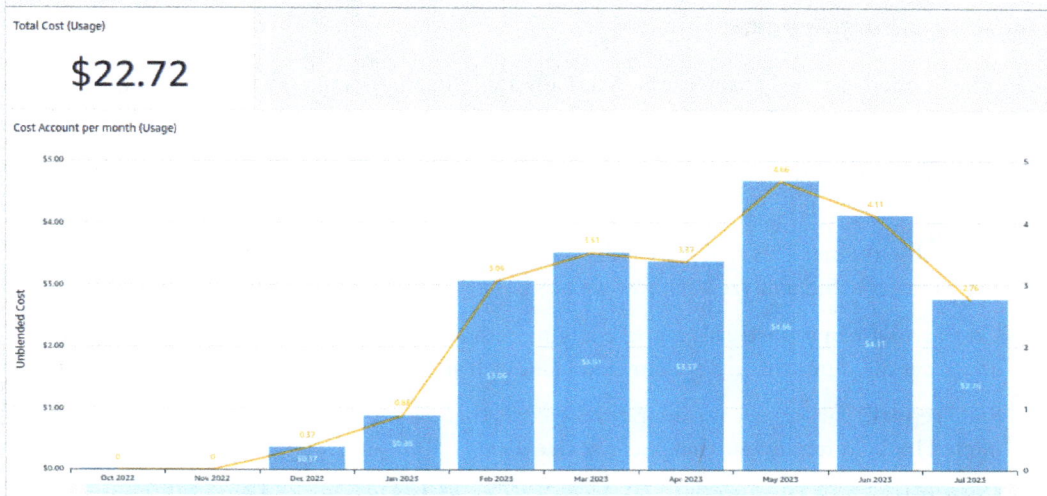

Figure 2.8 – AWS CUDOS dashboards

AWS Trusted Advisor

This tool is based on the Well-Architected pillars that were introduced earlier in this chapter. It provides recommendations on how to optimize our cloud resources in each of the six AWS Well-Architected pillars:

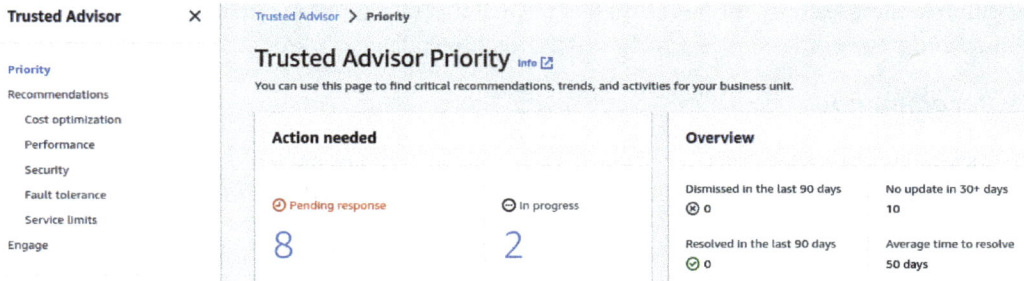

Figure 2.9 – AWS Trusted Advisor

From a cost optimization perspective, AWS Trusted Advisor offers some recommendations and the potential savings that we could obtain should we implement them. For example, it can show a database that has had no connections over the last few months and how much would we save if we deleted the resource.

A key thing to consider regarding this service is that the level of recommendations is tied to our AWS Enterprise Support plan. Having higher-level plans allows us to get the widest catalog of recommendations available, while basic plans, for example, only provide six security and service quota recommendations.

GCP Cloud Billing Reports

Google Cloud offers detailed cost reports in its Cloud Billing reports, mainly the following:

- **Billing report**: From this report, we can analyze costs using fields and filters, helping us answer key questions and discerning how our costs are distributed between different GCP services or any other constraint.

- **Cost table report**: This report includes all data that is issued on our GCP Cloud bills, so we can work on it and analyze whatever we need.

- **Cost breakdown report**: This report contains both the costs that are incurred as well as credits and other discounts that reduce our cloud bill, such as committed and sustained use discounts.

- **SKU report**: This report summarizes the costs that we incur on each of the cloud services offered in GCP, for anyone that needs such a view.

- **Committed Use Discounts (CUD) analysis report**: A dashboard with all the required information about CUD. From here, we can analyze whether we would get some additional discounts by sharing our CUD and other key KPIs to track CUD usage. Savings that we obtain by using CUD are also calculated automatically here.

- **Custom report**: We also have the possibility to move all this cost data to BigQuery, using other tools for data visualization and analysis in any way we want to.

In addition to these features, there is also a tool in GCP that recommends initiatives to improve cost optimization, among other things, which is GCP Recommender.

GCP Recommender

GCP Recommender is a tool offered by Google Cloud that provides recommendations and insights on how to optimize your cloud workloads, in a similar way to AWS Trusted Advisor and Azure Advisor.

The list of recommendations available per service can be checked out on this web page: `https://cloud.google.com/recommender/docs/recommenders`. There are recommendations for cost optimization, security, and the rest of Well-Architected pillars.

Recommendations are based on recent data gathered by Google Cloud, but they vary from service to service. For example, the Cloud SQL out-of-disk recommender analyzes the storage usage trends of the last 30 days, while the idle VM recommender only analyzes the previous 1–14 days.

As with other similar services, don't go and apply the recommendations without properly reviewing them. Keep in mind that recommendations are automatic and driven by an algorithm.

Exporting GCP Cloud Billing data to BigQuery

GCP offers a great feature out of the box that can ease the process of cost analysis in general, and that is the possibility of **exporting billing data directly to BigQuery**. While this is not directly a cost analysis or a recommender tool, it is a feature that can simplify greatly the process of cost analysis in general.

Instead of exporting to storage, as other cloud providers offer, and then processing the billing data, this feature directly exports this information to an existing dataset in GCP BigQuery. Once all this data is in BigQuery, we can use tools such as **Looker Studio** to create dashboards and other visualizations to perform cost analysis on our billing data.

In addition to these visualization tools, you can also use the BigQuery query engine to perform deep analysis of billing data. There are even some sample queries available to get you started on this analysis process.

This allows for much easier cost analysis beyond the tools that are offered as part of GCP Cloud Billing reports.

Market tools

Apart from solutions that come out of the box in each cloud provider, we also have market tools we can purchase to help us on our cost optimization journey.

These tools are not free; in fact, they can be really costly, as some of them even bill you based on a percentage of your actual cloud spend.

These tools can serve the following purposes in our view:

- When there are not enough resources or budget to properly assemble a FinOps team, these tools will do most of the heavy lifting for you, detecting non-optimized workloads and configurations that don't follow cost optimization best practices, and allowing your technical teams to look into them.

- If you have a proper FinOps team and FinOps maturity is in good shape, you should have budgets that can accommodate these tools. Having these tools helps FinOps practitioners avoid a lot of manual work, allowing them to focus on less technical details.

- If you want to avoid in-house FinOps developments and prefer to delegate them to a well-known vendor, these tools are ideal.

Without further ado, let's analyze some of the options offered by current industry leaders at the time of writing.

VMware Aria Cost by CloudHealth

Aria Cost by CloudHealth from VMware has been one of the cost optimization staples for some years now. It offers a wide set of features, such as the following:

- Cost allocation
- Budget management
- Cost analysis
- A resource optimization recommender
- Cost anomaly detection
- Resource tag management
- Customizable dashboards

One of the biggest selling points is that it supports different cloud providers and private/hybrid clouds, so you can effectively have a single pane of glass for cost optimization in multi-cloud or hybrid setups. At the time of writing, the public cloud vendors that are supported are Azure, AWS, GCP, and Oracle Cloud Infrastructure. It also supports a number of private and hybrid clouds, as well as on-premises data centers.

Additionally, if you are using VMware on-premises as your virtualization hypervisor, Aria Cost by CloudHealth can even help you calculate **Total Cost of Ownership** (**TCO**) for cloud scenarios and help with cost optimization before and after cloud migration.

You can find the complete information, such as detailed features and pricing, for this tool on the official home page, which can be found at the following link: `https://cloudhealth.vmware.com/es.html`.

NetApp Spot

Spot.io is NetApp's answer to cost optimization tools and is an automation-focused tool. It currently supports all the major public cloud providers – Azure, AWS, and GCP.

One of its biggest selling points is that it simplifies the management of different purchase models, such as Reserved Instances and on-demand and spot virtual machines. It also helps scale our cloud resources to meet user demand and is a great match for heavily containerized environments, especially in multi-cloud organizations where it is difficult to centrally manage and optimize workloads that are hosted in different clouds.

Spot.io offers four main products:

- **CloudChekr**: This product can be used for cost analysis across all three public clouds. Apart from this feature, it also provides visibility on user activities (i.e., an audit trail) in the cloud and eases the process of asset management for cloud resources, providing a full inventory of your services.

- **Ocean**: This is a container management tool that orchestrates compute provisioning for containerized workloads. It automatically scales our workloads to meet demand and minimize costs, while making use of Reserved Instances, Savings Plans, and Spot capacity. Additionally, it is able to provide metrics and insights on container usage, facilitating showback and chargeback exercises on containerized workloads.

- **Eco**: This helps you generate and track (this is important, as tracking savings is not an easy task) savings by efficiently making use of Reserved Instances and Savings Plans.

- **Elastigroup**: This is another orchestrator, and this one is designed to manage Spot instances across the public clouds. It solves one of the pain points of the Spot purchase model, which is recovery from cloud provider eviction process (which we will describe in depth in *Chapter 6* of this book), by distributing traffic between live virtual machines while discarding the ones that are reclaimed by the cloud provider. By using this tool, you are able to leverage this purchase model in a sustainable way, which will result in greater cost savings.

You can find the complete information, such as detailed features and pricing, for this tool on the official home page, which can be found at the following link: `https://spot.io/`.

Apptio Cloudability

Recently acquired by IBM, **Apptio Cloudability** is one of the first market products that appeared in the FinOps landscape. It covers Azure, AWS, and GCP as well as private clouds through its product brother, ApptioOne.

The main features that are offered are as follows:

- It enables chargeback processes by using its Business Mapping engine, which can be used to assign costs to different business units, projects, or any other relevant criteria. It also offers the ability to perform cost allocation in containerized workloads.

- It provides a module for Reserved Instance management and usage tracking.

- It has a dedicated module for rightsizing, recommending the most adequate size based on the usage of your virtual machines.

- Forecasting and budgeting for cloud costs.

- Cost dashboards that can be customized to adapt to your needs.

- Cost anomaly detection.
- Tagging Explorer to analyze the usage of tags across your resources.

You can find the complete information, such as detailed features and pricing, for this tool on the official home page, which can be found at the following link: `https://www.apptio.com/products/cloudability/`.

Other interesting tools

We wanted to dedicate a small section in this chapter to highlight some other offerings that bring something to the table.

There are a lot of FinOps professionals working on tools to make cost management and optimization easier for us practitioners, and they deserve a shout-out.

It is impossible to cover all the tools available in such a growing ecosystem as FinOps, so please consider this a small subset of tools that we have experience with or heard of and wanted to highlight.

CloudPouch

CloudPouch, developed by AWS Hero Paweł Zubkiewicz (`https://aws.amazon.com/developer/community/heroes/pawel-zubkiewicz/`), is an independent application designed to offer a streamlined, intuitive, and user-focused solution for AWS cost optimization. Crafted as a superior version of Cost Explorer, it helps users understand their costs while empowering them with great analysis tools, giving them specific cost optimization recommendations.

The key features of CloudPouch include the following:

- **Cost breakdown**: It provides visibility into AWS costs, breaking them down per service and usage type. Interactive dashboards allow for a more in-depth understanding of expenditures and make finding irregularities easier.
- **Cost insights**: Automated checks, within seconds, can find idle, misconfigured, or needless (waste) resources that generate unnecessary costs. Results are presented with explanations and actionable recommendations, empowering users to save money.
- **A desktop application**: Unlike its competition, CloudPouch is a desktop application, not a SaaS one. Billing data is directly downloaded from AWS to the user's computer, ensuring that no data is shared with third parties. This also means that it can work with any AWS account without any onboarding or setup needed.
- **AWS profiles**: The application supports AWS CLI profiles and SSO, making it a versatile solution for managing multiple accounts in enterprise environments.

- **A unique learning feature**: CloudPouch explains the meaning of cost components (usage types), thus saving users' time on external research.
- **Account switching**: Switching between AWS accounts has been simplified to a single click, a significant improvement over traditional logout/login methods.

Notably, CloudPouch offers a fully functional seven-day trial, with no credit card required. This allows you to test it before any commitment, which is always good with these products. We think that CloudPouch stands out as a straightforward, practical approach to AWS cost optimization, focused on saving users' time and, more importantly, their money in a simple but effective way.

You can find the complete information, such as detailed features and pricing, for this tool on the official homepage, which can be found at the following link: `https://cloudpouch.dev`.

Serverless360

Serverless360 is a product intended to be used as a support tool for cloud operations in Microsoft Azure, as it offers a lot of additional features, such as monitoring and automation, in addition to its cost management and optimization features.

The idea of this product is to solve the challenges that operating from the Azure portal presents, by offering a space where you can group your resources into applications and monitor their health and performance. This feature allows you to monitor applications that are composed of multiple resources as a complete service, which is really useful for technical teams in charge of operations. For these "composite" applications (as they are called in this product), you can also perform audit tasks and set up permissions at the composite application level.

Apart from these features, Serverless360 offers complete cost dashboards, where you can analyze costs using any cost driver or category to filter and select information, and it also detects workloads that are not optimized, providing you with recommendations on how much you can save by performing different optimization tasks.

You can find the complete information, such as detailed features and pricing, for this tool on the official home page, which can be found at the following link: `https://www.serverless360.com/`.

Kubecost

Kubecost is a tool intended for Kubernetes clusters and mainly focuses on cost allocation and optimization for Kubernetes.

It has two different versions:

- **OpenCost**: This is an open source tool maintained by the community. It provides cost allocation features for Azure, GCP, AWS, and on-premises Kubernetes clusters.

- **Kubecost commercial product**: On top of the cost allocation features of OpenCost, Kubecost offers optimization insights, budgeting, and cost alerts to detect budget overruns, cost anomalies, and underused resources.

Kubecost also offers interesting integrations with monitoring software such as Prometheus and Grafana.

The best thing about this tool is that you can test and use OpenCost without any monetary commitment or investment, installing it in your cluster in a matter of minutes, to help you on a task as complex as cost allocation in Kubernetes clusters, or even in cloud provider-managed Kubernetes clusters, such as **Azure Kubernetes Service (AKS)**, **AWS Elastic Kubernetes Services (EKS)**, or **Google Kubernetes Engine (GKE)**.

You can find the complete information, such as detailed features and pricing, for this tool on the official home page, which can be found at the following link: `https://www.kubecost.com/features`.

Infracost

Infracost is a software that offers cost estimation for Terraform.

By using this product as part of your pull request process, you can estimate how much impact your deployment is going to have on the overall costs of a solution, using a detailed breakdown that includes all costs per resource.

The tool connects to the different APIs from cloud providers (they also offer an API hosted by the Infracost team) and calculates the price of resources declared in Terraform code, integrating with the different phases of Terraform to incorporate cost changes into the plan and apply Terraform stages. This allows for improved cost visibility and helps technical teams become accountable for the impact that their changes make on overall costs.

As with other similar tools, Infracost offers a free, open source version that you can try and test in a matter of minutes.

You can find the complete information, such as detailed features and pricing, for this tool on the official home page, which can be found at the following link: `https://www.infracost.io/products`.

Summary

In this chapter, we have continued exploring the concept of FinOps, by understanding how it can adapt to different organizational situations and the methodologies and frameworks that are most used in the current industry.

We also reviewed all the base tools for cost management and optimization that are offered out of the box in each public cloud, as well as professional market tools that we can purchase to accompany our journey.

In the next chapter, we will begin to explore the Inform pillar by covering two key topics in cost allocation – naming conventions and tagging strategies.

Part 2: Inform – How to Increase Cost Visibility

This part describes in depth the **Inform** pillar, which is based on improving the visibility of cloud costs. We will cover naming conventions and tagging strategies, as well as how to estimate cloud costs and how to design and implement FinOps dashboards and reports.

This part has the following chapters:

- *Chapter 3, Designing and Executing the Tagging and Naming Convention Strategies*
- *Chapter 4, Estimating of Cloud Solutions Costs and Initiative Savings*
- *Chapter 5, Improving Cost Visibility with Dashboards and Reports*

3

Designing and Executing the Tagging and Naming Convention Strategies

Naming conventions and tagging strategies are two essential governance initiatives that every company should implement as one of the first steps in a cloud journey. They are also a fundamental piece of the *Inform* pillar, which will pave the way to work on other cost optimization initiatives.

In this chapter, we will explain why they are so important and all the benefits that implementing them provides, as well as the key differences between them.

Leaving theory aside, we will also propose how to get both naming convention and tagging started by providing some guidelines and key naming convention rules as well as a list of possible tags to be used, for you to be able to design a strong naming convention and tagging strategy.

To close the circle, we will also analyze how these initiatives can help in other domains, such as automation, cloud operations, and compliance or security policies.

Additionally, we will also review a key process for FinOps in a lot of organizations: cost allocation.

In this chapter, we will cover the following topics:

- The importance of naming conventions and tagging in FinOps
- Naming conventions for cloud resources
- Building a strong tagging strategy
- Tagging strategies
- Cost allocation

The importance of naming conventions and tagging in FinOps

Naming conventions are nothing new, as organizations have had their own rules for naming and organizing IT assets for a long time. But transitioning to the cloud also means adapting naming conventions to this new reality, as old rules won't apply anymore.

On the other hand, tagging is more like a new kid on the block. With cloud adoption rising worldwide, tagging is a great governance initiative to ease resource management, which is essential for cost allocation and cost analysis.

In this section, we will understand how naming conventions and tagging strategies are defined and the benefits that having them in place provides.

Why are naming conventions significant?

Having a proper **naming convention** in place can make an enormous difference. By having proper naming rules and enforcing them, we can identify essential information from a resource just by reading its name, such as the project this resource belongs to, the business unit, or any other important factor for our organization.

A naming convention is also key to ensure consistency and standardization across our cloud resources, making it so much easier to inventory and manage each resource's life cycle.

Let's use a figure to illustrate what a naming convention looks like:

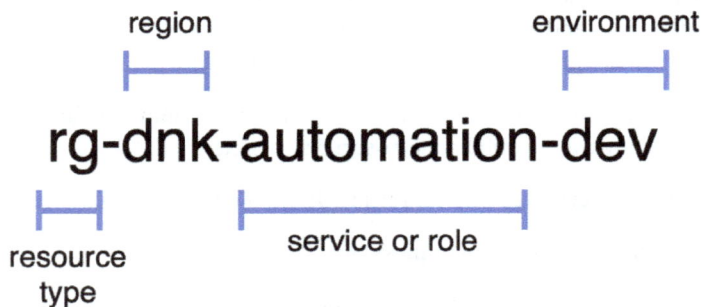

Figure 3.1 – An example of a naming convention

The main benefits of having a proper naming convention in place are as follows:

- Readable description of resources that includes all key information

- Consistency and standardization

- Easier tracking and filtering of resources across regions, business units, and even public cloud providers for easier governance

Having a naming convention in place means that no one will spend time thinking about which name to use for a specific resource again. This simplifies naming and eases governance processes in the future. In contrast, if we were to have no rules in place, it would mean that, depending on the person creating the resource, we could have different names because of different interpretations and opinions, which is not ideal for standardization.

> **Important note – AWS naming conventions**
>
> In **Amazon Web Services** (**AWS**), the process of naming resources is slightly different from Azure and Google.
>
> In Azure and **Google Cloud Platform** (**GCP**), part of creating a resource is to give it a name upon creation, and this is a requisite for creating almost any resource.
>
> In AWS, it does not work this way, as the name itself is optional and part of each resource's tags, which are also part of each cloud resource metadata. This means that, in AWS, naming conventions are considered a part of the tagging strategy, while in Azure and GCP, they are different governance processes entirely.
>
> This changes the approach in a big way. Why should we put all these fields in a complicated name tag following a naming convention, when we can just have as many tags as we need for these fields? Because of this, there is not much material and articles on the internet on AWS naming conventions as opposed to Azure and GCP.

As a reference, here are some key resources related to naming conventions from each of the cloud providers:

- **Azure**: `https://learn.microsoft.com/en-us/azure/cloud-adoption-framework/ready/azure-best-practices/naming-and-tagging` (part of the Well-Architected documentation)

- **AWS**: `https://docs.aws.amazon.com/whitepapers/latest/tagging-best-practices/tagging-best-practices.html`

- **GCP**: `https://cloud.google.com/compute/docs/naming-resources`

Now that we have an idea about naming conventions and how we can benefit from them for our cloud operations, let's move on to our next section, in which we will cover tagging strategies.

Why are tagging strategies significant?

Tagging is about adding metadata to resources, to add all relevant information to an object that cannot be part of the resource's name.

Tagging is widely used not only in cloud resources; it is also used for a lot of other use cases in day-to-day life, such as the following:

- **Photographs**: The geolocation where a photo was taken
- **Web page ads**: We are shown specific ads aligned with our interests when we visit any web page
- **Videos**: Tags are used to describe the topic and related keywords to match the video content to others, to simplify its search on platforms such as YouTube, Netflix, and so on
- **Music**: Genres are part of Spotify songs and album metadata

From these use cases, we see that tagging is nothing new, but the key point we want to focus on is to decide what information we need and how to present it.

Coming back to the cloud, tagging strategies in cloud operations are used as a governance initiative to ease management and complete missing information that is simply not present in resources without tags. Usually, we implement a tagging strategy that, once decided and set in stone, we should follow and even enforce across all our cloud resources.

Essentially, tags are key-value pairs that we add to an object.

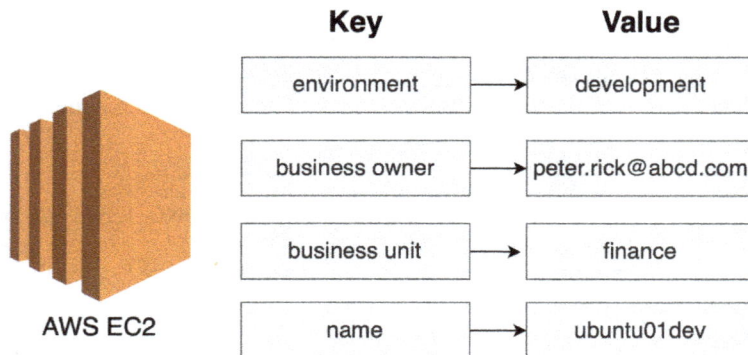

Figure 3.2 – An example of tagging

The main benefits of tagging are as follows:

- Tagging allows us to filter, group, and classify our cloud resources based on relevant terms, such as project, business unit, application, or environment, among others. It enables us to catalog and inventory our resources in different categories for easier management.

- Cost visibility and allocation. Having tags enables the possibility to analyze costs based on said terms that are important, so we can understand how each environment, project, or business unit contributes to our overall costs. By having tagging in place, we can uncover hidden truths.

- Having additional information as part of our resources metadata also simplifies operations, as key information for our resources for cloud operations can be included in the tags, such as the maintenance window, business or technical owner, and business criticality. We can also assign a special tag to critical business resources for our operations team to be aware of mission-critical workloads.

- Tagging can be used as a utility to mark specific resources that are subject to special company or security policies, such as resources that need to be compliant with ISO27001 standards, as well as who is accountable from an auditing perspective.

- Additionally, and not least important, tagging can be used for automation purposes. We can use tags to mark virtual machines that are to be shut down using a runbook that is run every night, or to set up a backup for specific virtual machines, for example.

- As organizations grow, information and knowledge about resources is lost over time and its management can get complex and messy. Tagging helps in getting back all that knowledge and information and getting a grip on your cloud resources.

To get all these benefits, we consider tagging a key governance initiative that should be in place in every organization should they want to increase maturity in the cloud and improve governance.

In further sections, we will analyze how we can define a tagging strategy starting from ground zero but, for now, let's jump to compare tagging and naming conventions.

Naming conventions versus tagging

It may be difficult at first to see the differences between tagging and naming convention strategies. They seem to work with the same purpose in mind, which is to standardize and provide additional information as metadata for our cloud resources.

However, the level required to implement each one varies a great deal.

For us, naming conventions should go first mainly for one reason, which is that naming a resource is not optional (except in AWS). We need to provide a name every time we create a resource, and if we use the wrong name, we most probably won't be able to correct it later on, as most cloud resources don't support being renamed. Also, creating a new correctly named resource and migrating data from the old one might have an impact on IT operations and should be planned, to change correspondent dependencies, connection strings, and so on.

On the other hand, tagging can be corrected and changed anytime during the life cycle of a resource. We can review tags periodically and, using automation or policies or any other method, detect and correct incorrect values in either keys or values of tags without any impact on IT operations. This means that we can evolve our tagging strategy over time, whereas the naming convention will remain mostly static.

In our view, a naming convention only tells a part of the story and should include the utmost essential information, while tagging can have both mandatory and non-mandatory tags to elaborate that information and extend it to other things that cannot be included in the name.

Also, there is some information that is not usually included in tags, such as the resource type, as it may seem like redundant information, but it is usually specified in the naming convention.

As a rule of thumb, we recommend putting the work into designing a proper naming convention first and completing this initiative before venturing into tagging strategies, which can be tackled in a later phase and improved iteratively. Applying a naming convention in a brownfield environment can be especially challenging as we will have a lot of resources in place with other (or no) naming conventions, but the sooner you start with it, the better.

In addition to these facts, a good naming convention can be the starting point for a tagging strategy, which adds to the idea that it may be a good choice to design a naming convention first. Although we don't recommend this, you could even begin by applying a naming convention in the management resources that act as containers for other resources (AWS accounts, Azure resource groups, or GCP folders) to start small and begin isolating for cost allocation, and then progress from there in your naming convention application by also enforcing it on all resources contained in these management resources, which will allow for much deeper granularity and possibilities.

Naming convention and tagging enforcement

Designing a naming convention and a tagging strategy for your resources can be a detailed task, but it is only a part of the story. For proper governance, once both strategies are defined, we should ensure that both naming conventions and tagging strategies are enforced across the organization.

With naming conventions, once we establish a naming convention, it should be used for all new resources, period. We cannot afford to have some resources that are compliant and some that are not. Resources with bad naming will make everything more complex, as we won't be able to fully trust our naming convention to be used for filtering and grouping resources.

As we discussed in the previous section, what we can be flexible on is the scope on which our naming policies apply. We can begin by enforcing naming conventions on specific resources, such as management resources or virtual machines, and then work from there.

However, if we enforce our full tagging policy at first, we may be blocking operations if we don't prepare the grounds enough (imagine technical teams denied from deploying resources because the resource is not compliant with the naming convention) or provide proper training to technical teams to understand tagging and how it can be used.

Instead of falling into analysis paralysis by trying to do everything at once, we recommend an iterative approach to naming convention and tagging policy enforcement. We can begin small – for example, using the simple tags that the cloud providers recommend and audit their use without enforcement, and over time, adapt them more to our needs while enforcing their use by using tagging policies.

A softer way of enforcing tagging can be auditing tags across our resources and setting up alerts when resources are not compliant or are missing essential tags. We can, for example, send a notification to the owner of the project that a resource belongs to whenever any resource that is part of that project is not fully compliant with our tagging strategy.

There are some guidelines to follow, in our view, regarding tagging enforcement:

- Go step by step. Begin enforcing some essential tags, such as `environment` or `project`, and work from there. As cloud maturity evolves and automation via **Infrastructure as Code (IaC)** and CI/CD progresses, having proper tagging may be simpler, in the long run, to fully enforce the set of tags that we consider mandatory.

- With this idea to begin small, we recommend enforcing tag policies, which means not allowing resources to be created if not compliant with tagging design, once tagging practices are mature, and not from the beginning.

- Use tags that use fixed values or a specific format for their values, so we know what to expect from the tag values when a specific tag key is present. A possible example can be to use a three-letter code for the environment, such as `{dev/pre/pro}`, and an email format for a business owner, such as `{name@company.com}`. It will be easier to track tagging compliance with these types of tags.

Naming conventions for cloud resources

In this section, we will explain the process of designing a naming convention step by step. Before doing so, we will introduce the basics, such as the format that we need to use in each cloud provider, as well as some general recommendations.

Please keep in mind the following guidelines before creating a naming convention:

- It should make resources unique through its name.

- As a general recommendation that we like to apply, information importance should go from left to right, with the right being the least important information (for example, we can have the numeral of a virtual machine at the right-hand side of a naming convention and the business unit at the left-hand side of the name). This is just a general rule to follow to be consistent with the information shown in the name.

One of the first points to keep in mind to design a proper naming convention is how long our names can be. For this, each cloud provider has its own rules and specificities. We will review these in the next section.

Style and format

To begin the exercise of creating a naming convention, we need to first understand the resource naming considerations in each cloud provider, from how many characters we are allowed to use to the uniqueness of the names across our cloud environments.

Azure

General considerations for naming conventions in Azure are as follows:

- Most of the resources have a maximum length of around 50–63 characters and resources are case-insensitive unless specified in the resource's specifications.

- Some resources, such as storage accounts, don't support hyphens.

- Some resources that enable a new URL by their name after creation (such as app services or function apps) have special restrictions.

- Some resources' names must be unique globally in Azure. Due to this, we may have trouble creating some resources with names such as `test1` or `virtualmachine`.

AWS

In AWS, naming a resource is achieved by using name tags, as we already explained. Due to this fact, the style and format for resource naming are the same as tagging considerations, which are mainly the following:

- Resources can have only one name tag with the same key name, as each tag key must be unique.

- The maximum key length is 128 characters, and their value is limited to 256 characters. Spaces are allowed and some special characters can be used.

- Tags keys and values are case-sensitive. Keep this in mind when using the AWS CLI or REST APIs.

- The `aws` prefix is reserved for automatic use. If you create a `User` tag using the `aws` prefix, you won't be able to change its value in the future.

- Some key operations on resources, such as stopping or terminating an EC2 virtual machine, cannot be performed using only tags as identifiers. For these key operations, you will need to specify the unique resource identifier.

GCP

In GCP, the name format should follow the following guidelines:

- Resource names should have a maximum length of 63 characters.

- The name should match the **regular expression (regex)** `^[a-z]([-a-z0-9]*[a-z0-9])?`, which means the following:

  - It should begin with a lowercase letter followed by any letters, hyphens, or digits

  - Resource names cannot end in a hyphen

- Google requires resource names to be compliant with RFC 1035 (ietf.org/rfc/rfc1035.txt), which is a **Domain Name System (DNS)** standard that was published in 1987 by the **Internet Engineering Task Force (IETF)** to be used for computers or any resource connected to a private network or the internet

- The names of resources must be unique across locations and projects

Now we know the naming rules that apply to different cloud providers, let's move on to a key component in any naming convention: separators.

Separators

Before going through the different key fields, let's try to answer the question of which separators we should use.

In our naming convention, we will have some fixed fields that will have a fixed length and format, so we know what to expect in that field. However, for some fields, it is going to be impossible for us to use a fixed length, and this is when separators come in handy.

Having separators allows us to get the information of our resources programmatically (for example, using a `split` operator with our separator as an argument) to filter them out without having to use fixed-length fields.

As an example, let's say we have the following naming convention in Azure:

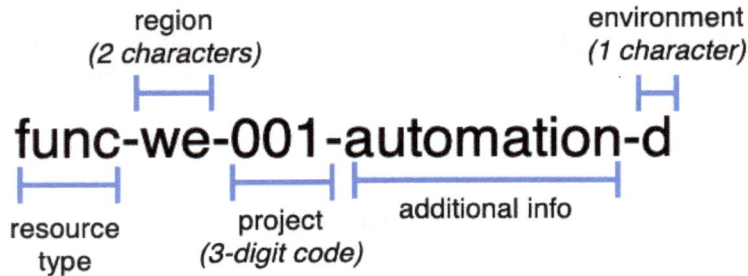

Figure 3.3 – An example of separators

The name shown in the figure stands for the following:

Function app in West Europe, part of Project 001 in the development environment

We could split the name in PowerShell programmatically and get the environment letter and project by using the following line of code and use it as part of any automation we want:

```
$environment=($resourceName -split "-")[-1]
$project=($resourceName -split "-")[2]
```

Our recommendation is to use a hyphen as a standard separator, as some cloud resources don't allow us to use underscores or other special characters that often. We also don't recommend using multiple separators as we want our naming convention to be as simple as possible.

Key fields to include

One of the things we need to address about a naming convention is that, as opposed to tagging, we cannot include all the information we want in the resources' names, as we always have a limit of maximum characters that our resources' names can have.

Creating a naming convention is about choosing the right fields to include and setting up how many characters each field can take.

In general, our recommendation is to limit each field length to a minimum when using fixed-length fields, so you will have more room for non-fixed-length fields.

In this section, we will also propose some key fields to include in your naming conventions.

Cloud provider

This field may seem redundant for organizations that only use one cloud, but it is important for multi-cloud organizations, or for companies that will adopt multi-cloud practices in the future.

This field will allow us to filter in the simplest manner and count how many resources we have in each cloud provider from a shared dashboard, for example.

This field is the perfect candidate for a fixed-length field, as the values here are really limited. If we limit its length to one character, we won't be able to tell Azure from AWS, as they both begin with the letter *A*, so our recommendation is to use two or three characters for this field, as in these examples:

- `az` or `azr` for Azure resources
- `aw` or `aws` for AWS resources
- `gc` or `gcp` for Google resources

Resource type

The resource type, in our view, is a key field to include in the naming convention. Having this field allows us to identify the type of each resource in a list of resources, which can be useful when checking out a long list of resources to get a high-level understanding of what is there.

Regarding the resource type, we could try to set a character limit (for example, three or four characters) to limit the length of a `Resource Type` field and then create a list of names per resource.

However, keeping this list up to date can be a tough job, as it will require an enormous administrative overhead. There are new cloud resource types being published every day while others are decommissioned.

Due to this, we don't recommend using a fixed length for cloud resource types or creating your own list of resource types. Instead, we recommend using the resource name abbreviations that each cloud provider recommends, if available. Using these lists, we can use a standardized field that is maintained by the cloud providers themselves, which will avoid any administrative overhead. The only downside of this is that we will need to accept a non-fixed-length field for our resource types. You can check out the following references on name abbreviations for more information:

- **Azure**:
  https://learn.microsoft.com/en-us/azure/cloud-adoption-framework/ready/azure-best-practices/resource-abbreviations
- **GCP**:
  - In *Naming Conventions*, section *3.4*, there are some recommendations:
    https://services.google.com/fh/files/misc/google-cloud-security-foundations-guide.pdf

- A good reference on the list of resources can be found in the Google Developer cheat sheet: `https://googlecloudcheatsheet.withgoogle.com/`

Part of the resource type is to create specific naming rules for resources that don't support standardized naming rules. For example, in Azure, we won't be able to use the same name format for resource groups and storage accounts, as storage accounts don't support hyphens and have specific name formats that are much more rigid than other resources.

Business unit/project

This field is especially important for allocating resources to projects or business units.

The format of this field depends on how business units and projects are named. Based on the naming, try to abbreviate business units and projects to reduce the characters used for this field to a minimum.

We recommend a fixed length for this field. As part of our name generator (we will speak about this later in this chapter), we can also publish our projects and business units' abbreviations and their corresponding project and business unit names.

Environment

The environment is another key piece of information to include in our resources' names. Different organizations and even business units may name their environments differently, so it is important to cover all the possibilities.

Having this information as part of the name is important, as different environments' resources are treated differently. For example, a virtual machine in *dev* can be subject to automatic shutdown during off-hours, while a production machine is not. You don't want to confuse one for the other when selecting which one to power off programmatically.

Following our idea of using as few characters as possible, we could use a single-letter approach:

- `d` for development
- `t` for testing environment
- `u` for user acceptance testing
- `p` for production
- `s` for staging/preproduction

We can also use a three-letter approach if we want the information to be easily readable:

- `dev` for development
- `tst` for testing environment
- `uat` for user acceptance testing

- `pro` for production
- `stg` for staging
- `pre` for preproduction

In cloud operations, we almost always provide some shared resources that are used from resources across different projects and environments. We consider that these shared resources don't have a specific environment assigned to them, so we can use a special environment to designate those, such as `hub` or `ccc`:

- `h` for hub/shared services with a single-letter environment tag
- `hub`/`ccc` for hub/shared services with a three-letter environment tag

Location/region

The location of resources is another key piece of information we should specify, especially to keep resource names unique. We could have different cloud resources that are used with the same purpose and name in different regions.

Additionally, the location of our resources is also important for resource grouping, for high availability purposes, and even for cost considerations (some regions are cheaper than others).

The `Location` is an easy field to include in our naming convention, as the list of locations available by each cloud provider is well-known information, subject to be included in a fixed-length field.

Additional information

The `Additional information` field is one of the most important fields to include in a naming convention, so we can have some wiggle room to add additional information that the rest of the fields cannot provide.

This field is ideal for explaining the role of the resource in a project, which identifies multiple resources of the same type in the same project and location as each other.

Some considerations about this field are as follows:

- If there are multiple resources with the exact same role, we can also add digits at the end to differentiate resources from each other
- We should avoid putting redundant information, such as the project, in this field if it's already present in the rest of the name
- It is difficult to fix the length for this field, as it will limit the possibilities of this field, but it can be done if it's needed

Parent and child resources

Some resources can be considered child resources from a parent resource, such as disks or network adapters and virtual machines.

For these resources, we should use a specific naming convention rule so we can refer to the parent resource name, which should already contain all the information we need.

Having a specific naming convention for these resources helps us to tie them together. These resources should have the same life cycle and we should be able to obtain this fact from how they are named as well.

For these child resources, we often follow a specific naming convention, such as the following:

```
{resource type}-{parent resource name}
```

Here are a few examples of names following this convention:

- `vm-ccc-it-adfs1-pro` (virtual machine parent resource)
- `nic-vm-ccc-it-adfs1-pro` (virtual machine network adapter child resource)
- `nsg-vm-ccc-it-adfs1-pro` (network security group child resource)

With these essential fields covered, we now have a grasp of what information to include in a naming convention. In the next section, we will learn how name generators can help us apply and implement a naming convention.

Creating a name generator

It is difficult enough to design a naming convention by itself. But after the job is done, we have another challenge to face, and that is to implement what we just designed.

We could prepare a 20-page document with all the explanations and all the details and rationale behind our naming convention. However, having this document does not guarantee that everyone is going to read it and that no mistakes are going to be made interpreting our naming convention.

To address this issue, we always recommend publishing an internal name generator that implements the naming convention of an organization.

This naming generator can be prepared in Excel, or it can be an in-house-developed simple web application with an input form on which users provide the information for each field of the resources, with pre-fixed value fields and others that are on free text. It can even be a simple script that implements the logic behind our naming convention.

As an example or starting point to picture how these tools can help in this process, Microsoft provides a web application called **AzNamingTool**, under their Well-Architected framework, that can exemplify how these naming generators work. You can find it here: `https://github.com/microsoft/CloudAdoptionFramework/tree/master/ready/AzNamingTool`. AzNamingTool is a simple .NET application with some customization options. It also provides a RESTful API that can be used to create names from other resources such as Terraform.

Under the same concepts, there is another interesting tool that can be used for this purpose that follows the Microsoft Cloud Adoption Framework, which is the **azurecaf** Terraform provider (`https://github.com/aztfmod/terraform-provider-azurecaf`). By using this provider as part of your IaC process, you will be able to generate names on the fly for the resources you are going to deploy programmatically from Terraform.

It essentially acts as a black box; you provide as input the information of the resource, such as the resource type, and it will output a resource name compliant with your naming convention strategy. Make sure to check it out, as it may be useful for your naming journey.

As an additional reference, we would also like to recommend the work of Tao Yang (`https://www.linkedin.com/in/tao-yang-988325a/`), who has developed some PowerShell modules that can be used as a name generator as well and can be integrated with CI/CD IaC pipelines. You can check them out by visiting the following links:

- `https://blog.tyang.org/2022/09/10/programmatically-generate-cloud-resource-names-part-1/`
- `https://blog.tyang.org/2022/09/10/programmatically-generate-cloud-resource-names-part-2/`

Now that we have grasped the process of defining a naming convention in detail, let's move to the next section, where we will go over some ideas to create a tagging strategy.

Building a tagging strategy

Following the same train of thought that we used for our naming convention, before we try to decide upon which tags we should add to our tagging strategy, we need to analyze what possibilities we have with regard to the format and style of tags in each cloud provider.

Style and format

As we did for our naming convention, let's review the different rules that apply to each cloud provider regarding tagging.

Azure

The following rules apply on Azure when tagging resources:

- A resource can have a maximum of 50 tags. If we need more, we can always resort to using JSON blocks as values (compound tags), which we will analyze further in an upcoming section.

- Tag keys are limited to 512 characters while tag values are limited to 256 characters. There are some resources, such as storage accounts, that support fewer characters for tag values.

- Tag keys cannot contain the following special characters: <, >, %, &, \, ?, and /.

- Some resources don't support tags. You can check which ones on this web page: `https://github.com/tfitzmac/resource-capabilities/blob/main/tag-support.csv`.

AWS

In AWS, there are some specific rules we need to keep in mind for our tagging strategies:

- In AWS, there are two types of tags: a set that is user-defined and other tags that are set automatically by AWS. The user cannot modify tags created by AWS and you cannot use the `aws:` prefix as it is already reserved.

- An object can have a maximum of 50 user-created tags.

- Tag values are limited to 128 characters while tag values are limited to 256 characters.

- There are some disallowed characters in tags that vary from one resource to another. You can check this in the **Tags** section in each AWS service's documentation. In general, in tags, you can use letters, numbers, spaces, and the _, ., :, /, =, +, -, and @ characters.

- In AWS, tags are case-sensitive.

- As a safety measure, you cannot delete or terminate a resource based solely on its tags. Such operations will ask for the identifier to make sure you are targeting the right resource.

GCP

Last but not least, these are the rules that apply in GCP regarding resource tagging:

- In GCP, tags and values are limited to 63 characters.

- Allowed characters are UTF8-encoded characters except ', ", \, and /.

- You cannot add a tag unless it has both a tag key and value.

- Tag inheritance is activated by default in GCP. This means that child resources will get the tags of parent resources. We will analyze this later in this chapter.

One key point regarding tags in general in all three cloud providers is that their information is fully visible and not encrypted. Due to this, tags should not contain any kind of **Personal Identifiable Information (PII)** to avoid GDPR and compliance conflicts.

Simple and compound tags

With everything that we have covered so far, it should already be clear how we can tag resources and the possible keys and values that we can have as part of our strategy.

There is one key feature that is currently only offered in Azure resources, which is **compound tags**. Compound tags consist of using a JSON block as a value for a tag, instead of a single string of text and numbers. By using this JSON block, we can nest additional information under one tag key, which we can access programmatically in any way we want.

For example, we can define the following JSON in a text file:

```
{
    "companyname"    :    "imagineinc",
    "businessunit"   :    "finance",
    "city"           :    "madrid",
    "region"         :    "spain"
}
```

Then, we can use it as a value for a tag key named `companyinfo`:

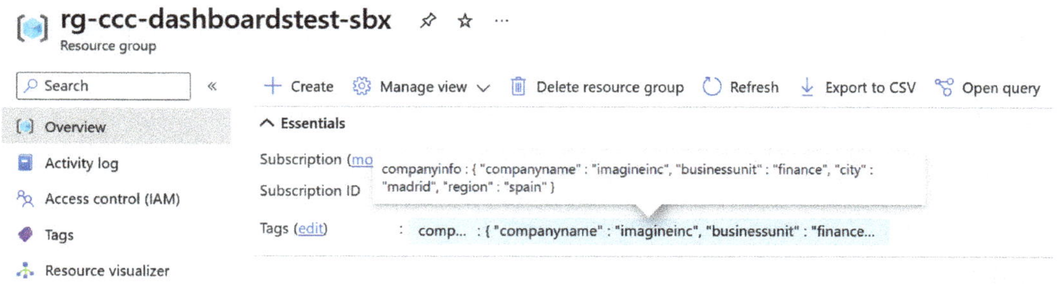

Figure 3.4 – Azure compound tag example

Using compound tags and values in JSON can help to overcome the limit of 50 tags per resource. They can also be used as the foundation to have more elaborate tags if we want to reduce our tag keys to the minimum while providing the most information possible.

Remember that these tags are also included in our billing data, so they can be a good option to enable cost allocation, a process that we will describe in detail at the end of this chapter. **Tags** is a column that is included in our billing data in Azure, so we can use these special tag values to include much more information than we could by using a simple string.

All in all, using this feature will make your tags more elaborate and complex and it should only be used for programmatic scenarios, as information won't be as readable in JSON compound tags, but it is a nice feature, in our view, that can suit some organizations.

Creating a tagging strategy

Creating a tagging strategy is about defining which tags should be used in cloud resources and which values can they take. It may seem like a simple task, but it is not, as it has so many implications.

When we think of the purpose of tags, which is to add additional information as metadata to our cloud resources, we may think that more is always more. We want as much information as possible.

However, this approach is not ideal, as having a lot of information and tags will make our tags really difficult to manage and maintain over time.

This is why, in our view, part of creating a tagging strategy is to find the sweet spot between providing as much information as possible and selecting a simple set of tags to be used that are easily readable, understandable, and maintainable for our technical and operations teams.

Another principle to consider when creating a tagging strategy is that we should not provide redundant information on resources, which is information that we can get directly from a resource configuration. An example of redundancy can be indicating the region of a resource by using tags.

In this section, we will propose some useful tags that can be used to provide valuable information, in our view.

Before going further, we want to propose a format to define our tags, the information they should contain, and in which format.

We can define our tags in the following manner (using an Azure tag as an example):

Characteristics	Value
Tag name	`environment`
Tag type	Technical
Defined by	Infrastructure
Mandatory?	Yes
Scope	All resource groups
Uniqueness	Shared across resources in a resource group
Application	Manual
Value format	Three letters
Possible values	`pro/pre/dev/sbx`

Inheritance	Yes
Description	This tag is used to specify the environment a resource belongs to, which can be useful for many use cases such as cost allocation, automation, and even company policies

Table 3.1 – Tag specification

Using these specification tables for our tags, we define not only the format and the value but also how our tags are applied and to which resources they apply. Let's analyze the different fields we propose for this specification sheet:

- **Tag name**: The tag key to be used. Should be case-sensitive.

- **Tag type**: We can classify our tags in categories based on their purpose. We can recommend some categories, such as the following:

  - Hierarchy, business, and cost allocation tags

  - Technical tags

  - Automation tags

  - Resource classification tags

  - Security tags

- **Defined by**: Who defined this tag and included it in the tagging strategy? This one is optional and can be useful especially when there are multiple teams contributing to the tagging strategy.

- **Mandatory?**: This field indicates whether our tag is going to be enforced by a policy or not.

- **Scope**: This field indicates to which resources this tag should be applied.

- **Uniqueness**: This field is more abstract, and it describes how unique a tag is across multiple resources. It can help understand how tags are going to be used and, depending on the tag, which differences between resources will be highlighted by each tag.

- **Application**: This can be **Manual** or **Automated**. We will review automated tagging in one of the following sections of this chapter.

- **Value format**: We can specify the format that will be required for our tag values. In this field, if we feel comfortable with it, we could even have a regex to clarify the format of each tag, which will be useful for enforcing the tag.

- **Possible values**: For tags that use fixed values, we can specify the values that should be expected here.

- **Inheritance**: This field indicates whether or not this tag is inherited by child resources or resources contained inside a management resource.

- **Description**: A description of the tag and the rationale behind selecting this tag.

Using specification sheets as a reference, we can select a set of tags that can be useful for our organization and complete all required information to prepare our own tagging strategy aligned to our needs.

In the following section, we will propose some tags to be used, from mandatory tags that every resource may have to non-mandatory or optional tags that can provide useful information in some cases.

We won't go through fulfilling a complete tagging strategy, but at least we want to provide both the guidelines and some possible tag ideas for your future tagging strategies.

Please feel free to choose the best name or abbreviation for your tag keys in your view. What we want to highlight in the following section is the information that each of our proposed tags can provide.

Mandatory tags

These sets of tags are important enough for us to consider them mandatory. This means that we should enforce their application throughout all our cloud resources and environments.

These are the tags proposed:

- **Cost Center**: The cost center or department where the costs should be allocated for a specific resource. This tag value can be either a number, an identifier, or the name in a fixed values list we can prepare beforehand.

- **Project Name**: The project that the resource belongs to. The project can use either a unique code or plain text for the name. This tag is essential as well for cost allocation.

- **Business Unit**: A cost center is not the same as a business unit in some organizations, so we may need to know both of them. It is also important to know which department has ownership of resources.

- **Business Owner**: This field usually refers to the product owner or the senior management person ultimately accountable for a resource. We recommend putting the value in email format, to make it easier to contact such a person in case it's needed.

- **Technical Owner**: This field refers to the technical owner of the resource, which usually is the responsible party as per the RACI matrix. We recommend putting the value in email format, to make it easier to contact such a person in case it's needed.

- **Environment**: The environment that a resource belongs to. The same ideas that we covered in the naming convention section apply here as well. For the tag, we can use the non-abbreviated name of the environment if we want to.

- **Operating System**: It may seem redundant, as we can usually get the underlying operating system in use from a virtual machine. However, in some special cases, this data is not available and should be clarified, which adds to the value of having this tag. Some examples of this may be containers (in which, due to using Dockerfiles or previously built container images, the operating system in use may not be fully transparent) or virtual machines using custom images. This data is really valuable from an architecture perspective, to be able to calculate the operating systems in use across all of our infrastructure.

- **Database Engine**: This field can be redundant in PaaS resources (as the resource type almost always includes this information) but extremely useful in IaaS resources where we install the database from scratch. As with operating systems, this tag is invaluable for architecture and cost allocation as well.

- **Compliance or Security standards**: There are times when organizations need to be compliant with specific security policies or audit policies, such as ISO27001. We can also use tags to mark resources that are subject to special treatment or that need specific processes to be implemented. Another example of this could be resources that, as part of the Business Continuity Plan, need to be subject to annual contingency tests.

- **Asset ID**: This field can be useful when we have Change Management and a **Configuration Management Database** (**CMDB**) in place with all our assets. A CMDB is an independent database on which all IT assets used to provide IT services are inventoried, and it is part of IT service management systems and tools.. Using this tag, we would be able to correlate the CMDB asset with the real cloud resource. If we pair this tag with some automation to keep the CMDB updated, this tag can create a lot of value by improving integration between the cloud and the CMDB.

- **Management Level**: This tag is a utility one that can complement the resources information. The idea of this tag is to indicate the level of management, either IaaS, PaaS, CaaS, FaaS, or SaaS resources, so we can oversee how many resources we have of each type. This information can be key to creating some modernization KPIs. Adding this information by hand can be a lot of work, so our recommendation is to try to automate the application of this tag to avoid administrative overhead. We will review this specific topic in an upcoming section.

- **Business Criticality**: Business critical resources should be treated with utmost care by operational teams, to avoid causing any impact on running workloads and, therefore, on our organization's business. On the other hand, we can use this to mark non-important resources that we can shut down without any expected impact, for example. Our proposal on this tag is that you could create different levels of criticality, from **0** (most critical) to **4** (less critical), for example, to show how critical each resource is at a glance.

With all these ideas, let's move on to other tags that may not be as important as to be considered mandatory, but that can offer really interesting information on some use cases.

Non-mandatory tags

Apart from the tags that, in our view, should be mandatory, we would also like to cover additional tags that can add other valuable information to our resources metadata:

- **Expiration Date**: This field can show the expected expiration date of a resource if we know it beforehand. It can be useful to track the expected life cycle of a resource.

- **Maintenance Window**: For operations teams, we can indicate the maintenance window that can be used in a resource such as a virtual machine. This is especially useful in organizations where there are some pre-established maintenance windows used for all the resources.

- **Resource Role**: There are some cloud services on which the name is not enough to explain the role a service is playing. For these cases, we could use an additional tag to clarify this information. For example, we could have multiple domain controllers, but by using a tag, we can specify which ones are also providing DNS services.

- **Schedule of Use**: Similar to the **Maintenance Window** tag but reversed. In this tag, we can indicate the schedule when a resource is used. For example, in a web application, we could indicate the schedule that the users connecting to the application follow.

With these tags at hand, you should already have a good foundation to build your own tagging strategy. Keeping this in mind, let's move on to the next section, where we will describe how we can leverage automated resource tagging in different ways.

Automated tagging

If tagging was a manual task, it would be almost impossible to keep it up to date across all of our resources, especially in big environments.

Fortunately, it is not, and this is why, in this section, we are going to review the different ways that we have available to fully automate tag applications, from inheritance to other interesting features, such as AWS-generated tags.

Tagging inheritance

It is possible to implement inheritance when tagging resources. Tagging inheritance means that a parent resource can pass on its tag keys and values to child resources.

Each cloud provider has a different take on how tags can be inherited and this feature works in a different way from one cloud provider to another. Let's analyze the different ways in which this feature is implemented in Azure, AWS, and GCP.

Azure

In Azure, we can set up tagging inheritance by using specific Azure policies that transfer tags from the resource group or subscription to all the resources contained inside. This simplifies tag management as tags are often shared inside a resource group (environment or project tags, for example)

We don't even need to create an Azure Policy by ourselves, as this policy is already built in Azure out of the box and we just need to assign it to our subscriptions (more information can be found at `https://learn.microsoft.com/en-us/azure/azure-resource-manager/management/tag-policies`).

Let's illustrate how tagging inheritance works in Azure with a simple diagram:

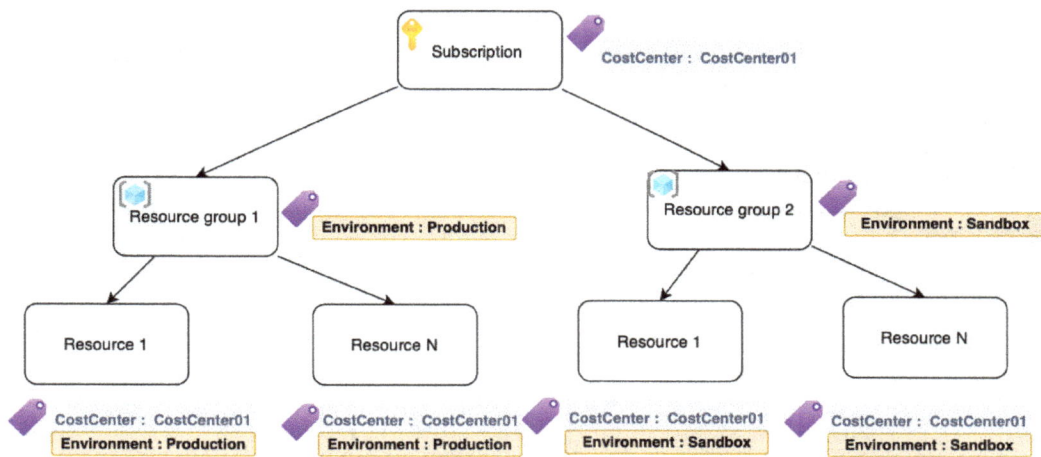

Figure 3.5 – Azure tagging inheritance

These are the policies we can use that are available on Azure:

- *Inherit a tag from the resource group*

- *Inherit a tag from a subscription*

- *Inherit a tag from the resource group if missing*

- *Append a tag and its value from the resource group*

We can also use these policies as the foundation to create our own policies to apply inheritance to specific tags upon certain conditions.

AWS

In AWS, there are two ways to implement tagging inheritance from AWS Organizations to the child resources:

- We can use an **AWS tag policy**, where we define a tagging rule that will be inherited top-down, from AWS Organizations down to underlying AWS **organizational units** (**OUs**) and AWS accounts. Using tag policies, we can combine parent policies coming from the parent resources (parent policy) with the policies that we define at the resource level (child policy), to define policies that will be inherited by child resources.

- By using **Service Control Policies** (**SCPs**), we can establish organizational policies to manage the permissions in the different AWS accounts that we have. Using SCPs, we can require some tags to be present in all resources, as well as block existing tags to be modified after creation to service principals, to ensure consistency. SCP policies also allow for inheritance in the same way as tag policies, so we can apply SCP policies coming from the parent level and combine them with other SCP policies that will be inherited in child resources.

Let's illustrate how tagging inheritance works in AWS with a simple diagram:

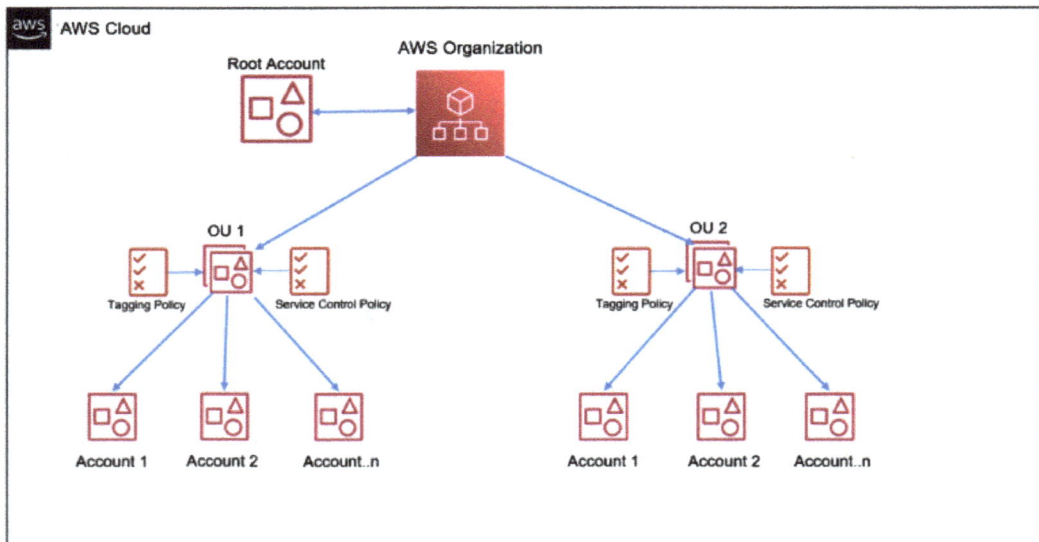

Figure 3.6 – AWS tagging inheritance

To summarize, we can create both tag policies and SCP policies in AWS Organizations, at the OU or account level, to define the tags and their values and enforce their use in a very flexible manner, allowing for complex hierarchies and any tagging strategy that we want.

GCP

In GCP, tags are inherited by default throughout the resource hierarchy (company, folders, projects, and then resources) by child resources.

We can override any inherited tag from a child resource if needed. We need to be careful when planning tagging because of this fact.

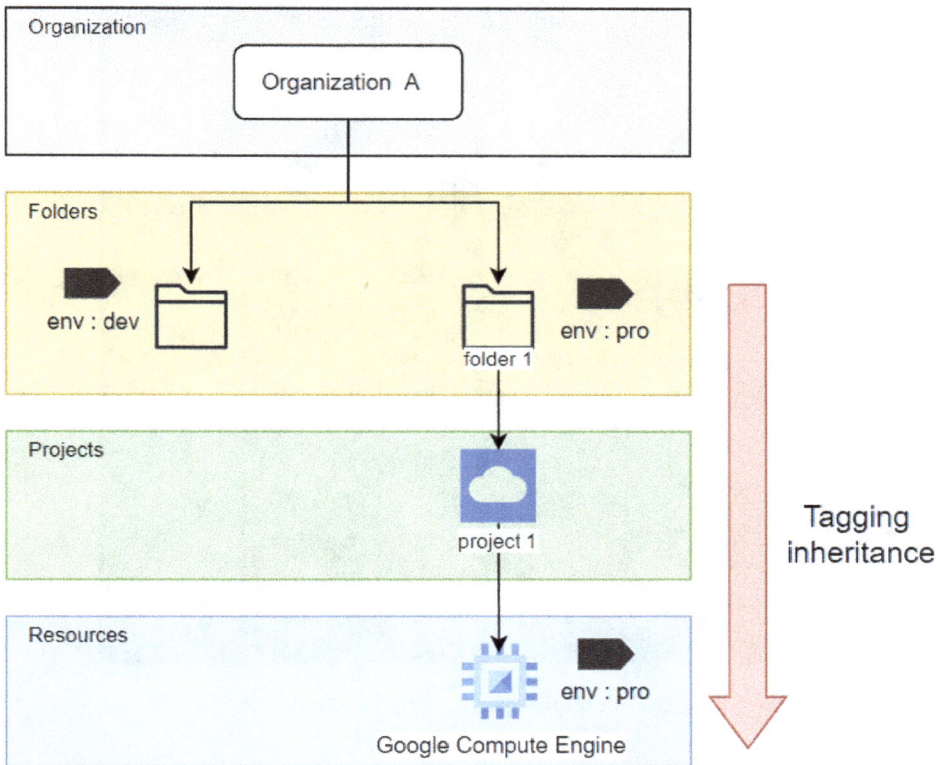

Figure 3.7 – GCP tagging inheritance

Given how tagging inheritance works in GCP, our recommendation is to tag key information at the highest level in the hierarchy possible, which will streamline and simplify the process of tagging a great deal, avoiding repetition and manual application resource by resource.

This is a great approach by Google in our view, and it should be taken as a model by other cloud providers.

Tags and IaC

IaC and tagging go hand in hand. We can include, in our IaC, the specific tags for each resource, which helps with consistency across resources. Here's an example:

```
# Create a resource group
resource "azurerm_resource_group" "myRG" {
  name     = "myRG"
  location = "North Europe"

  tags     = {
    Environment = "pro"
    Application = "superwebapp"
    TechnicalOwner = "alfonso.san.miguel@abc.com"
  }
}

# Create a virtual network within the resource group
resource "azurerm_virtual_network" "myVNET" {
  name                = "example-network"
  resource_group_name = azurerm_resource_group.myRG.name
  location            = azurerm_resource_group.myRG.location
  address_space       = ["10.0.0.0/16"]

  tags     = {
    Environment = "pro"
    Application = "superwebapp"
    TechnicalOwner = "alfonso.san.miguel@abc.com"
  }
}
```

Figure 3.8 – Resource-specific tagging in Terraform

We can also apply here a good coding practice that is based on the DRY concept. **DRY** stands for **Don't Repeat Yourself**, which proposes avoiding repetition as much as possible in our code. For this use case, it means that if we need to specify the tags one by one for all the resources we are going to declare, we are going to end up with a lot of lines of code. We should try to avoid this as it is not the best way to go and will increase the risk of human error.

To avoid repetition, we can declare tags as a JSON block, then store it in a variable or even declare it as a parameter, and just assign it to all the resources instead of repeating the code again and again. Here's an example:

```
locals {
  common_tags = {
    Environment = "pro"
    Application = "superwebapp"
    TechnicalOwner = "alfonso.san.miguel@abc.com"
  }
}

# Create a resource group
resource "azurerm_resource_group" "myRG" {
  name     = "myRG"
  location = "North Europe"

  tags     = local.common_tags
}

# Create a virtual network within the resource group
resource "azurerm_virtual_network" "myVNET" {
  name                = "example-network"
  resource_group_name = azurerm_resource_group.myRG.name
  location            = azurerm_resource_group.myRG.location
  address_space       = ["10.0.0.0/16"]

  tags       = local.common_tags
}
```

Figure 3.9 – Common tagging in IaC

> **Important note – AWS default tags**
>
> In Terraform, AWS has made available a new feature called **default tags** from AWS provider version *3.38.0* onward that lets you declare tags at the provider level, which will be shared with all the resources declared in that Terraform project.
>
> You can get more details from the HashiCorp documentation: `https://developer.hashicorp.com/terraform/tutorials/aws/aws-default-tags`.

By using IaC, we can also help enforce and audit tagging usage. There are two main ways to do this:

- We can use tools such as **HashiCorp Sentinel** (`https://docs.hashicorp.com/sentinel/intro`), **Aqua Security's tfsec** (`https://aquasecurity.github.io/tfsec/v1.1.5/getting-started/configuration/custom-checks/`), or any other tool that is capable of analyzing the code beforehand – some sort of tagging unit test before our infra is deployed that audits the tags present there to ensure that the tags declared in code are compliant with our tagging policy. We could call this a pre-deployment strategy.

- In our IaC code, we can declare the policies to enforce and audit tagging usage across our resources by declaring and assigning the policies using code (Azure Policy policies, GCP organization policies, and AWS Organizations policies using AWS CloudFormation, for example). This would be a post-deployment strategy, even though our policies will block resources that are not compliant if we are enforcing tagging standards.

AWS-generated tags versus user tags

AWS has a slightly different way of applying tags than other cloud providers, as it offers some tags that are generated automatically and others that can be created by users.

There are two types of tags:

- **AWS-created tags**: These tags have the prefix `aws:` before the tag name and they cannot be modified once they are assigned a value
- **User-created tags**: These tags have the prefix `user:` before the tag name and these are fully managed by the user

Let's illustrate how this works by providing two examples, one for each tag type:

- `aws:createdBy`: This tag is automatically created by AWS and it includes the user ID and username of the user who created the resource. This information can be really valuable for housekeeping and to understand when and why a resource was created.
- `user:Environment`: This is a user-defined tag in which we can put the environment this resource belongs to in any way and format we want.

These tags are also relevant to cost optimization, as we can use them as filters in AWS billing services such as AWS Cost Explorer or AWS Cost and Usage Reports.

Having these AWS-generated tags provides really useful information on our resources, and they are offered out of the box with no user interaction, which is a plus in our view.

Be mindful that we need to activate or deactivate these tags depending on our needs so we have them available for cost analysis and reporting.

Some considerations regarding activation and deactivation are as follows:

- We can only have a maximum of 500 tags in a resource
- It takes some time (up to 24 hours) for the tags to appear or disappear after a tag is enabled or disabled

- Tags should be activated from the AWS payer account
- If the AWS payer account, changes we must activate the tags again so information is not lost during the process

With the differences clear between user- and AWS-generated tags, let's move now to a key topic that is essential for cost management in the cloud in general and is especially related to tagging, which is cost allocation.

Cost allocation

Cost allocation is a financial process that identifies, allocates, and assigns costs – in this case, the cloud spend – across regions, departments, or business units in an organization.

Cost allocation is a key process for organizations that transition to the cloud, as it often needs to be built from ground zero. It is also one of the most common challenges that big organizations, or organizations that provide services to others, need to face somewhere in their journey to the cloud.

Before the cloud, costs were usually distributed on the basis of simple variables such as volumes (number of virtual machines), users, or any other constraint we had at hand.

The pay-as-you-go model in the cloud changes everything. Costs are now more visible (and painful) than ever and, as we are only charged for what we use, we should then distribute the costs in a much more precise manner to avoid wrongly assigned costs.

Cost allocation is especially important in big organizations where we usually have one contract with our cloud providers, so we can benefit from volumes and get bigger discounts, which are used to host cloud resources from different business units, departments, and regions within the same company.

As we described during the introduction, cost allocation is often followed by showback and chargeback processes. It should be also crystal clear that for us to perform showback or chargeback, tagging is definitely a requirement:

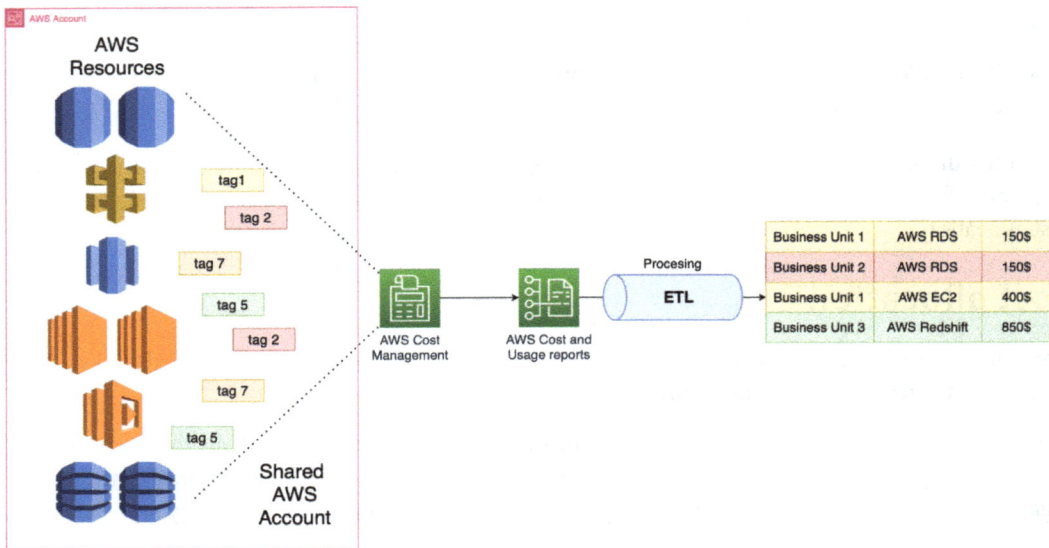

Figure 3.10 – Example of showback/chargeback in a shared AWS account

There are two main types of costs when doing cost allocation exercises:

- Costs from resources that can be assigned fully to a business unit, project, and so on
- Shared costs that we need to distribute between different business units or projects

We can achieve the first point easily by having proper tags or naming conventions, but the second point requires specific processing of cost data. Some examples of the second point can be licensing costs or the cost of shared cloud services that are used by multiple business units (e.g., network services such as firewalls or VPNs).

There are different ways to distribute costs, as follows:

- They can be distributed equally between all business units or projects that make use of a resource. This is the easiest approach but not the best way, as there can be differences in how much each party uses a resource.
- They can be distributed using a factor to divide our resource usage, to assign different weights to different business units or projects. This factor can be anything depending on the resource. The ideal approach would be to have sufficient information to identify how much each department has used a shared resource, but this is complex as it requires much more technical detail and processing. A good starting point, if such information is not available right away, is to use another factor, such as the number of users per business unit using a shared resource.

One way or another, we should process billing data to provide the allocated costs for each business unit or department as output. Part of this exercise is to leave no cost unallocated due to a lack of proper tags or naming conventions.

After the cost processing is done, we must set up a proper review process to ensure that there are no mistakes, especially in the first steps of cost allocation, to create trust in the process and foster technical teams' accountability on cloud costs. A joint review of the distributed costs with all the different business units present can be really beneficial.

Summary

In this chapter, we reviewed naming conventions and tagging strategies for cloud resources in depth. We now understand their key benefits as well as their differences, and how they synergize with other processes such as IaC, organizational policies, and cost allocation. In addition to this, we also looked at some recommendations on how to build a naming convention and a tagging strategy from scratch, which can be a good starting point for FinOps practitioners to create their own.

We also analyzed the differences between different public clouds in their approach to naming resources and using tags.

We hope that this content has proven to be useful for you.

In the next chapter, we will advance and deepen our review of the *Inform* pillar by explaining how we can estimate cloud savings to be used as a driver, prioritizer, and selling point for our FinOps initiatives going forward.

4

Estimating Cloud Solution Costs and Initiative Saving

Estimating the costs of your cloud solutions is one of the pain points for every organization transitioning to the cloud, from both technical and financial perspectives. It is also a key part of the Inform pillar, as it enables a lot of visibility on possible cost optimization opportunities, as well as projected costs of future projects not yet delivered.

When our infrastructure is hosted in the cloud, we need to be able to estimate the cost of new projects, to make this factor into our decision-making, as costs can make or break a project if not justified.

Apart from being able to estimate the potential cost of new projects, we need to be able to analyze the impact of our cost optimization initiatives beforehand. For example, if we know that an initiative is going to yield big savings, we should prioritize it over others.

In this chapter, we will review the different tools that we have at our disposal to estimate cloud costs, as well as analyzing the process that we should go through for proper cost estimation.

In this chapter, we will cover the following topics:

- How to calculate the TCO for cloud solutions
- Pricing APIs from cloud providers and how to work with them
- Estimating potential savings of cost optimization initiatives
- How to automate cost estimation

Technical requirements

In this chapter, we are going to mix non-technical information, such as an introduction to the TCO and its concepts, with more technical information, such as how to use pricing APIs from different cloud providers.

Nevertheless, the technical information covered here should be easily understandable as we are not going to dig deep into such technical aspects.

How to calculate the TCO for cloud solutions

As mentioned in *Chapter 1*, calculating the **Total Cost of Ownership** (**TCO**) is one of the key processes we need to go through when considering moving to the cloud.

We make estimations in our daily lives, such as when we are going to buy a car or a house. When considering these purchases, we usually first address its purchase price, but we also keep in mind indirect costs, such as electricity in the case of the house or proper insurance in regard to the car. Maintenance costs of small house repairs or expendable parts in the car should also be considered to consider this exercise complete.

Without these costs, our estimation is just not going to reflect reality. The same thing applies to the TCO but from a technological perspective.

In this section, we will go through a complete review of what the TCO is and the factors that weigh into this key cost analysis process.

TCO introduction

TCO was already introduced as a key concept in FinOps in the first chapter of this book. In this chapter, we are going to deep dive into what the TCO is and how we can calculate it, as well as understanding how important it is in designing cloud solutions and planning for cloud migrations.

The TCO for a solution should include all direct and indirect costs of hosting such a solution using cloud services. There should be complete transparency in costs for all relevant parties by providing clear and exact information, hence avoiding falling for the old iceberg of hidden costs (as described in *Chapter 1*).

When we transition to the cloud, there are new, indirect costs that make an appearance, such as the following:

- Migration costs (service providers that will help us migrate workloads)
- Data transfer
- Cloud training and enablement for our teams
- Licensing costs (billed through Azure)

With the same idea, there are costs that disappear and won't be part of the TCO, such as the following:

- Virtualization licensing
- Hosting and housing costs
- Electricity

As you can imagine, these parameters are unique for each organization. Transitioning to the cloud means *shifting CAPEX costs to OPEX costs*, so we need to expect a lot of changes in how costs work and the main cost drivers of organizations.

Estimating a proper TCO is essential to define cloud budgets. One of the most common mistakes that organizations make is trying to set up the budget without creating a proper TCO analysis, which often leads to misalignment between expected costs and reality, often due to indirect costs not being taken into account.

Due to this dependency between the TCO and cloud budgets, it is essential for our estimations to be as precise as possible.

As using cloud services implies the adoption of a pay-as-you-go model and delegating management of different areas of IT operations to the cloud providers, our TCO will vary depending on the specific configurations of our resources and will vary over time.

As we advance in using less managed services in the cloud, we will also need to keep up with better and better estimations that include small details, such as data transfer costs or service uptime in PaaS services. During this journey, estimating cloud costs will get increasingly more complex and require more technical expertise.

As an additional idea, we should always consider **vendor lock-in** as a risk that could potentially result in additional costs. Once we commit to using fully managed SaaS or PaaS solutions in the cloud, we are locked to a specific technology or service that is only offered by one vendor. Moving to another cloud vendor from that point can be hard or expensive or require different degrees of transformation work.

Our recommendation is to avoid using special purchase models as part of our TCO exercise, such as **Reserved Instances** or **Spot virtual machines** (which we will analyze in depth in *Chapter 6*, of this book), to simplify the estimation process. We will use these as optimization initiatives that we will propose later as part of our project's life cycle and promote accordingly, but we should not consider these as part of our initial estimation.

We also want to note that estimating the TCO is an iterative process that will improve in time with practice, as teams that are involved in it grow more proficient in cloud services and more knowledgeable in FinOps practices.

The TCO should represent reality. For example, if we have a workload that is not optimized on-premises, we should highlight the impact that this can have on cloud migration, to use the projected cost as a driver for optimization and modernization.

Let's analyze the different costs that we need to take into account in our TCO analysis. To avoid too much complexity, we are going to focus on migration costs, direct cloud costs, and indirect cloud costs, leaving aside other indirect costs, such as training for technical teams. We are not going to delve further into this area for the purpose of this book, but we want to provide an additional article whose approach we like and share as an additional reference on this specific topic, by *Network World*: `https://www.networkworld.com/article/3164444/how-to-calculate-the-true-cost-of-migrating-to-the-cloud.html`.

Migration costs

As part of our TCO exercise, we must consider the costs of our migration to the cloud. Migrating our solutions to the cloud is going to take time, resources, and money, to put in motion the required projects to perform the different migrations. The first thing that our management needs to understand is that it is not going to be cheap in any way, and there are no shortcuts to go to the cloud.

Organizations need to invest in the cloud if they want to fully unlock its power, including training, migration, and hosting costs. Additionally, cost optimization can only happen after proper monetary and resource investment.

There are different ways to migrate applications to the cloud. We often refer to the different alternatives as the **six Rs of cloud migration**, which are as follows:

- **Rehost or lift and shift**: This approach entails migrating what is hosted on-premises to the cloud as an exact copy without any changes. We need to consider that on-premises solutions are usually not fully optimized as, due to traditional capacity management, this was never needed. This means that if we take this migration approach, inefficiencies in our solutions will be carried over to the cloud. From a cost perspective, this approach is usually the one that results in bigger costs but less transformation effort. There are specialized tools to help us in this process, such as **Azure Migrate** or **AWS Application Migration Service**.

- **Replatform**: In this migration approach, we may change small parts of our workloads to be optimized to the cloud, such as replacing a database hosted in virtual machines to PaaS. This migration approach usually doesn't require much migration work and offers great benefits, such as less administrative overhead. This approach usually does not result in big cost savings.

- **Refactor or rearchitect**: This consists of reimagining an application or workload to be cloud native, transforming its components deeply to fully leverage cloud advantages and newer architecture paradigms such as **microservices** or **event-driven**. This approach requires a lot of work on transformation, but it can result in greater cost optimization.

- **Repurchase**: This approach is basically purchasing another product, usually a SaaS solution that reduces the operational work and complexity of our solutions, to replace an existing application hosted on-premises.

- **Retire**: If we have an application in a big organization that is used by two or three users and is redundant with other applications, does it really make sense to migrate it to the cloud? Usually not, and this is why sometimes the best choice is to retire applications that are no longer used.

- **Retain**: We need to accept that sometimes the best approach possible is to do nothing. Due to dependencies and business factors, it may be impossible to migrate some applications, and that is OK.

Let us summarize the different approaches with the help of the following table:

Method	Cloud costs	Transformation costs	Administrative overhead	Example
Rehost	High	Almost none	High	Migrate an application hosted on-premises to the cloud using IaaS services
Replatform	High	Low	Medium	Migrate an application hosted on-premises to the cloud using IaaS services, but with the database hosted in PaaS
Refactor	Medium-low	High	Low	Change a monolithic application to microservices architectures making use of serverless and PaaS cloud services
Repurchase	Depends on solution	Depends on solution	Low	Replace ITSM ticketing tool hosted on-premises to Jira SaaS
Retire	None	None	N/A	Eliminate an application that is only used by a handful of users
Retain	None	None	Depends on solution	Decide to retain on-premises a mainframe application written in COBOL for business and compliance reasons

Table 4.1 – Six Rs of cloud migration

We should choose one over the other mainly based on these factors:

- **Company strategy**: Organizational strategy may play a part in deciding which choice we can make. Some organizations, for example, have decided to follow a cloud-first approach, where all new solutions should be hosted in the cloud.

- **Value**: How much value are we creating by migrating an application to the cloud? In some cases, it can be a big factor.

- **Cost and budget**: This factor may seem obvious, but it is only part of the story. We could choose to dedicate a lot of budget and resources to a cloud migration if the outcome is worth it.

- **End of life/application or contract life cycle**: Out-of-support applications and servers are always a risk for organizations, but they are commonly used, despite it being a bad practice. As another example, we may have a hosting contract with our data center service provider that is expiring soon, so we can use the opportunity to save some costs by moving to the cloud. Both of these reasons can be a great driver for cloud migration.

- **Impact**: How much impact is our migration going to have, either for good or for bad, on users?

- **Administrative overhead**: Some applications, even if small and not widely used, can require a huge administrative and operational overhead. Reducing this administrative overhead, to manage more applications with less technical resources, can be a big selling point for cloud migrations as well.

- **Cloud readiness and technical expertise**: It is not the same going through a migration with a fully cloud-knowledgeable team than with a team that has never worked in the cloud. Due to this fact, the expertise and technical level should also be considered when deciding which approach to take for cloud migrations.

Choosing one of these scenarios is going to have an impact on our TCO one way or the other. The important takeaway is that the decisions we make should be smart. If we decide to refactor an application over a six-month project with a lot of budget and resources, it needs to be worth it, from a business, cost, and operational perspective.

With all this information in mind, let's move on to the part of the cost that is more straightforward, which is cloud services monthly costs.

Direct cloud costs

As we already know, having our solutions hosted in the cloud means adopting a pay-as-you-go cost model, which is often offered in two different flavors by cloud providers:

- In a **monthly payment model**, each month we are billed for the services that we used throughout the month. At the end of the month, we get the corresponding cloud bill with all the services we used and their cost.

- In a **commitment or advance pay model**, we pay in advance to get cloud credits that we use until there are no credits left. This model is used by companies that want more control over costs or during the first years of the journey to the cloud. This payment model can also be used with Reserved Instances and Saving Plans, which will be covered in *Chapter 6*, of this book.

One way or the other, the outcome is the same, as we will end up paying for the services used in the cloud regardless of the payment model.

At first glance, it may seem easy to estimate the costs of a solution hosted in the cloud. We can review the publicly available pricing tables and make our own cost analysis to calculate the costs of a solution. For these purposes, we can check either the **public pricing tables** available from cloud providers or **calculators** that the cloud providers have made available for us to use. We will analyze how to get this information in the next section of this chapter.

However, the price sheet shown publicly is most probably not the same as the one applied to your organization, especially if it's a big one. Organizations often have some degree of **discounts** on all or specific services, mainly based on volumes (you get greater discounts as organizations grow bigger and invest more in the cloud).

These discounts are often set when a contract or agreement is signed with the cloud provider, and the conditions can be reviewed periodically. Our recommendation is to always use discounted prices when possible, to aim for the best estimations possible.

In addition to this, the prices vary over time, and things such as the effect of **inflation** over cloud prices can have an impact on the prices, going up or down depending on global market trends, as well as price adjustments to remain competitive with other cloud providers.

As we covered in *Chapter 1*, we need to also remember that pricing varies from region to region, the cheapest region being in the US, followed by European regions, with Asia regions being the ones with the highest cost.

Due to these facts, estimating cloud costs is a challenging exercise, and we need to know and anticipate in advance possible factors impacting our estimations.

Indirect cloud costs

We consider **direct cloud costs** as the ones that can be traced to specific resources and can be seen clearly in our cloud bills and thus estimated in advance using calculator tools.

However, these costs don't tell the whole story, as there are additional costs that are much harder to calculate, estimate, and trace in our cloud bills, such as data transfer costs of traffic going out of services in the cloud. We call these additional costs **indirect costs**.

These costs should not be dismissed, and we should aim to include indirect costs in all our estimations, even in a rough form.

As an example, let's say we need to move some files, around 1 TB, to a storage service in the cloud. The cost of 1 TB in the cloud is around 20-25 dollars in non-archive tiers in Azure, AWS, and GCP. However, the storage capacity does not tell the whole story. We must also ask ourselves the following questions:

- *How much does it cost to copy our data into the cloud?* This type of traffic coming into the cloud is called **ingress traffic**. Fortunately, cloud providers do not charge for such traffic.
- *How much are we going to read or write on this storage platform in the cloud?* We will be charged for all operations on our data hosted in the cloud.
- *Will we need any kind of replication to other cloud regions or to move data to other cloud services?* If that's the case, we will be billed with additional costs as well.

In most use cases, indirect costs are not going to be a big percentage of your total costs. However, in some special cases, indirect costs can raise the cost enormously if they are not considered.

Cloud pricing calculators

Before we migrate to the cloud, we need to assess the potential cost of hosting our solutions in the cloud. No organization is going to commit without an estimate, as it is too high a risk to do so.

Because of this, there are different tools in all three cloud providers to simplify cost estimation. In this section, we are going to describe the different tools available and how they work.

Before going into these tools, we need to understand that, in order to use such tools, we need to have basic knowledge of the cloud provider of our choice should we want our analysis to be fruitful. Especially with calculators, we need to know the different services offered in the cloud and all possible configurations before jumping in to make estimations, as well as answers to some other key questions, such as the following:

- *How many virtual machines does the workload need?*
- *Which configuration is needed for virtual machines?*
- *Does the solution need high availability?*
- *Do we need to implement backup?*
- *Do we need a Load balancer?*

In other words, we need to have a **proper solution architecture design** as a starting point to iterate around before using these tools.

With this said, let's jump into the different tools that Microsoft offers for the Azure cloud.

Azure

In Azure, we have at our disposal the following tools to estimate the cost of solutions hosted in this cloud:

- Azure Calculator
- Azure TCO Calculator

Azure Calculator

Azure Calculator is a tool designed to estimate the cost of our solutions running in the Azure cloud. You can select which resources are going to be part of your solution and its configuration, as well as other additional inputs, such as backup storage or data transfer costs.

Essentially, it lets you select the different services that are going to be part of your estimation with its specific configuration, and it does the cost calculations for you, including totals and cost per resource and service. You can also save your estimates to be used at a later time.

This tool does not require authentication. However, if you already have a contract with Microsoft such as an **Enterprise Agreement (EA)** you can use your username and password to authenticate, which will enable you to see the services pricing from your current contract, including discounts. This option allows much more exact estimations and saves us the hassle of adding discounts manually after the calculations with tools such as Excel.

Azure Calculator is an essential tool for FinOps practitioners in Azure, so make sure you know how to work around it inside and out. Keep in mind that this tool is not aimed to estimate the cost of a complete cloud environment, but to estimate the cost of specific solutions or workloads.

You can find more information and details related to this tool with the following link: `https://azure.microsoft.com/en-us/pricing/calculator/`.

Azure TCO Calculator

Azure TCO calculator is a tool intended for organizations that are beginning their transition to the cloud. Essentially, it helps estimate the TCO for our cloud workloads.

It works by requesting key information from the user, such as the following:

- Number of servers, databases, storage volumes, and others to be moved to the cloud
- Assumptions for the estimate, such as current on-premises costs

With this information, it creates an estimate on how much it would cost to store these workloads in the cloud as opposed to keeping them on-premises.

The tool shows the changes in costs that will happen after moving to the cloud, such as shifting to OPEX costs from CAPEX+OPEX costs.

Compared to Azure Calculator, this tool is more focused on helping to create a business case to go to the cloud, and therefore it is way simpler than Azure Calculator and requires way less cloud knowledge. For more detailed information or to elaborate specific solution estimates, we should always resort to Azure Calculator.

To use this tool, you don't need to be authenticated or have a contract with Microsoft.

You can find more information and details related to this tool with the following link: `https://azure.microsoft.com/en-us/pricing/tco/calculator/`.

AWS

In AWS, the tools that are available for cost calculations and migration scenarios are the following:

- AWS Calculator
- AWS Migration Evaluator

AWS Calculator

AWS calculator is the tool that is offered as the standard for cost estimation in AWS. It includes all available services with all their cost drivers that we can configure as estimation parameters, such as region, disk size, and some indirect costs such as data transfer.

It offers a simple interface that lets you add the resources you want to your estimation and obtain the totals, which can be exported or shared between teams.

One thing to keep in mind is that estimations are in dollar currency and there is no possibility to change this.

It also lets you add a description for individual services, custom names for your estimations, and names for groups of services, which is always useful for adding additional information to our estimations, especially if we are going to share them across teams.

You can find more information and details related to this tool with the following link: `https://calculator.aws/#/`.

AWS Migration Evaluator

AWS Migration Evaluator is a tool offered by AWS for migration scenarios. It enables us to evaluate the possible migration of applications and workloads, providing us with the translated costs of running them in the cloud after an assessment process.

This tool is not available right away. To be able to access the application, a form must be filled in with contact information and details about the business case. After this, AWS will send you an invitation to use the application.

One thing to keep in mind regarding this tool is that it requires setup, with the installation of software on-premises such as agents and certificates. Once the software is installed, information will be collected in regard to our hardware specs and usage over time, to create a proper estimation that will include cost optimization initiatives such as Reserved Instances or specific virtual machine instance families, among others.

You can find more information and details related to this tool with the following link: `https://aws.amazon.com/migration-evaluator/`.

GCP

In GCP, there are different tools offered, such as the following:

- GCP Cloud Pricing Calculator
- GCP TCO Assessment
- GCP Migration Center

GCP Cloud Pricing Calculator

GCP Cloud Pricing Calculator is GCP's equivalent of the Azure and AWS calculators. This tool can help us estimate the cost of GCP solutions.

Using this calculator, we can add to our estimation different services with specific configurations to get the total cost of the solution.

It can be a great tool to estimate the costs of future projects and solutions before their implementation, or to estimate how potential changes in existing solutions may impact overall costs.

This tool does not require authentication and is free.

You can find more information and details related to this tool with the following link: `https://cloud.google.com/products/calculator`.

GCP TCO Assessment

In GCP, currently there is not a dedicated tool to be used to calculate the TCO.

What you can do, though, is fill in a form (`https://inthecloud.withgoogle.com/tco-assessment-19/form.html`) and request a **TCO assessment** from Google. A Google representative or Google Partner will contact you and will help you in the process of gathering all the required information with tools dedicated to migration scenarios.

After the data collection is complete, they will create your customized report within one week, including all key information needed to add to your cloud business case.

You can find more information and details related to this tool with the following link: `https://inthecloud.withgoogle.com/tco-assessment-19/form.html`.

GCP Migration Center

GCP Migration Center is a platform that allows us to perform end-to-end cloud migration scenarios, from the initial estimation or asset discovery to the migration.

Basically, it follows a process that is divided into the following steps:

1. **Estimate**: In this phase, we can make estimates of running our workloads in the cloud. Essentially, we provide the specifications, and the tool will come up with a configuration that matches those specifications while selecting the optimal way to host the workload in the cloud.

2. **Discover**: In this phase, we can either install some agents in our infrastructure to enable automatic discovery or enter all required data manually. The idea is to clearly delimit the scope of this migration exercise.

3. **Assess**: Once information about assets to be migrated is in GCP Migration Center, it generates a TCO analysis based on the migration settings that the user inputs, with the potential end solution hosted in GCP. As part of this phase, we may identify additional dependencies for our workloads that we may need to take into account for the migration. In this phase, we will use specific migration products such as StratoZone.

4. **Plan**: Once cost estimation is done, it is time to plan for the migration of resources. In this phase, GCP Migration Center can help by providing best practices and helping us create a proper roadmap, dividing the migration in waves depending on dependencies, criticality, or other key considerations.

5. **Migrate**: Once a plan is clearly defined and agreed upon, the only point left in the process is to migrate. In this phase, we will be presented with migration tools that can help us in whatever flavor of migration we have chosen to follow (lift and shift, replace, refactor, and so on).

You can find more information and details related to this tool with the following link: `https://cloud.google.com/migration-center/docs/migration-center-overview`.

With all these migration tools at hand for our future migration exercises, let's now move on to the next section, where we will review how to work with pricing APIs from different cloud providers and how they function.

Pricing APIs from cloud providers and how to work with them

REST APIs nowadays are the most common programming interfaces for application integration.

REST APIs, also known as RESTful APIs, are one of the gold standards that modern applications have adopted to get information and interact in other ways with all applications. Cloud providers are no exception, and, apart from the cloud console and the CLI and SDKs that are available, all cloud providers offer a complete set of APIs to interact with cloud services.

Before explaining the different APIs that we can use, we want to recommend some tools to ease the process of making queries and getting data from different APIs.

If we are going to be working with APIs and even use automation to query REST APIs, we must be comfortable working with them. The best way to get familiar with and test the different methods that we have available thoroughly is to make use of these helper tools.

Off the top of our heads, we can recommend the following:

- **cURL**: Stands for **Client URL** and is a command-line tool that can be used to make HTTP requests and communicate with multiple web services and applications. We can use cURL to easily test REST APIs as it is usually installed in most Linux distributions (https://curl.se/).

- **Postman**: A software application that can be used to create REST API requests for testing and troubleshooting purposes (https://www.postman.com/).

- **Thunder Client (Visual Studio Code extension)**: Using this extension, we can have similar functionality as what we get with Postman from the Visual Studio Code IDE (https://www.thunderclient.com/).

- **HTTPie**: Another command-line tool that also offers a web browser version and a desktop version that are similar to Postman (https://httpie.io).

- **Rapid API (Visual Studio Code extension)**: An alternative to Thunder client that works in a similar way (https://rapidapi.com/guides/categories/rapidapi-client-vscode).

These tools are essential to try out the different APIs that we are going to cover here, as they make working with them much simpler, easing things such as authentication, file attachments, and the request's structure and content.

Pricing APIs overview

In this section, we want to go through the different pricing APIs offered by the cloud providers Azure, AWS, and GCP. We also consider this API as one of the basic tools that every FinOps practitioner should include in their toolset, and one of the starting points to become familiar with FinOps practices.

Microsoft, Amazon, and Google offer a wide range of APIs we can use to unlock the full potential of their clouds and to work with every single service. For the purpose of this book, we are going to limit the scope to pricing APIs for the sake of simplicity.

Let's begin by discussing the options available in Azure.

Azure

In Azure, we need to leverage the following REST APIs if we want to work on cloud costs:

- Azure Retail prices
- Azure Price Sheet

One of the things we need to know about Azure APIs in general is that when there are too many results as a result of an API call, we will get a pointer to get the rest of the results.

As an example, this is how this pointer looks in the API results when we make a simple call with the HTTPie web browser version querying for virtual machine pricing. If we scroll to the end of the response, we will find the pointer:

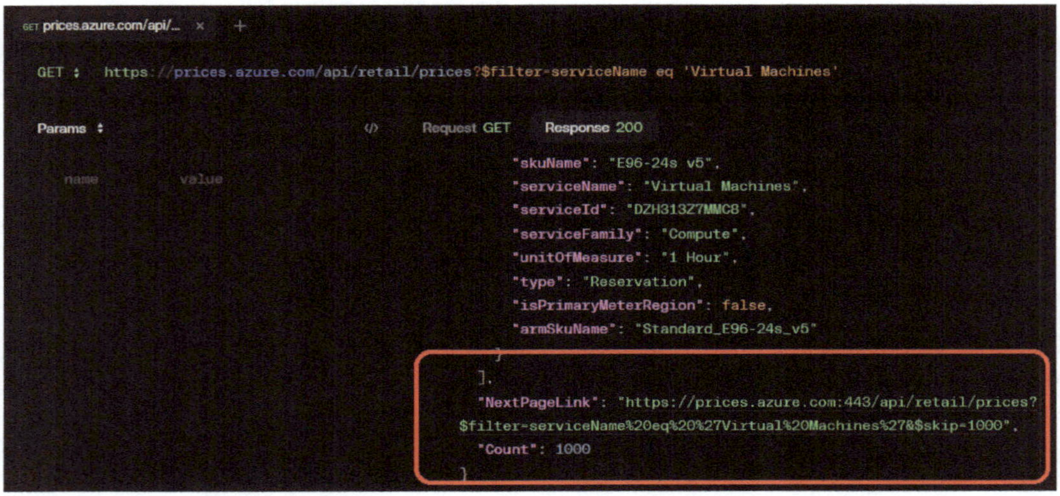

Figure 4.1 – Azure Retail prices REST API example

Using this `NextPageLink` pointer, we can query for the next set of results; in this case, we got 100 results from the first API call, and we can get the next 100 by using this pointer.

If we are querying for the complete list of results, then we need to iteratively query the REST APIs in the following way:

- Process current results

- Iteratively call `NextPageLink` until the value is not present in the API call

We want to recommend this documentation, `https://learn.microsoft.com/en-us/rest/api/azure/`, before going deep into Azure REST APIs, as it can get complex if the caller is not experienced in using REST APIs, especially on APIs that require authentication.

Azure Retail prices REST API

This RESTful API is one of the APIs offered by Microsoft on Azure. Essentially, it provides the retail list price of all services offered in Azure. It allows simple filtering, so we can query for specific services such as *Virtual machine prices in West Europe region*.

It may seem like a simple API, but in our experience it is not. Azure uses a complex data structure for this API, so it may be hard to navigate the response format and how information is presented. Constructing the specific filter that you need can take multiple test calls. You will need to test it thoroughly to get the exact result you need, as there are some things that need to be considered:

- We cannot use all fields for filtering.

- Using the `currencyCode` setting, we can get the prices in any of the supported currencies the API offers.

- We need to be aware of the format of each field and know it in advance.

- When we query the price of a service, it will give us the pricing in all purchase models, including Reserved Instances, Spot, and Dev/Test. We must make sure to filter it out to select the one we want.

Keeping these considerations in mind, we will most probably need to do postprocessing after we get information from the API call, as it is not the most versatile interface to query. The good news is that in the official documentation, we will find an example of how to query data from Python, which can save a lot of time using the API programmatically.

As a small example, this is how we can query the API. Let's say that we want to get the previous query that we referenced, which is *Virtual machine prices in West Europe region*.

To do so, we need two different filters:

- One to filter `serviceName` to be equal to `Virtual Machines`
- Another to filter the region to `West Europe`

After working a little bit on the API and making some test calls, we can get the final call, which will look like this:

```
prices.azure.com/api/retail/prices?$filter=serviceName eq %27Virtual
Machines%27 and armRegionName eq %27westeurope%27
```

This API requires no authentication, and it is publicly available for everyone to use.

You can check the API reference for the current version here: `https://learn.microsoft.com/en-us/rest/api/cost-management/retail-prices/azure-retail-prices`.

Azure Price Sheet

This API generates a CSV with a complete report of the prices for all services under our current contract with Microsoft, which includes specific discounts and offers.

This feature is only available for customers that have an EA, **Microsoft Customer Agreement** (**MCA**), or **Microsoft Partner Agreement** (**MPA**) license. It does not work with the **Cloud Service Provider** (**CSP**) agreement contract model, which is really inconvenient.

Unlike Azure Retail pricing API, this API requires authentication. This means that, in order to get the information from the API REST, you should have acquired a token first from Azure Active Directory and added it to your HTTP call as part of the header.

To simplify obtaining this file, we can also check this price sheet in the following ways:

- From the Azure portal, with Azure Cost Management.
- Using Azure PowerShell, once we have installed and imported the `Az.Consumption` module.
- Using the Azure CLI, through the `az consumption pricesheet show` command. As opposed to other methods, we will get the information in JSON format, which may be easier to process and query through the Azure CLI.
- For customers with an EA contract, the price sheet is also available from the EA portal (`https://ea.azure.com/`).

This file should be the reference used for our cost estimation, so make sure to process it properly and have it at hand. A word of advice, though: the file is pretty large and may be hard to process.

Keep in mind that the price sheet will vary over time, so it is generally a good idea to keep historical pricing of our services in a database or Azure Data Lake Storage Gen2 and feed a report with this information. Having this clear view of cost fluctuation for Azure cloud services may answer a lot of questions about cost variation month by month.

You can find the complete API documentation with this link: `https://learn.microsoft.com/en-us/rest/api/cost-management/price-sheet/download?tabs=HTTP`.

AWS

In AWS, we have the following APIs available:

- AWS Price List Bulk API
- AWS Price List Query API

AWS Price List Bulk API

This API allows us to get historical pricing data directly from AWS. The API will return the prices that are queried. This API does not require any kind of authentication as the information shown in this API is public, and the result of the calls is a JSON or CSV file.

AWS makes available an endpoint to query this data, which is as follows:

`https://pricing.us-east-1.amazonaws.com/`

Using this API as our base URL, we can build more elaborate URLs that will enable us to query for specific services. To have the complete information, we will need to use two types of files:

- **Offer index file**: This file is only available in JSON format, and its purpose is to list all available services with an associated URL that we can use to download its current pricing.
- **Offer file**: This file contains the complete price list for the service that we are referencing. For example, if we are in AWS RDS, this list will include all the instance types available for AWS RDS and their corresponding price.

We will begin the process to query for specific services by accessing the offer index file, which contains the references to all available services, and then go forward to the specific offer file using the link that we got from the offer index file.

Accessing the Offer index file will lead us to the following information:

- `versionIndexUrl`: A link to different pricing versions over time. From this link, we could get the pricing that AWS EC2 virtual machines had, for example, in 2020.

- `currentVersionURL`: This URL is a reference to the offer file, which includes the current price list in all regions for this service.

- `currentRegionIndexUrl`: This URL is a reference to the Offer index file containing all the services that are available in the current region.

- `savingsPlansVersionIndexUrl`: This is a reference to all the different pricing versions for Saving Plans on this service. This reference will take you to a list of all regions' Saving Plans details. These lines will only be available in services that support Saving Plans.

- `currentSavingPlansIndexUrl`: This URL is the offer file in which all the Saving Plans available in all regions are included. These lines will only be available in services that support Saving Plans.

As an example, let's say we get the Offer index file through its endpoint, as follows:

`https://pricing.us-east-1.amazonaws.com/offers/v1.0/aws/index.json`

In this file, we can search for the service we want, for example, Amazon EC2.

```
"AmazonEC2" : {
  "offerCode" : "AmazonEC2",
  "versionIndexUrl" : "/offers/v1.0/aws/AmazonEC2/index.json",
  "currentVersionUrl" : "/offers/v1.0/aws/AmazonEC2/current/index.json",
  "currentRegionIndexUrl" : "/offers/v1.0/aws/AmazonEC2/current/region_index.json",
  "savingsPlanVersionIndexUrl" : "/savingsPlan/v1.0/aws/AWSComputeSavingsPlan/current/index.json",
  "currentSavingsPlanIndexUrl" : "/savingsPlan/v1.0/aws/AWSComputeSavingsPlan/current/region_index.json"
```

Figure 4.2 – AWS Price List Bulk API Amazon EC2 on offer index file

From this point, let's say we want to check the current pricing. We should use `currentVersionUrl`:

`https://pricing.us-east-1.amazonaws.com/offers/v1.0/aws/AmazonEC2/current/index.json`

On this web page, we will get the pricing of Amazon EC2 in all regions, following this data structure:

```
"QUMEF4UK3NPT4MN3" : {
  "sku" : "QUMEF4UK3NPT4MN3",
  "productFamily" : "Compute Instance",
  "attributes" : {
    "servicecode" : "AmazonEC2",
    "location" : "US East (N. Virginia)",
    "locationType" : "AWS Region",
    "instanceType" : "c3.xlarge",
    "currentGeneration" : "No",
    "instanceFamily" : "Compute optimized",
    "vcpu" : "4",
    "physicalProcessor" : "Intel Xeon E5-2680 v2 (Ivy Bridge)",
    "clockSpeed" : "2.8 GHz",
    "memory" : "7.5 GiB",
    "storage" : "2 x 40 SSD",
    "networkPerformance" : "Moderate",
    "processorArchitecture" : "64-bit",
    "tenancy" : "Shared",
    "operatingSystem" : "Windows",
    "licenseModel" : "No License required",
    "usagetype" : "UnusedBox:c3.xlarge",
    "operation" : "RunInstances:0002",
    "availabilityzone" : "NA",
    "capacitystatus" : "UnusedCapacityReservation",
    "classicnetworkingsupport" : "true",
    "ecu" : "14",
    "enhancedNetworkingSupported" : "Yes",
    "gpuMemory" : "NA",
    "instancesku" : "7MS6E9W2YWKJZRX5",
    "intelAvxAvailable" : "Yes",
    "intelAvx2Available" : "No",
    "intelTurboAvailable" : "Yes",
    "marketoption" : "OnDemand",
    "normalizationSizeFactor" : "8",
    "preInstalledSw" : "NA",
    "processorFeatures" : "Intel AVX; Intel Turbo",
    "regionCode" : "us-east-1",
    "servicename" : "Amazon Elastic Compute Cloud",
    "vpcnetworkingsupport" : "true"
  }
}
```

Figure 4.3 — AWS Price List Bulk API Amazon EC2 current offer file

This API may seem complicated to use, but in our view, it is a great highlight that it offers all historical information about prices in AWS. This information can be key to understanding AWS pricing fluctuations and we can make it visible to cloud teams through dashboards or reports.

AWS Price List Query API

While this tool may seem similar to the one we just covered, as its objective is to get cloud services pricing, there are some slight differences between them that are worth mentioning:

- We need authentication to use this API.
- To use the AWS Price List Query API, we need to use either the AWS Console or the AWS SDK.
- We can use filters to select the specific data that we want to get. For example, we can filter by region, tenancy, or specific configurations.
- As of August 2023, there are three different APIs available.

We are going to exemplify how to use this API to illustrate how it works, in this case, using the AWS CLI.

The first step to keep in mind is that we will need to update the AWS credentials file, which usually is located in `~/.aws/credentials` in Linux. Before directly querying for the prices, we can get the different attributes that we can use for our queries using a specific command: `describe-services`.

We can use the following command to get Amazon EC2 pricing:

```
aws pricing describe-services --service-code AmazonEC2
```

As an example, this is the response we'd get after executing this query:

```
"Services": [
    {
        "ServiceCode": "AmazonEC2",
        "AttributeNames": [
            "regionCode",
            "instanceCapacityMetal",
            "volumeType",
            "maxIopsvolume",
            "instance",
            "classicnetworkingsupport",
            "instanceCapacity10xlarge",
            "fromRegionCode",
            "locationType",
            "toLocationType",
            "instanceFamily",
            "operatingSystem",
            "toRegionCode",
            "clockSpeed",
            "LeaseContractLength",
            "ecu",
            "networkPerformance",
            "instanceCapacity8xlarge",
            "group",
            "gpuMemory",
            "maxThroughputvolume",
            "ebsOptimized",
            "vpcnetworkingsupport",
            "maxVolumeSize",
            "gpu",
            "intelAvxAvailable",
            "processorFeatures",
            "instanceCapacity4xlarge",
            "servicecode",
            "groupDescription",
            "elasticGraphicsType",
            "volumeApiName",
            "processorArchitecture",
            "physicalCores",
            "fromLocation",
            "snapshotarchivefeetype",
            "marketoption",
            "availabilityzone",
            "productFamily",
            "fromLocationType",
            "enhancedNetworkingSupported",
            "intelTurboAvailable",
            "memory",
            "dedicatedEbsThroughput",
            "vcpu",
            "OfferingClass",
            "instanceCapacityLarge",
            "capacitystatus",
            "termType",
            "storage",
            "toLocation",
            "intelAvx2Available",
            "storageMedia",
            "physicalProcessor",
            "provisioned",
            "servicename",
            "PurchaseOption",
            "instancesku",
            "productType",
            "instanceCapacity18xlarge",
            "instanceType",
```

Figure 4.4 — AWS Price List Query API describing Amazon EC2

Note that we have selected the attribute used to specify the region.

Once we know the attributes that we want to use, we can use these attribute names as references for adding filters in the next step, which is to directly query the pricing APIs.

Let's say that we want to get EC2 pricing for the `eu-west-1` region using an `m5.xlarge` instance:

```
aws pricing get-products --filters Type=TERM_
MATCH,Field=regionCode,Value='eu-west-1' Type=TERM_
MATCH,Field=instanceType,Value='m5.xlarge' --max-results 1 --service-
code AmazonEC2
```

This is the result that we get using this command:

```
"product": {
  "productFamily": "Compute Instance",
  "attributes": {
    "enhancedNetworkingSupported": "Yes",
    "intelTurboAvailable": "Yes",
    "memory": "16 GiB",
    "dedicatedEbsThroughput": "Up to 2120 Mbps",
    "vcpu": "4",
    "classicnetworkingsupport": "false",
    "capacitystatus": "Used",
    "locationType": "AWS Region",
    "storage": "EBS only",
    "instanceFamily": "General purpose",
    "operatingSystem": "Red Hat Enterprise Linux with HA",
    "intelAvx2Available": "Yes",
    "regionCode": "eu-west-1",
    "physicalProcessor": "Intel Xeon Platinum 8175",
    "clockSpeed": "3.1 GHz",
    "ecu": "16",
    "networkPerformance": "Up to 10 Gigabit",
    "servicename": "Amazon Elastic Compute Cloud",
    "gpuMemory": "NA",
    "vpcnetworkingsupport": "true",
    "instanceType": "m5.xlarge",
    "tenancy": "Dedicated",
    "usagetype": "EU-DedicatedUsage:m5.xlarge",
    "normalizationSizeFactor": "8",
    "intelAvxAvailable": "Yes",
    "processorFeatures": "Intel AVX; Intel AVX2; Intel AVX512; Intel Turbo",
    "servicecode": "AmazonEC2",
    "licenseModel": "No License required",
    "currentGeneration": "Yes",
    "preInstalledSw": "SQL Ent",
    "location": "EU (Ireland)",
    "processorArchitecture": "64-bit",
    "marketoption": "OnDemand",
    "operation": "RunInstances:1110",
    "availabilityzone": "NA"
  },
  "sku": "25KZHD4N5DK4EY7Y"
},
"serviceCode": "AmazonEC2",
"terms": {
  "OnDemand": {
    "25KZHD4N5DK4EY7Y.JRTCKXETXF": {
      "priceDimensions": {
        "25KZHD4N5DK4EY7Y.JRTCKXETXF.6YS6EN2CT7": {
          "unit": "Hrs",
          "endRange": "Inf",
          "description": "$1.822 per Dedicated RHEL with HA and SQL Enterprise m5.xlarge Instance Hour",
          "appliesTo": [],
          "rateCode": "25KZHD4N5DK4EY7Y.JRTCKXETXF.6YS6EN2CT7",
          "beginRange": "0",
          "pricePerUnit": {
            "USD": "1.8220000000"
          }
        }
      },
      "sku": "25KZHD4N5DK4EY7Y",
      "effectiveDate": "2023-08-01T00:00:00Z",
      "offerTermCode": "JRTCKXETXF",
      "termAttributes": {}
    }
  }
},
"version": "20230823155509",
```

Figure 4.5 — AWS Price List Query API pricing get-products

In the case of this API, the prices that we get are current prices, and it is not possible to get historical data.

GCP

Google offers the following pricing APIs:

- **GCP Cloud BIlling Catalog API**: This shows public prices for GCP services
- **GCP Custom Pricing API**: This offers customer-specific pricing under a GCP agreement or contract, including discounts and special offers

Unlike its counterparts in Azure and AWS, as a prerequisite to use this API, we need to do the following:

- Enable the APIs for our project
- Create an API key to use alongside our requests, even for the public APIs that Google offers

Let's dive into how to use them in detail.

GCP Cloud Billing Catalog API

This API shows the public price list of services offered by GCP. The process to query the API is as follows:

1. We get the list of services available in GCP.
2. We look for the service we want more information on and get the identifiers.
3. We query for the price list of the exact service we want to search for, which will get us the list of SKUs with their pricing and characteristics.

Once the key has been created, we can make a request to this API using the following endpoint:

```
https://cloudbilling.googleapis.com/v1/services?key={API_KEY}
```

Here, `API_KEY` is the key we created previously. Remember that this `API_KEY` value is unique to your project and should not be shared, as it allows for anonymous access to different public Google APIs. As an example, this is part of the result that we get when we call the API:

```
{
  "name": "services/24E6-581D-38E5",
  "serviceId": "24E6-581D-38E5",
  "displayName": "BigQuery",
  "businessEntityName": "businessEntities/GCP"
},
{
  "name": "services/251D-E8E2-38A4",
  "serviceId": "251D-E8E2-38A4",
  "displayName": "MySQL 5.6 on Ubuntu 14.04 LTS",
  "businessEntityName": "businessEntities/GCP"
},
```

Figure 4.6 — GCP Cloud Billing Catalog API services list

From this first call, we will be able to get from the `serviceId` field the identifier that we will use in subsequent steps. Once we have the `serviceId` field, we will make another request with this format:

```
https://cloudbilling.googleapis.com/v1/services/{serviceId}/
skus?key={API_KEY}
```

This API will give us the list of SKUs or possible configurations for this service alongside its pricing.

As an example, here is a screenshot showing an example of a response from this API:


```
{
  "name": "services/6F81-5844-456A/skus/48EB-F1F3-223B",
  "skuId": "48EB-F1F3-223B",
  "description": "Spot Preemptible N2D AMD Instance Core running in Singapore",
  "category": {
    "serviceDisplayName": "Compute Engine",
    "resourceFamily": "Compute",
    "resourceGroup": "CPU",
    "usageType": "Preemptible"
  },
  "serviceRegions": [
    "asia-southeast1"
  ],
  "pricingInfo": [
    {
      "summary": "",
      "pricingExpression": {
        "usageUnit": "h",
        "displayQuantity": 1,
        "tieredRates": [
          {
            "startUsageAmount": 0,
            "unitPrice": {
              "currencyCode": "USD",
              "units": "0",
              "nanos": 8940000
            }
          }
        ],
        "usageUnitDescription": "hour",
        "baseUnit": "s",
        "baseUnitDescription": "second",
        "baseUnitConversionFactor": 3600
      },
      "currencyConversionRate": 1,
      "effectiveTime": "2023-08-25T15:34:24.519849Z"
    }
  ],
  "serviceProviderName": "Google",
  "geoTaxonomy": {
    "type": "REGIONAL",
    "regions": [
      "asia-southeast1"
    ]
  }
},
```


Figure 4.7 — GCP Cloud Billing Catalog API response final result

Keep in mind that the price shown here uses the **units and nanos** format, which essentially means that whole numbers are represented using units while decimals are represented using the `nanos` field in nano format.

As an example, *$2.25 = 2 units* and *250,000,00 nanos*.

You can find more information about this format with this URL: `https://cloud.google.com/recommender/docs/reference/rest/Shared.Types/Money`.

This API also uses pagination as it delivers a maximum of 5,000 results in each API call. **Pagination** works in a similar way as Azure APIs, but in this case, the field to look for is called. To get the results for subsequent pages, we need to add the value of `nextPageToken` to our next API call like this:

```
https://cloudbilling.googleapis.com/v1/services/{serviceId}/skus?key={API_KEY}&pageToken={nextPageToken}
```

We can also add as a parameter a custom page size, adding the `pageSize` parameter to the call: `https://cloudbilling.googleapis.com/v1/services/{serviceId}/skus?key={API_KEY}&pageSize={pageSize}`.

GCP Custom Pricing API

This API, unlike the GCP Cloud Billing Catalog API on which we query for public prices, will show the current prices including any discounts under our agreement with GCP, which can help us get more exact cost estimations.

It works in the same way as the GCP Cloud Billing Catalog API, which basically involves going through the following steps:

1. We get the list of services available in GCP.

2. We look for the service we want more information on and get the identifiers.

3. We query for the price list of the exact service we want to search for, which will get us the list of SKUs with their pricing and characteristics.

The endpoint to be used in this case is different: `https://cloudbilling.googleapis.com/v2beta/services`.

But there is one key difference, and that is that when listing the SKUs, we need to use a filter to select the proper `serviceId`, like this:

```
https://cloudbilling.googleapis.com/v2beta/skus?key={API_
KEY}&filter=service="service/{serviceId}}
```

As an example, this is a possible result if we search for the BigQuery service in `asia-southeast1`:

```
{
  "name": "skus/0033-9FD2-2500",
  "skuId": "0033-9FD2-2500",
  "displayName": "BigQuery BI Engine Flat Rate for Singapore",
  "service": "services/650B-3C82-34DB",
  "productTaxonomy": {
    "taxonomyCategories": [
      {
        "category": "GCP"
      },
      {
        "category": "Analytics"
      },
      {
        "category": "GBQ"
      },
      {
        "category": "BigQuery BI Engine"
      },
      {
        "category": "Flat Rate"
      }
    ]
  },
  "geoTaxonomy": {
    "type": "TYPE_REGIONAL",
    "regionalMetadata": {
      "region": {
        "region": "asia-southeast1"
      }
    }
  }
},
```

Figure 4.8 — GCP Custom Pricing API BigQuery

From this endpoint, we will get the SKU ID that we want to query to be able to make the final call that will get us the current pricing:

```
https://cloudbilling.googleapis.com/v1beta/skus/SKU_ID/price?key=API_
KEY
```

In this last step, we can also specific a currency code using another parameter, `currencyCode`, which should follow the ISO-4217 currency code format:

```
https://cloudbilling.googleapis.com/v1beta/skus/SKU_ID/price?key=API_
KEY?currencycode={CURRENCYCODE}
```

As we can see, it works similarly to the public API but with some specifics to keep in mind. We have not described in detail the fields that are returned as a result of calling this API, as it is way beyond the scope of this book.

With this practical understanding of how to use APIs to query service pricing, let's now analyze how to properly estimate the potential savings as a result of adopting FinOps initiatives.

Estimating potential savings of cost optimization initiatives

Once we have a basic understanding of how to estimate solutions hosted in the cloud, we are prepared to go one step further and begin analyzing the potential savings brought about by FinOps initiatives.

In this section, we are going to analyze how we can leverage proper cost estimation to drive our FinOps initiatives.

As part of FinOps practices, FinOps practitioners or other cloud teams propose a series of possible cost optimization initiatives, to make the most out of the cloud budget.

Imagine we go through a lot of work to apply some initiatives and, after all the work and effort, we find out that the savings we have created are minimal. Not ideal, is it?

To avoid this, our proposal is to use **potential savings estimation** to score our FinOps initiatives. What we are going to do is calculate in advance the impact of FinOps initiatives on our solutions.

To implement this analysis process, we like to use the **as is, to be, gap approach**. This approach uses three concepts:

- **As is**: The current solution
- **To be**: The potential solution with different initiatives implemented
- **Gap**: What we need to do to go from our current solution to our potential solution

The process can be illustrated by the following figure:

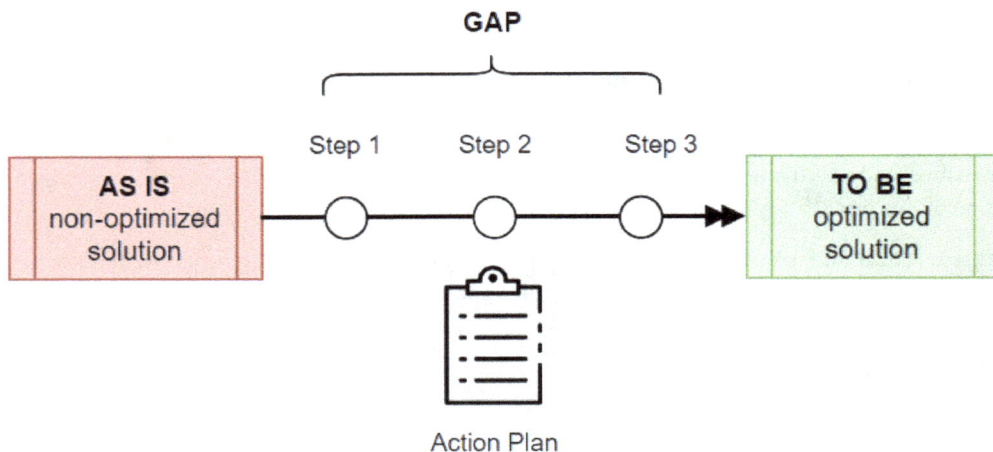

Figure 4.9 – As is, to be, and gap approach

After we do the analysis, our final decision to prioritize some initiatives over others will depend on the following:

- Costs

- How big the gap is, which represents how much effort it will require to apply the initiative

We can consider either small projects that only entail one step as the action plan or bigger projects with a multi-step action plan, to conform to a plan with multiple stages that will yield even more cost optimization.

Using this process, we can go through the following steps to calculate potential savings:

1. Analyze the current status of our cloud services; that is, we will consider our as-is status.

2. Using the list of possible FinOps initiatives that apply to this workload, create multiple potential to-be scenarios, including short-term and long-term initiatives with different effort levels required.

3. Calculate our as-is TCO analysis using the cloud bills as references, or using cloud calculators.

4. Once we have our as-is specification and the to-be scenarios, we should thoroughly calculate, one by one, the impact that each initiative will have on total costs. We can then calculate the savings for each initiative, including both the percentage and the actual figure.

5. Include a proposed action plan per initiative and try to evaluate the effort that each initiative will require, either in man days or scored with a number from 0 to 5 or 0 to 10.

6. With the savings at hand, begin considering the ones that yield greater savings with less effort.

We will use a simple example to illustrate what this process will look like.

Our *as is* can be as follows.

Let's say that we have migrated some virtual machines from non-productive environments (development and sandbox) to AWS from our data center as the first step of our cloud journey. Because our teams are not cloud experts, we are running our instances 24x7, and we don't want to purchase Reserved Instances as we are not sure whether our environment is correctly sized – so we are using a pay-as-you-go or on-demand payment model.

As part of our as-is information, we can add that CPU and memory consumption never surpasses 50% of use.

As the first step, we can use the AWS calculator to obtain the total price of our virtual machines:

Number of instances	Service	Availability	vCPU	RAM	Virtual machine size	EBS disk size (OS + data disk)	OS	Price ($/mo)
10	AWS EC2	24x7	4	16	m5.xlarge	GP3 (128 GB)	Linux	$1,674.84

Table 4.2 — As is stage of AWS example

Now let's move to our *to be* proposal. We are going to propose two initiatives here:

1. **Step 1**: Shutdown of virtual machines when not in use
2. **Step 2**: Virtual machines rightsizing, to try to find smaller virtual machines better suited to our computing needs

For step 1, let's consider that we choose to only use the virtual machines on a 12x7 schedule, or 12 hours a day every day. If we make the proper calculations in the AWS Calculator, we have the following:

Number of instances	Service	Availability	vCPU	RAM	Virtual machine size	EBS disk size (OS + data disk)	OS	Price ($/mo)
10	AWS EC2	12x7	4	16	m5.xlarge	GP3 (128 GB)	Linux	$837.42

Table 4.3 — To be stage of AWS example (step 1)

Moving on to step 2, we can propose a change to the T3 virtual machine family, keeping the same amount of vCPUs and RAM, to try to find a virtual machine better sized for our needs.

Combining step 1 with step 2 results in the following:

Number of instances	Service	Availability	vCPU	RAM	Virtual machine size	EBS disk size (OS + data disk)	OS	Price ($/mo)
10	AWS EC2	12x7	4	16	t3.xlarge	GP3 (128GB)	Linux	$722.08

Table 4.4 — To be stage of AWS example (step 1 + step 2)

The result of our potential savings before and after steps 1 and step 2 will then be as follows:

Original cost before optimization	Step 1	Step 2
$1,674.84	$837.42	$772.08
Savings	50%	53.90%

Table 4.5 — To be stage of AWS example overall results

As we can see, proposing these two simple initiatives may result in *more than 50% savings* on the total cost of this solution, which can be a strong factor or driver for this initiative to be implemented as soon as possible.

Don't worry if the technical part of steps 1 and 2 are not yet clear, as we will be discussing the technical aspects of these initiatives in depth, as well as many others, in *Chapters 6, 7, 8,* and *9* of this book.

Of course, making these decisions requires proper expert analysis and FinOps experience. Microsoft Excel or similar may be a good tool and starting point to make these estimations, as there is plenty of data processing we can do with this kind of product.

There is one key downside to this process, and that is that this process is totally manual. These exercises will take a lot of time from our FinOps practitioners, as these analyses require both knowledge and thorough work.

Due to this fact, we should aim to automate these estimation processes as our organization gets more mature in our FinOps practice. We will try to cover how to automate these analyses in the next section.

How to automate cost estimation

So far, we have learned how to calculate the cost of cloud solutions as well as how we can properly estimate the outcome of our cost optimization initiatives before their implementation.

To fully close the circle, we are going to describe in this section how we can automate the end-to-end process of potential savings estimation, to be able to always have this information available in a dashboard or similar, which will improve our FinOps posture a great deal. Having this means that we will be able to choose first the initiatives that will result in the best outcome from a savings perspective, as long as the effort required for such an initiative is reasonable enough.

The complete process will look like this:

1. Data sources selection
2. Data consolidation
3. Estimation calculation
4. Change notifications
5. Data update

The first point we need to keep in mind before diving into this process is that even though cloud services pricing does not change regularly, this does not mean that it does not change at all. These changes will impact everything, from our cost estimations of future projects to the current cost of running infrastructure, as well as our cost savings estimations as a result of FinOps initiatives.

Apart from these slight cost changes, we need to keep in mind that services offered by cloud providers are in constant change. We need to be up to date at all times on the latest instances offered and published if we want to stay informed on cost optimization strategies. Due to this, we always recommend regularly reading the different web pages on which cloud providers publish news and updates regarding cloud services, such as the following:

- Azure updates: `https://azure.microsoft.com/es-es/updates/`
- What's new on AWS feed: `https://aws.amazon.com/new`
- What's new on GCP Cloud: `https://cloud.google.com/blog/topics/inside-google-cloud/whats-new-google-cloud`

With this in mind and without further ado, let's jump in to explain what we can do in these phases.

Data sources selection

Choosing the right data sources for our automated estimation process may not be easy, as they will be the foundation that we will build from to get the most exact data possible for our FinOps initiatives and estimations. We now know that we can get this data from multiple sources, but how can ensure that the information we'll get will be robust and trustworthy?

We also need to understand the limitations of each data source to be able to build something on top of it. For example, the Azure Pricing API does not include the specifications of virtual machines or databases in detail. If this information is needed, we will need to add another data source to provide it and correlate and aggregate data from different sources. This does not mean that one data source is better than the other, just that different data sources may require different approaches.

Another point, which we've covered already and need to take into account, is whether the APIs offer historical data or not. For example, the public Azure Pricing API does not offer historical data. Due to this, if historical data is needed, we will need to provision an automated process that queries the API and gets data from it from time to time to store it in a database or data lake, to overcome this existing limitation.

As an example, if we wanted to have a dashboard for cost optimization in Azure that calculates the potential savings for rightsizing and scheduled shutdown of virtual machines, we would need information from the following sources:

- **Azure Cost Management** to get the price sheet for the current contract with Custom Pricing (if we have a contract with Microsoft that includes some discounts).

- **Azure Monitor** to get the current CPU and memory consumption of the virtual machines we are going to analyze. From Azure Monitor, we can also get the hours of use or uptime for all virtual machines.

- **Azure REST API**, **PowerShell**, or the **Azure CLI** to get specifications and configuration of each virtual machine (number of vCores, RAM, and so on).

Once we've selected our data sources, we are ready to move on to data consolidation.

Data consolidation

Before we embark on the data consolidation phase, our recommendation is to devise a data model or data structure that fulfills our current and future needs. We know that considering possible future needs now may be challenging, but we need a data model to be flexible and adaptable to change.

This data model may change over time, as nothing in the cloud stays static. The service provider may add new fields or change existing ones, or financial or technical teams may request new data to be processed and included in our calculations.

One of the decisions we also need to make in this phase is how frequently we should refresh pricing data from the cloud provider. As we already explained, pricing does not change frequently but it does change. We consider a monthly or weekly data refresh as a good starting point, after an initial full data ingestion.

Another key point to decide upon in this phase is the currency that will be used. Some cloud providers offer pricing data in dollars, which means we will need another processing layer to convert pricing to our currency (if it's not dollars, of course) using current currency exchange pricing, which means another data processing layer.

After the data is processed, we need some kind of storage or database platform to host the resulting data. Depending on the use case and size of the data, we will choose a **storage service**, **database**, or **data lake** to host our data. From a cost optimization perspective, a storage service or a data lake may be the best choice. We will explain the differences between storage services, data lakes, and databases in *Chapter 8*.

Now that we have our data in place, let's describe what we can do with all this information.

Estimation calculation

A **cost estimation** is a projection of how much a specific solution would cost using current pricing and specific conditions and configurations. When a cost estimation is made by FinOps teams, it is usually optimistic and tailored to result in maximum savings. Due to the nature of estimations, they may be affected by things happening in parallel in the organization, or even other FinOps initiatives implementations, such as Reserved Instances being purchased or the use of Spot instances. On top of this, indirect costs can have an unexpected impact on the overall cost, as we have already explained. Because of these external factors, we won't always be able to nail our estimations, and we need to accept that an estimation is a starting point or a reference, not something that's written in stone.

Additionally, when we calculate estimations manually, **human error** may be an important factor as well. If the calculations are automated, the risk of mistakes is lower, but then we will need our automation to work like a well-oiled machine, which takes time and effort.

Now we have the million-dollar question, which is, what we will use to automate calculations? This point is crucial as our calculations need to be refined and precise and not depend in any way on manual steps and calculations. Building automation processes always requires work and effort to set up, but once the process is done and properly tested, it does not require much maintenance.

There are a number of services available for such calculations, such as Databricks or AWS Glue. A word of advice, though: these tools require specific knowledge of technical teams, so a **cost-benefit analysis** should be considered for this.

Let's say that we create an estimation with some new cost optimization improvements on a current solution with virtual machines. If the cloud provider publishes a new instance type that is cheaper and performs better, it would be good to be notified of this and, therefore, update our estimation. This is what we'll cover next.

Change notification

Once a big enough change is detected in our data sources data, such as pricing increases, we need to automatically notify our cloud and financial teams when pricing changes occur accordingly using any way of notification (Slack, Microsoft Teams, ITSM ticket, email, and so on), and our estimations and dashboards should be updated accordingly. Alongside this update process, we also recommend having a field to specify when data was last updated, so we have visibility over changes that occur over time.

Notification processes are so important in this case, as it is impossible to manually look at all this data gathered from APIs to detect changes. If this data is lost, we will be losing cost optimization opportunities, which is not desirable at all.

Data update

Once our data is updated at the end of this process, it is also good practice to notify teams of these changes, as they can have an impact on running initiatives. Updated data can deprioritize current initiatives to focus on others that may offer more potential savings with less effort, feeding us valuable information to make key decisions in each step of this long cost optimization iterative journey.

Updating this data usually impacts the following assets:

- Cost estimates for future projects
- Operational dashboards
- Financial dashboards
- FinOps dashboards

Once our data is updated, we may choose to keep history information for reference if we want to keep full traceability of past scenarios and initiatives, which can be useful along the way to explain the reasoning behind decision-making.

As a small example to illustrate the whole process, this is how it would look in AWS to update the pricing data used for three estimates created for future projects that need to be up to date:

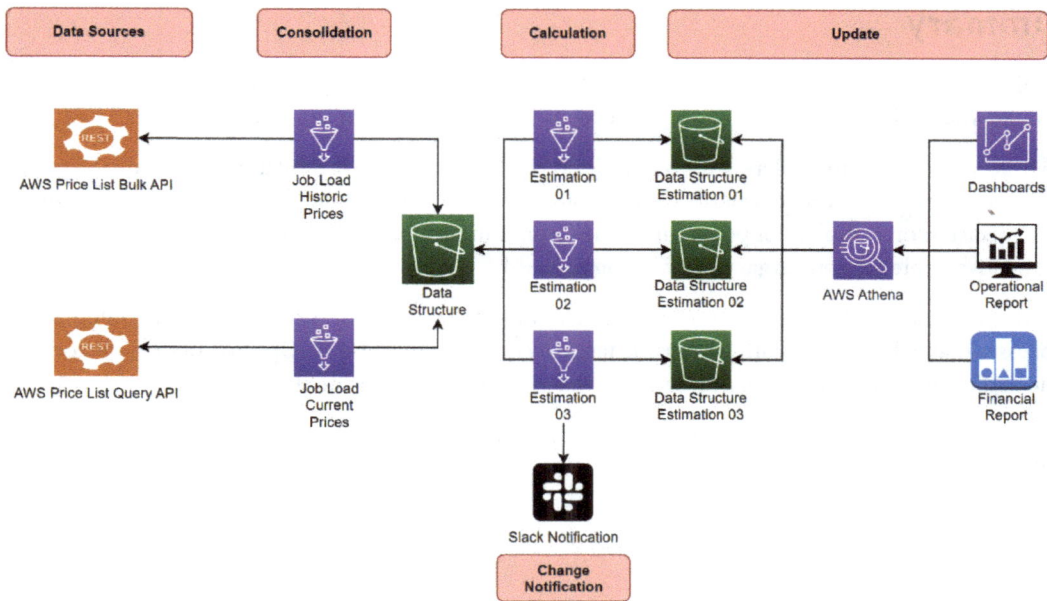

Figure 4.10 — Example of fully automated estimation in AWS

In this example, we have used the following AWS services:

- The **AWS Price List Bulk API** to load historical pricing data

- The **AWS Price List Query API** to load current prices

- **AWS S3** as our storage service, on which we will keep the files that result from our automation process

- **AWS Athena** as our analysis service to show the result of our estimations

- **AWS QuickSight** and other dashboards that feed on Athena information

With this, we have come to the end of the chapter.

Summary

In this chapter, we have learned about the different means and concepts that we need to take into account when calculating the TCO and the cost of solutions hosted in the cloud.

We have also covered the different tools that cloud providers have made available to estimate the total cost of solutions and to find out the pricing of cloud services. Understanding these tools and increasing your precision on estimates of potential savings from FinOps initiatives and costs of future projects will be invaluable for your organization going forward.

To bring this chapter to a close, we also explained how to go from manual estimations to fully automated processes that will avoid repeating the same tasks over and over, saving FinOps practitioners precious time and evolving the quality of our data and the information we can get from it.

In the next chapter, we will analyze how to build FinOps dashboards and reports to keep improving visibility and control of cloud costs.

5

Improving Cost Visibility with Dashboards and Reports

Understanding our cloud costs comes before any cost-saving initiative. To understand our costs, we must comprehend the key cost drivers that shift costs up and down and how the costs are distributed across different services present in our cloud environments.

Dashboards and reports are key assets that we can use to ease this process, and let us check the current status and track the development and implementation of FinOps initiatives. The purpose of this chapter is to learn how to use the built-in dashboards and reports available from our cloud providers, as well as how to create our own, and understand the information that we can select and highlight to increase cost visibility and optimization.

To be able to work on reports and dashboards, we will also cover the financial basics required for this exercise, such as calculating trends and understanding financial key concepts that can help us throughout FinOps practices.

Making use of the available tools before considering creating our own is one of the key points that will be covered as a way to minimize the effort required.

In this chapter, we will cover the following topics:

- Understanding cloud invoices and billing data
- Dashboards and reports
- How to prepare cost evolution reports and their importance
- How to prepare FinOps dashboards

Understanding cloud invoices and billing data

One of the first steps to understand cloud costs is to understand how cloud services are billed. In this section, we will learn how to read invoices and cloud bills and understand how billing in the cloud works.

Understanding cloud billing data is not easy at all, and to do so, we must also understand the challenges we might face:

- It often requires a great deal of knowledge about cloud services and their cost drivers.

- Cloud providers constantly make changes, such as changing how a service is billed and publishing or retiring cloud services from their portfolio, which will impact billing as well.

- Billing data and pricing have a strong dependency on our agreement and current contract with our cloud provider. Prices can change over time depending on our contract.

- There are special purchase models, such as **Reserved Instances** and **Spot virtual machines** (which are priced differently from pay-as-you-go models). Using these purchase models will also increase bill complexity.

Each cloud provider has a different way of billing us for our cloud services, with similar yet slightly different takes. We are not going to delve into detail on the specifics as it goes beyond the purpose of this book and it is ever-changing, but we would like to provide some links where we can look up the different fields and their purpose in each cloud provider's billing data. In these links, you will find the fields that are used in some services that allow for billing data to be visualized and queried:

- **Azure**: `https://learn.microsoft.com/en-us/azure/cost-management-billing/automate/understand-usage-details-fields`

- **AWS**: `https://docs.aws.amazon.com/cur/latest/userguide/data-dictionary.html`

- **Google Cloud**:

  - Standard billing data usage: `https://cloud.google.com/billing/docs/how-to/export-data-bigquery-tables/standard-usage`

  - Detailed billing data usage: `https://cloud.google.com/billing/docs/how-to/export-data-bigquery-tables/detailed-usage`

Any cloud invoice has a standard structure and common features to describe which services we have used in what amount, as well as the rate applied for these services.

Any bill or invoice, not just cloud ones, should at least include the following elements:

- An itemized breakdown of services and products used
- The rate or price paid for these services
- The quantity, which refers to how many products have been used of each type

The costs are calculated with the help of the following formula:

Cost = Rate x Quantity

When working in the cloud, *Quantity* can be both the number of services used and, depending on the cost driver, can also refer to how many hours or minutes a service has been used. This fact introduces another key variable that we need to check in the bill, the **unit of measure**, which will be different for each cost driver. When purchasing storage services capacity, for example, the unit of measure will be *GB/month*, while in a **Virtual Machine (VM)**, it can be the *hours* for which the VM has been used.

As we already explained at the beginning of this section, even though there are differences in each cloud provider's take on cloud billing, we can find some common ground between them in the fields that are commonly used for cloud invoices, which are the following:

- **Billing Period Date**: The time range for the scope of the invoice.
- **Invoice ID**: The unique identifier for the invoice for accounting controls. Usually, if the billing period is not yet closed, it may appear blank.
- **Account ID**: The unique identifier for the contract or agreement with the cloud provider on which this invoice is under.
- **Currency**: The currency used for the invoice. The currency can have a lot of impact on our cloud costs if the currency exchange rateis not stable over time.
- **Location**: The geographic region that each invoice line belongs to. This parameter also has an impact on the rate we pay for services.
- **Service ID**: The unique identifier of the service that corresponds to each invoice line.
- **Service description or service name**: The common name of the service that corresponds to each cost line.
- **Resource ID**: The unique identifier of the resource (if there is one) that this invoice line cost belongs to.
- **Resource name**: The common name of the resource (if there is one) that this invoice line belongs to.
- **Usage period**: The time range in which the service was functioning and for which we are billed.
- **Meter**: This field goes side by side with **Unit**, and it describes what is measured.

- **Unit**: A metric that will be used to calculate each service cost based on units that measure the **Meter** field.

- **Amount**: The number of units that are billed for the usage of services.

- **Rate**: The cost of each unit.

- **Cost**: The final cost of the service considering all units. This price is calculated by multiplying the amount per unit rate.

As an example, here is what an Azure invoice looks like:

Category	Subcategory	Unit	Consumed	Cost	UnitPrice	Currency	Meter	Resource Group	Resource Locati
SQL Database	SQL Database Single/Elastic Pool General Purpose - Compute GenS	1 Hour	76,00	129,06	1,70	NOK	vCore	rg-fin-ok4-pro	westeurope
SQL Database	SQL Database Single/Elastic Pool General Purpose - SQL License	1 Hour	76,00	77,05	1,01	NOK	vCore	rg-fin-ok4-pro	westeurope
Virtual Network	Virtual Network Private Link	1 Hour	648,00	65,72	0,10	NOK	Standard Private Endpoint	rg-fin-ok4-pro	westeurope
Storage	Standard HDD Managed Disks	1/Month	0,90	53,75	59,72	NOK	S10 Disks	rg-fin-shared-hub	westeurope
SQL Database	SQL Database Single/Elastic Pool PITR Backup Storage	1 GB/Month	17,04	25,75	1,51	NOK	ZRS Data Stored	rg-fin-ok4-pro	westeurope
SQL Database	SQL Database Single/Elastic Pool PITR Backup Storage	1 GB/Month	16,74	25,30	1,51	NOK	ZRS Data Stored	rg-fin-ok4-pro	westeurope
SQL Database	SQL Database Single/Elastic Pool PITR Backup Storage	1 GB/Month	15,97	24,13	1,51	NOK	ZRS Data Stored	rg-fin-ok4-pro	westeurope
SQL Database	SQL Database Single/Elastic Pool PITR Backup Storage	1 GB/Month	15,71	23,74	1,51	NOK	ZRS Data Stored	rg-fin-ok4-pro	westeurope
SQL Database	SQL Database Single/Elastic Pool PITR Backup Storage	1 GB/Month	14,72	22,24	1,51	NOK	ZRS Data Stored	rg-fin-ok4-pro	westeurope

Figure 5.1 – Example of an Azure invoice

Note the different services and how they are measured using different units and meters.

We should keep in mind that our invoices will contain direct costs as well as indirect costs such as data egress or data transfer, which can generate a lot of billing information, making it harder to decipher our billing data at first sight. It is important to also understand how these costs work and how they are measured to avoid unwanted surprises.

Cloud providers allow us to export billing information to plain files that we can store in cloud storage services. Keep in mind, though, that these exports can generate large volumes of data that will need to be processed and curated. This process requires a great deal of knowledge and technical expertise in order to get the most out of this data, which not all companies can afford.

Due to this fact, our recommendation is to make use of billing tools, as well as dashboards and reports, starting with the ones that are offered by cloud providers out of the box, to navigate billing data without added complexity. This will allow our technical and financial teams to have a centralized and complete view of cloud costs and billing data since the first days of FinOps practices.

Now that we know the basics of our cloud billing data, let's move on to the differences between a dashboard and a report and what each one brings to the table.

Dashboards and reports

Dashboards and reports are invaluable tools to show information and are often used to showcase information about cloud services such as monitoring, operations, or KPIs for all kinds of organizations. They can both increase cloud cost visibility and help us understand how our costs are distributed across services, projects, or any other logical grouping that we need. However, there are some differences between them that we want to highlight before going forward.

The main differences between a report and a dashboard

The main differences between a report and a dashboard are detailed in the following table:

Concept	Dashboard	Report
Data refresh rate	Real time	Captures the status at a specific moment in time, even though it can be done regularly
Purpose	Broader scope, covering multiple areas of interest in the same domain	Specific purpose that is aligned with the data shown
Accessibility and audience	Permissions are often required to access a dashboard. Multiple teams can use the same dashboard for their day-to-day work.	Delivered directly to stakeholders and interested parties. Access to reports is usually more limited.
Data visualization and format	Interactive data visualization using filters, dimensions, drilldowns, and other data visualization mechanisms	Fixed visualization that contains the exact data needed with a specific format, as agreed previously with the report audience
Effort required and creation process	More generic and faster to develop. It often begins with a baseline of information that can be expanded further as the dashboard matures.	The audience is involved in its creation, defining and selecting the data required, as well as validating the contents of the report

Table 5.1 – Dashboards versus reports

Now the differences between them are clear on a conceptual level, let's proceed to review how they provide value to organizations.

Key benefits

Dashboards and reports provide some key benefits that make a difference, by visibilizing key organizational insights, which can help organizations improve their processes in many ways. Reports show summarized key information to stakeholders that takes a lot of effort to gather and calculate, while dashboards are a tool used to accompany daily operations.

Both empower teams to make informed decisions using the information they provide. In the FinOps domain, they answer fundamental questions such as the following:

- Are we using the correct technology?
- How much does each of the projects we run cost?
- How much does each environment cost?

They also provide many other key insights to make technical teams accountable for cloud costs and the consequences of each action and change in monthly bills.

In addition to these points, dashboards and reports centralize information, and multiple teams can draw on them as the central source of information, which helps align teams in many ways. Despite this fact, we recommend that FinOps information for each team is shown separately by segregating permissions or using mechanisms such as **row-level security**, which limits the visibility of a user to a specific scope depending on their permissions. This avoids unhealthy competition between teams, leaving the centralized view for the main FinOps functions.

Dashboards from another view – simulators

To close this introduction to dashboards, we want to provide an additional way in which they can be used that we have used for our FinOps work, which has proved to be really useful.

The standard way of using dashboards is to view information and visualize metrics, or any kind of data we want. They provide a highly interactive way of querying data by using filters, selectors, and other data-filtering mechanisms.

Dashboards can also be used as *What if?* simulators, where we can test different scenarios and understand the outcomes. This has a clear application on different parts of FinOps practices, to accompany initiatives such as Reserved Instances and disk and VM rightsizing, which will be described in depth later in this book.

Creating a dashboard with this idea in mind can require different degrees of effort, as you need to build up a custom dashboard from ground zero to act as your simulator, but in some situations, it can provide great insights and prove to be very useful. Cloud provider calculators are examples of these scenario simulations.

Example – disk simulator dashboard

Disk rightsizing is a key initiative that is part of the *Optimize* pillar and will be thoroughly described later on in this book. It consists of choosing the most adequate disk for our VMs, ensuring it is properly sized to each specific VM without falling short or having too much throughput and disk speed.

During our FinOps experience, we needed to find the sweet spot for our disks, so we decided to build a simulator to help us in this process.

Concept and main goal

The simulator should help us find the break-even point where a specific disk tier stops being optimal and aid us in simulating scenarios to choose a new disk tier that is properly sized and adequate.

Disks have many parameters that we can configure, such as capacity, size, and throughput, and it can be challenging to navigate through all the different tiers. Some disk tiers offer fixed parameters and others let you customize these parameters, such as **Azure Ultra Disks** or **AWS EBS io1/io2**. Having this wide range of options can be overwhelming and may make estimating costs and choosing the best option difficult.

The objective of this simulator is to tackle these challenges by presenting a tool similar to a pricing calculator that only focuses on disks, and that is able to find break-even points between tiers and simulate different scenarios. It can be used both to determine the best tier for existing disks and to correctly size disks in new projects.

In this case, we will exemplify how to build this simulator in **Amazon Web Services (AWS)**.

Technologies used

To begin this exercise, one of the first points is to choose the right platform or technology to build this dashboard. There are many ways to build this dashboard but, in this case, we chose AWS tools that we were familiar with and knew well.

We can describe the different stages of processing and showing our data by using the following diagram:

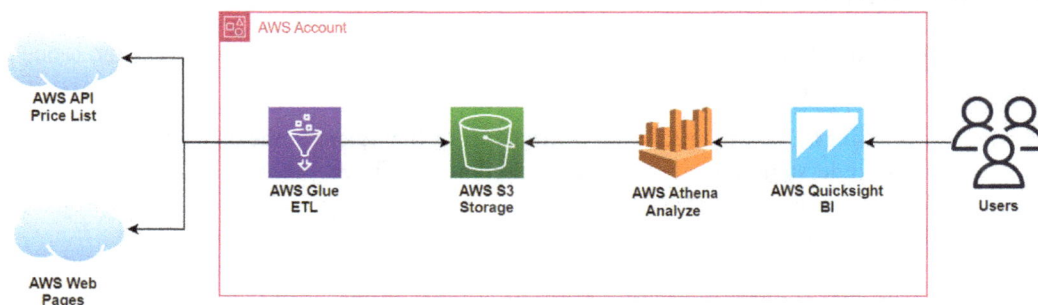

Figure 5.2 – Example of disk simulation dashboard

In summary, this is what we did:

1. We chose **AWS Glue** to implement our **extract, transform, and load** (**ETL**) process. This service allows us to collect all the required information for our dashboard (which is mainly the different disk types of cost drivers and their characteristics) and build our dataset. Once the dataset is built and its dimensions clear, we can use Spark processes to calculate all possible permutations with the different parameters and their corresponding cost. The idea is to create a complete data product with this information, which we will show later on in this process.

2. To store our results after data processing, we will use the **AWS Glue Data Catalog**, where we can define a schema for our data and define the tables we need after the database has been created in the service.

3. All this information needs a storage platform where we can put our data during the different phases of our ETL process. We opted for **AWS S3 buckets** as our storage platform.

4. As our analytics tool, we selected **AWS Athena**, which allows us to perform all the required analyses on our data, such as creating specific views on top of the different tables or filtering and processing the data that we collected from AWS Glue.

5. As the last point of our process to build this dashboard, we chose **AWS QuickSight** as our business intelligence tool. AWS QuickSight allows us to create a presentation layer that will show all the required data for our dashboard and let us arrange the information in any way we want, creating calculated fields, letting us choose the format of our data, and adding filters and selectors for our simulator.

This is the train of thought that we followed when we created this dashboard, which provided useful information for technical teams more interested in getting the most out of AWS EBS disks.

Having explained how we work and create a dashboard, we will now move on to cost evolution dashboards and reports, which are invaluable tools that help us work on the *Inform* and *Optimize* pillars.

How to prepare cost evolution reports and dashboards and their importance

As we have explained time and time again, understanding how our cloud costs are distributed is one of the first points we need to tackle for any FinOps practice.

But as important as this is, we also need to add a fundamental ingredient to the mix, which is to have clear visibility and understanding of how our costs shift up and down, growing or decreasing over time, which should be aligned with the usage we make of cloud resources, how many new projects are developed, as well as how many cloud resources are created or retired over time.

This alignment between costs and cloud demand is of the utmost importance, as misalignment can quickly lead to cloud waste and overspending. This misalignment can happen for many reasons outside our control, such as misconfiguration, changes in cloud services or how they are billed, and changes in how we make use of our resources, among others, and it is on FinOps practitioners to detect these unwanted patterns and solve them as soon as they appear.

This view is essential for any organization to understand how much they spend in the cloud, to properly adjust cloud budgets, and to try to estimate the costs that will be incurred in the future.

In essence, we need a complete view of our costs over time. We refer to these kinds of views as **cost evolution dashboards and reports**.

Such assets should show the evolution of our cloud costs and allow us to filter by key dimensions such as the following:

- Environment
- Business unit
- Project
- Region

These styles of reports are the ones provided by tools such as **Azure Cost Management**, **AWS Cost Explorer**, and **GCP Cloud Billing Reports**.

Such tools are a great starting point for cost analysis, but if we want to dig deeper and deeply analyze costs from a financial perspective (such as calculating trends), these tools will fall short, so we may end up exporting data to CSV files and doing a manual analysis by ourselves.

These manual analyses, of course, should turn into full-blown reports and dashboards once the needs of our clients, stakeholders, and technical teams have been assessed and are clearly defined and agreed to.

As an example, this is what a cost evolution report in Azure can look like:

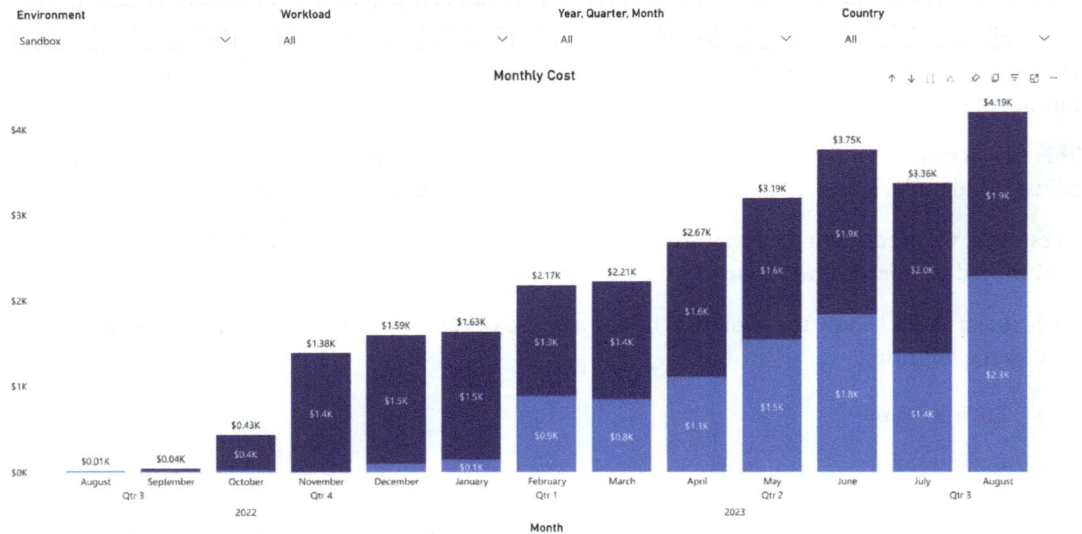

Figure 5.3 – Example of cost evolution dashboard

In this case, we have used Power BI as our reports platform, and we are extracting billing data from Azure Cost Management REST APIs on a monthly basis.

Here, the report is dynamic, and the report user can filter information and get really interesting cost insights, such as filtering per business unit (country, in this case) or per environment or application.

Now you have seen an example of what a cost evolution report looks like, let's move on to our next topic, which is a summary of some financial basics that can help us in our work as FinOps practitioners.

Financial basics

FinOps is about bridging the gap between teams – mainly between technical teams, FinOps practitioners, and financial teams.

One of the problems that we have experienced first-hand is that technical teams may not be familiar with even the most basic financial concepts. FinOps is about costs, so part of the preparation work that needs to be done is to reduce this gap by helping our teams comprehend at least the most basic financial concepts.

In this section, we will go through some of these concepts and how we can use them as part of our FinOps practices in our daily work.

Savings and waste introduction

FinOps is about cost reduction, among other things. Cost reduction, which can also be referred to as **savings**, can come from multiple sources, such as deleting resources that are not used, rightsizing current resources, or moving to other platforms or technologies that are more cost-efficient. Savings usually come from reductions in both *Quantity* and *Rate* (remember that *Cost = Quantity x Rate*), which comes through cost avoidance and rate reduction, respectively:

- **Cost avoidance**: Generating savings before actual costs are incurred. This means paying just for what we use – nothing more, nothing less – with our goal being to make the most out of our cloud resources.
- **Rate reduction**: Generating savings by reducing the rate we pay for services. This can involve changing parameters such as resource configuration, region placement, or any other parameter that can have a direct impact on rates.

Savings must be tracked and quantified. You could be doing top-notch FinOps work, but without clear results for stakeholders and interested parties, FinOps practices will always be up to the eye of the beholder. To demonstrate results, you need objective terms that can be measured. To measure FinOps success, savings are one of the first starting points. Our cost evolution dashboards and reports ideally should include the savings we have generated alongside the costs, as they are equally important.

But we also need to consider the other side of the coin that can be measured, which is cloud waste. **Cloud waste** is the excess costs we are paying for services that are not optimized, in the form of underused or unused resources, badly designed solutions, and non-optimal cloud usage in general. This waste can also be measured, as with savings, if we know our environment well enough.

As FinOps practices evolve, we should see cloud waste being reduced while savings are generated. If we have no cloud waste, then there are no savings to be made. This is why there will come a point where no more cloud savings can be generated if our cloud environment is fully optimized. This is fundamentally why FinOps is not all about savings. A small word of advice though: it can be really challenging to get to that point, as it will require a lot of work to measure and identify how much money is wasted by non-optimal configurations.

Trend lines and tendencies

A **trend line** is a superimposed line that can be added to a graph to highlight the overall direction of data. Trend lines can help us see the hidden high-level tendency of our costs, which eases the process of future cost estimation. This is an example of a trend line:

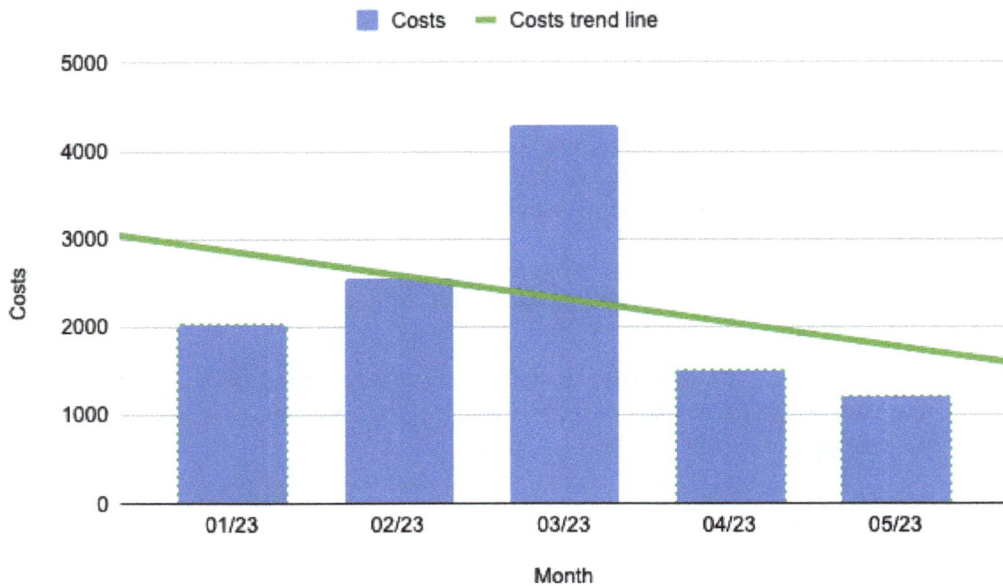

Figure 5.4 – Example of a trend line in a cost plot

In addition to trend lines, we can accompany our cost analysis with the following parameters that can help us abstract day-to-day data into tendencies and patterns behind our cost data:

- **Year-to-Date (YTD)**: This refers to the period of time from the first day of the fiscal or calendar year to the present.

- **Month-to-Date (MTD)**: This refers to the period of time from the first day of the month to the present.

- **Year-over-Year (YoY)**: This refers to the financial comparison between the costs of the previous year to the costs of the current year. This KPI can help us see the big picture and gauge our cloud journey evolution in objective terms.

- **Month-over-Month (MoM)**: This refers to the financial comparison between the costs of the previous month to the costs of the current month. This KPI can help us see the big picture and gauge our cloud journey evolution in objective terms.

Purchase options and amortization

Cloud services are known for having adopted a pay-as-you-go payment model, in which you are billed for the services you use.

But for some services, cloud providers offer a purchase model in which you can prepay or commit to a certain capacity or cloud services usage in exchange for some additional discounts. We call this payment model a **purchase option**.

To analyze the impact of purchase options on our cloud financials and measure the discounts and cost benefits we get from it, we can use **amortization**.

Amortization is the act of spreading an upfront or unique payment cost over time. Amortization can be used for upfront paid services, such as Reserved Capacity, Reserved Instances, or Savings Plans, which we will cover in detail later on in this book.

Tracking savings to initiatives and adding milestones

As we already explained, tracking the savings we generate is essential to help us visualize the success of our FinOps practices.

Now that the different types of savings are fully understood, we should aim to include savings and waste as new plots that we can show in our cost evolution dashboards and reports.

Having these views will also allow us to interpret and analyze trend lines and add this valuable information to cost tendencies. Having this complete view will enable us to gauge the impact of our cost optimization initiatives.

Let's imagine that we have all this information at hand. To close the circle, we want to tie savings to ongoing and completed cost optimization initiatives. In some cases, savings can come from different sources, so this exercise can be really complex.

A simple way to begin tying savings with initiatives can be to just add each initiative's milestones to our reports, which will help us explain the cost trends and their shifts up and down. In addition to this, we can also enrich our cost evolution dashboards and reports with other major project milestones, even those not relevant to the FinOps domain, and changes that can impact our cloud costs, such as a new application deployment or decommissioning of cloud resources.

As an example, this is how our initial cost evolution dashboard could look after we incorporate all this information:

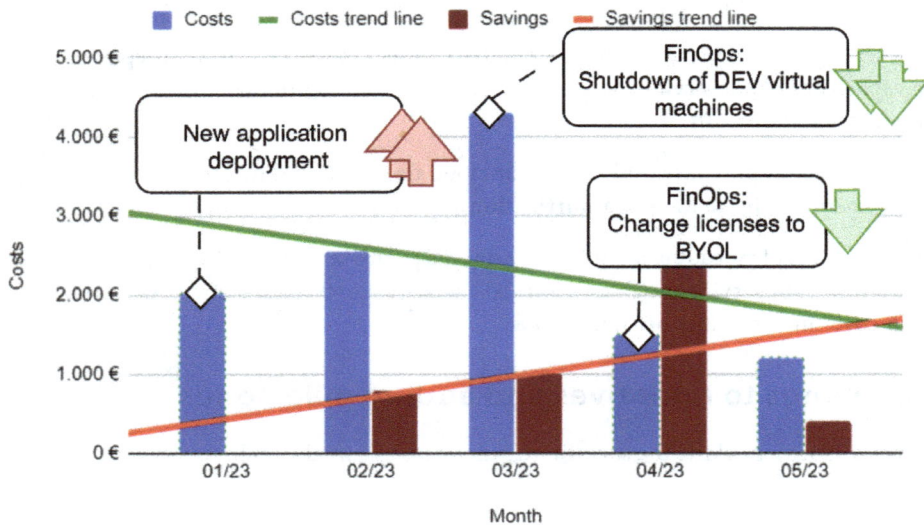

Figure 5.5 – Extended information shown in the cost plot

Note that savings and costs have their own trend lines, and we have incorporated project milestones (**New application deployment**) with FinOps initiatives, which helps in explaining cost variation month to month. This figure represents just a suggestion of how the information could be shown.

We could also incorporate the *YoY/MoM* and *YTD* financial concepts into savings as well once we have the savings information there.

Now that we have a basic grasp on how to prepare cost evolution reports and dashboards, let's introduce a key concept that can set these reports and dashboards apart while providing great value to the business, which is the concept of unit economics.

Unit economics

To close this section, we will introduce the idea of **unit economics**, which is a key concept that aims to create value by aligning cloud financial information with objective business parameters that are expressed on a per-unit basis.

Let's imagine that we are working in an organization that develops web applications that are offered to our customers. After some work, we have our cost evolution report and our cloud costs are more or less stable, without major ups and downs, over a one-year period. If we only take the costs into consideration without any more insights, an external observer (who is not present on a day-to-day basis in the cloud operations) may think that there is not much effort on cost optimization or any major projects.

The cost evolution report may look like this:

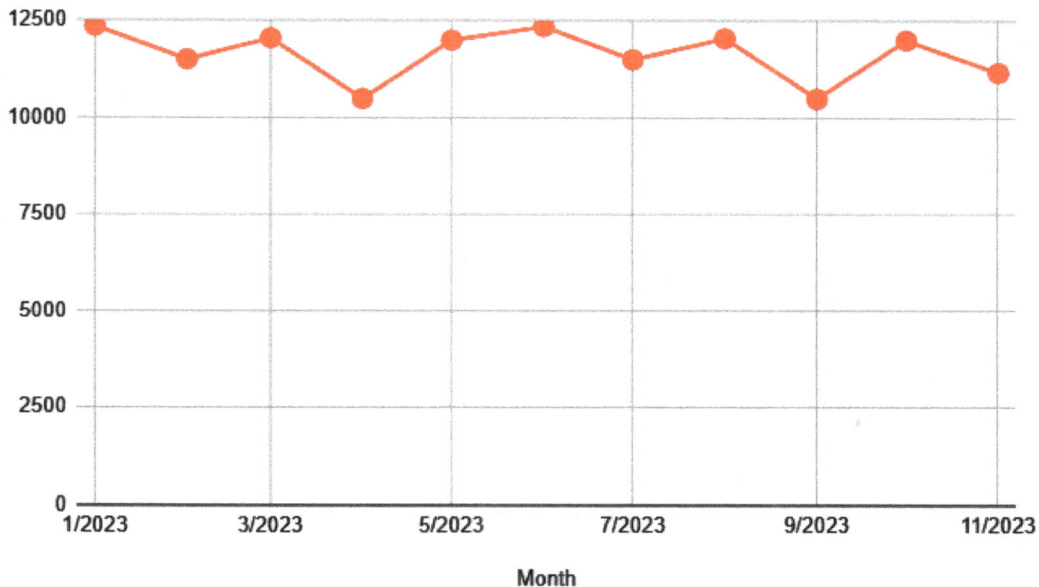

Figure 5.6 – Unit economics cost evolution example 1

It is clear that this information is not enough to interpret how we are performing and the value we are creating with the cloud. Let's consider two scenarios:

- **Scenario 1**: Even though our cloud costs are stable, our application has performed incredibly well, triplicating its users over the same one-year period. From a business perspective, the organization is thriving and maximizing its revenues.

- **Scenario 2**: Even though our cloud costs are stable, our application is losing users over time and it's not being well received, to the point that after one year, our users have halved in number. This resulted in great losses in revenue for this organization.

Let's imagine that we add a KPI for our business to this representation, which could be the *number of registered users in our application* for both scenarios, displaying scenario 1 with square points and scenario 2 with triangle points on top of the cost information (even though they can be in completely different scales of magnitude).

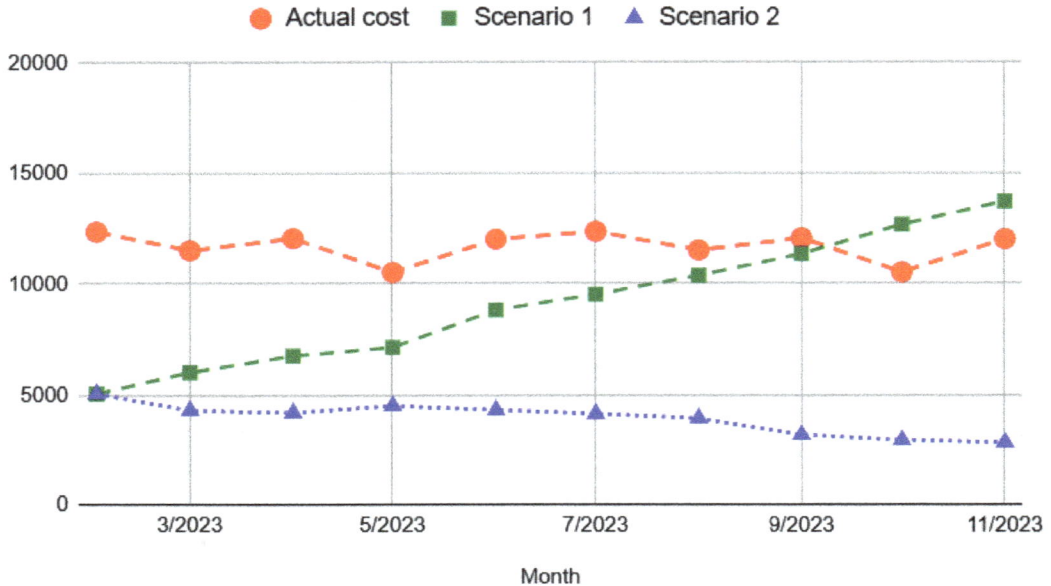

Figure 5.7 – Unit economics cost evolution example 2

Regardless of which scale is represented and used on the *y* axis, it is clear from the figure that our costs represent really good news in scenario 1, while scenario 2 paints a much darker and more challenging picture for our imaginary organization.

How could we measure this performance using this set of data? This is where unit economics enters the picture, by making use of a unitary parameter that we can objectively measure to analyze our performance. For this example, we could define a new KPI that represents *cloud costs per user registered*. This KPI will measure how much we are paying for each client that makes use of our application. This KPI can be interpreted in the following manner:

- Higher values on this KPI will mean that we are paying a lot of money for each user, which could be translated as we are not fully optimizing our cloud costs

- Lower values on this KPI will mean that we are making the most of our cloud resources and being optimal

If we just represent these KPI values instead of cost and user data, we get the following figure:

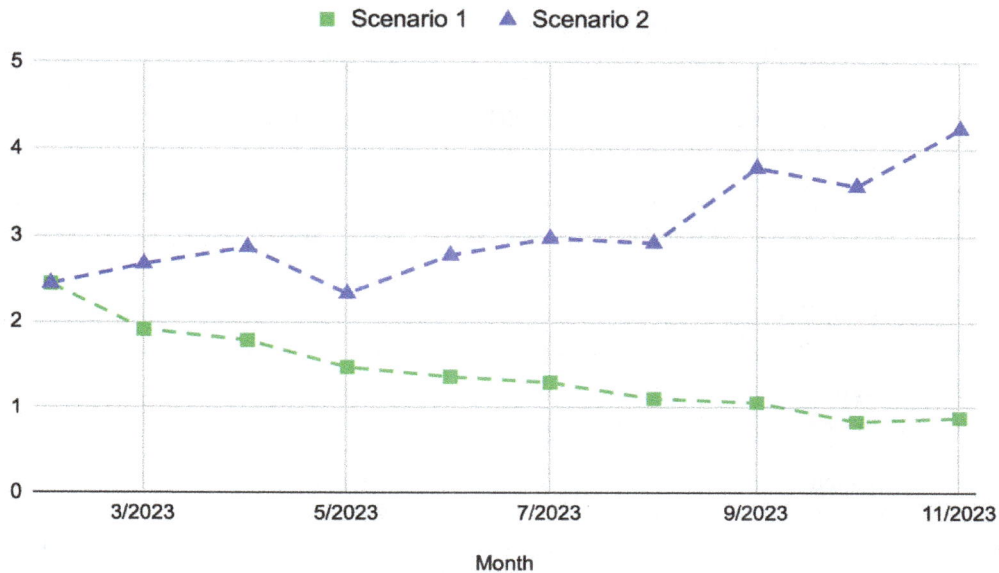

Figure 5.8 – Unit economics cost evolution conclusion

From the figure, we obtain clear information in both scenarios, which start from the same place, while having a KPI that we can try to maximize as much as possible to strive for cost optimization.

In this final figure, we provide key insights for our business by being able to measure success and performance while increasing the visibility of cloud cost optimization. This is what unit economics is about.

Measuring unit economics is not a simple task, as it requires a great deal of knowledge about cloud services and their key cost drivers, but the final outcome clearly outweighs the effort required to set this concept up. It also requires a great deal of information beforehand, as we won't be able to do any analysis without it.

Unit economics bridges the gap between technical teams and business and financial teams by tying KPIs from each of these domains together, helping them speak the same language.

As an idea to get started, we have used this concept to measure parameters such as the following:

- We could continue the example that we have used of cost per customer and divide the costs into the different purchase models in the cloud, which are IaaS, PaaS, and SaaS. This will result in IaaS, PaaS, and SaaS unit costs. Having this information can also help in measuring the degree of **modernization** in an organization, which, over time, should shift from IaaS to PaaS and SaaS if we are evolving into using fewer managed services.

- To go for a more business-focused KPI, we could directly **correlate organizational revenues over cloud costs**, which could help for deep analysis to seek any kind of impact between these key metrics.

- **Cost per vCPU (IaaS computing)**: This KPI offers really interesting information, as it represents how big (equals costly) the VMs used throughout an organization are.

- With the same idea as the previous point, we could measure cost per GB in our IaaS disks, cost per GB in storage, and other similar unitary KPIs.

We can use any metric we want, even metrics that are not related to costs, as long as they provide some useful information. There won't be a unique metric to rule them all (sorry for the *The Lord Of The Rings* reference; we could not help it), but instead, a lot of metrics that will help us build a case or a message that is fully supported by data. A specific metric can be invaluable for one organization and not useful at all for another, so our recommendation is to research and propose as much as possible, as we will get better from iteration (remember the **crawl, walk, and run** process?). Remember to invite financial and business teams, as well as technical teams, to participate in this process and encourage collaboration, as they may offer new points of view or ideas that you will all benefit from.

As some closing thoughts, these are the benefits of unit economics:

- It creates value by providing new key insights
- It improves decision-making through objective and measurable metrics
- It helps in measuring cost optimization, as well as performance and success
- It fosters cloud cost accountability and team collaboration

With this explanation of unit economics, we close our section on cost evolution dashboards and reports. In the next section, we will highlight some reports and dashboards that we can use as a starting point for our analyses, as well as provide some recommendations and ideas to build your own custom dashboards from scratch.

How to prepare FinOps dashboards and reports

When setting up a FinOps practice, there may be a point where we need to begin creating our own dashboards and reports. It can be difficult to choose whether we should use existing solutions or build our own. Both of these will have an impact on costs, either by paying for a product license or for the development that is required.

In this section, we will try to provide as many tools as possible to ease this process. We will describe some built-in resources that we could use from ground zero, as they are already available from cloud providers, and also give recommendations and ideas for the process of creation itself.

Existing dashboards and reports

The first recommendation we want to make is to start using what is already available, especially in the first steps of FinOps practices. As organizations mature, their needs can be also clearer and more defined, and it can be worth investing in developing a custom solution. Until that moment comes, we can use several tools as a source of inspiration and information, which will reduce the time required to create value from the first steps of the practice.

In this section, we will analyze some assets that can be used throughout this process. Keep in mind that some of these assets may incur additional costs.

Azure Cost Management Power BI report

This is a dashboard offered for Azure that allows us to analyze our cloud costs from Power BI, which is the **business intelligence (BI)** offering from Microsoft.

This dashboard is currently only available for **Enterprise Agreement** customers, which leaves out many customers using other contract models such as **Cloud Solution Providers (CSPs)**. We will discuss the different contract models later in this book.

Windows Server AHB usage report

AHB vCPUs consumed (last 30-days)

This Windows Server Azure Hybrid Benefits usage report shows how many have VMs have AHB enabled, and of those VMs how many AHB vCPUs or cores are being utilized.

For more information around Azure Hybrid Benefit see: https://azure.microsoft.com/en-us/pricing/hybrid-benefit/

Windows Server AHB enabled (last day)

6

Windows Server AHB vCPUs used (last day)

64

AHB enabled SKUs with less than 8 vCPU (last day)

	VCPUs
Standard_DS1...	3
Standard_D4s...	1

SKUs with 8+ vCPUs where AHB not enabled (last day)

	VCPUs
Standard_B8ms	2 (8)
Standard_DS13...	2 (20)
Standard_F72s...	2 (24)
Standard_D15...	1 (72)
Standard_NC2...	

Resource Details (last day)

Date	SKU	VCPUs	AHB vCPUs	Azure Hybrid Benefit	SubscriptionId	SubscriptionName	ResourceGroup	ResourceName
1/27/2020 12:00:00 AM	Standard_F72s_v2	72	72	Not enabled	73c0021f-a37d-433f-8baa-7450cb54eea6	Trey Research Finance	JUMPER	Jumper3
1/27/2020 12:00:00 AM	Standard_F72s_v2	72	72	Not enabled	73c0021f-a37d-433f-8baa-7450cb54eea6	Trey Research Finance	JUMPER	Jumper4
1/27/2020 12:00:00 AM	Standard_NC24s_v3	24	24	Not enabled	b85af93a-5507-4886-b799-fec6d505b988	Contoso	USAGE	SpotTest2
1/27/2020 12:00:00 AM	Standard_D15_v2	20	24	Not enabled	73c0021f-a37d-433f-8baa-7450cb54eea6	Trey Research Finance	JUMPER	Jumper1
1/27/2020 12:00:00 AM	Standard_H16_Promo	16	16	Enabled	b85af93a-5507-4886-b799-fec6d505b988	Contoso	ADFCOSTTEST	AzureEAtest3
1/27/2020 12:00:00 AM	Standard_D5S_v2	16	16	Enabled	1caaa5a3-2b66-438e-8ab4-bce37d518c5d	Cost Management Research	formatica	DVM
1/27/2020 12:00:00 AM	Standard_DS13_v2	8	8	Not enabled	9be7120a-cf89-46c8-b091-d341776575e9	TestEA Towboat Customer2	GunnarcDC	sql-vm-A
1/27/2020 12:00:00 AM	Standard_DS13_v2	8	8	Not enabled	73c0021f-a37d-433f-8baa-7450cb54eea6	Trey Research Finance	Gekko	GekkoPower
1/27/2020 12:00:00 AM	Standard_DS3_v2	4	8	Not enabled	d5bcaab0-0add-4d7f-b41d-0f2b8e829540	ISO27001	FABRIKAM-SHAREDSVCS-ADDS-RG	fabrikam-shared

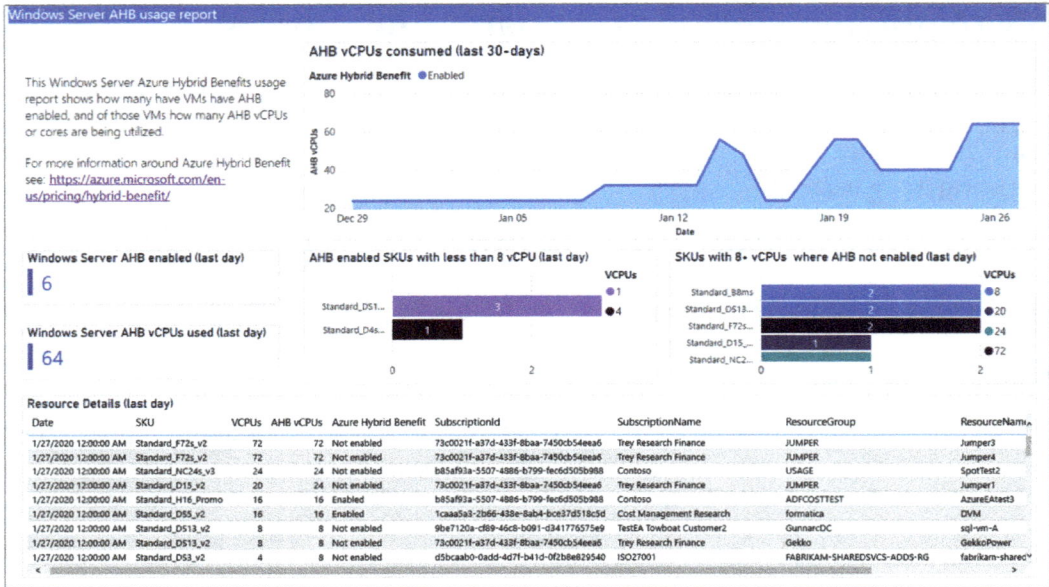

Figure 5.9 – Azure Cost Management Power BI report

This dashboard offers a lot of interesting insights relative to our cost data, including a complete drilldown of our costs, as well as information on Reserved Instances and Azure Hybrid Benefit, which are cost optimization initiatives in Azure that we will cover in *Chapter 6* of this book. You can find more information related to this Power BI dashboard at the following link: https://learn.microsoft.com/en-us/azure/cost-management-billing/costs/analyze-cost-data-azure-cost-management-power-bi-template-app.

This dashboard is a good starting point for our cost analysis, but how can we customize it?

This is a non-customizable report. Despite this fact, we can make use of a Power BI connector specific to Azure Cost Management to load Cost Management data into Power BI and present it in any way we want. In the documentation of the connector, we can find the complete dataset available as well as its format.

This connector has some limitations, and the main one is that it will only be able to retrieve 1 million rows as a maximum. For bigger subscriptions or accounts with a lot of resources, you will surely reach this limit, especially with queries that span over many months, which can result in a lot of invoice lines.

This dashboard can be a good starting point to explore and gather ideas to create your own dashboards and reports for cost optimization in Azure. The good thing is that it does not need any requirements apart from the license, as it lets you directly query the service without a database or a data warehouse to store the cost data, which can work really well for limited scopes or smaller environments.

You can find more information about this connector at `https://learn.microsoft.com/en-us/power-bi/connect-data/desktop-connect-azure-cost-management`.

Keep in mind that in order to use the Power BI report and this connector to create your own dashboards and reports, you will need a Power BI license.

Azure FinOps Workbook (Work In Progress)

This is an Azure Log Analytics workbook created by Sam Bell, a cloud solution architect from Microsoft (`https://www.linkedin.com/in/sambellau/`).

In our experience, **Azure Log Analytics workbooks** are a great platform to create cost optimization dashboards, as they can be connected to multiple sources of data, such as Azure Log Analytics and Resource Graph, and are easily customizable. They use **Kusto KQL** as a querying language to format and transform the data shown in our workbooks and allow for many transformations.

This workbook is a great place to start to develop your own FinOps dashboards. Not only does it give a lot of useful information for cost optimization, such as orphaned resources and optimization insights, but it also represents a really good starting point to begin implementing your own workbooks. You can just download this workbook, import it into your Azure subscription, and begin customizing it to your specific needs, which will save a lot of time.

You can find the workbook in its GitHub repository here: `https://github.com/ms-sambell/azure-finops-workbook`.

AWS CUDOS

As we already introduced during *Chapter 2*, **AWS CUDOS** dashboards are amazing reports that contain a wide range of information regarding cloud costs and cost optimization in general.

In our opinion, AWS CUDOS is an amazing example of what a FinOps dashboard can look like, and AWS has put a lot of work and thought into it. Any cloud practitioner who wants to get into cost optimization should use these dashboards as a starting point to get deeper into AWS cost optimization.

On top of all the information they offer, these dashboards can also be customized further to adapt them to each organization's needs. You can change how data is visualized and filtered, or even add new KPIs and initiative tracking if you are versed in **AWS QuickSight** and **AWS Athena**.

In our view, this tool is essential for AWS FinOps practitioners and sets the quality bar really high for other cloud providers, whose built-in cost optimization dashboard offerings pale in comparison.

Custom dashboards and reports

When our needs regarding dashboards and reports are not met after having exhausted built-in resources, and if the tooling does not provide the required information, sometimes, the only way forward is to create our own custom dashboards or reports.

In this section, we will propose a process to create such assets and make the most of them. Apart from proposing a simple process that we can follow, we will also recommend some cloud services that can be used for each part of the process, to try to extend the FinOps toolbox as much as possible.

The first thing we must keep in mind is that creating these assets will require an investment in both time and resources, as these developments require specific FinOps knowledge as well as technical proficiency with cloud services. The investment will pay off, as having these assets will help cost optimization, and therefore, help organizations optimize and get the most out of their cloud budgets. Due to the need for this investment, it needs to be taken into consideration and accepted by stakeholders and all interested parties.

As with any other report and dashboard, we should follow a series of steps to ensure that our developments will be successful.

End goal analysis

Before putting any work into the development process itself, we need to think about the end goal for this dashboard or report.

The first point we need to consider is whether we need to develop a dashboard or a report. We have already covered the main differences between them, and this first point will impact the rest of the process. Please make sure that the differences are understood and that the use case is aligned with the answer to this question before going forward.

Once this first point is clear, we need to ask ourselves questions such as the following:

- What is the value that this asset brings to the table?
- What information does it offer?
- How many data sources will it use?
- From a high-level perspective, what kind of processing do we need?
- How will the information be presented?

We should have thought about all of these questions to have a good basis to begin development work properly. All this analysis and business case information should be shared with stakeholders and interested parties, as in any other project.

Ingestion

The first point to be tackled is where to get the required information that we require for our project.

More often than not, our data sources will be already available, though some may require some form of authentication. We could directly use available data sources, such as REST APIs or the CLI, to get the information we need. However, reports and dashboards usually require presenting historical information as well as current.

In addition to this fact, *REST API responses are limited in size by definition*. Some sources, such as billing data or VM CPU usage, can generate large volumes of information that we won't be able to get in full from API calls due to this size limit. Also, the services that different cloud providers offer have limited cost data retention, as shown in these examples:

- **Azure Cost Management** is limited to **13 months**

- **AWS Cost Explorer** is limited to **12 months**

- In GCP, the period is longer but there are some specifics that you can check at this link: `https://cloud.google.com/billing/docs/how-to/reports#data_availability`

Due to this, our recommendation is always to ingest data when possible into a storage platform or even a database service, which we will use as the foundation for later steps in the process. We can ingest data on a daily or even shorter basis if needed, making the calls much smaller and easier to manage.

Regarding billing information, cloud providers have different services that allow billing data to be exported to storage, as we covered in *Chapter 2*, which matches perfectly our needs, such as **Azure Cost Management exports**, **AWS Cost and Usage Reports**, or **GCP Billing BigQuery exports**.

If the information we need is related to resource usage, such as CPU or memory usage, we can resort to monitoring services that also have APIs that we can use to get any data we require. Services such as **Azure Monitor**, **AWS CloudWatch**, and **GCP Cloud Monitoring** offer REST APIs that we can query to get our data for this ingestion process. This information, in addition to billing and cost information, will be the cornerstone for our rightsizing exercises that will be introduced in *Chapter 6* of this book.

After the information is gathered from different data sources, our dataset will be done and we are ready for the next step of the process, which is data processing.

Processing

Depending on our needs, we may require additional processing for our data to make specific transformations in our dataset that will make our information more readable to be presented in reports or dashboards.

We are not going to go really deep in this part as it goes way past the purpose of this book, but we can recommend some cloud services that can be used for this purpose, such as **Azure Data Factory** or **Azure Synapse**, **AWS Glue** and **AWS Athena**, and **GCP BigQuery**.

Presentation

The last step in our process is to present the information we have prepared, selected, and transformed. Nowadays, there are a myriad of products and technologies that can be used for this purpose, such as **Microsoft Power BI**, **BigQuery BI Engine**, or **AWS QuickSight**, if we limit the scope as to what the cloud providers themselves offer.

If we want less complex solutions, we can also use **Azure Log Analytics workbooks** (which we already presented briefly), **Azure Dashboards** (which have limited functionalities), **AWS CloudWatch dashboards**, and **GCP custom dashboards**. Keep in mind that some of these products may not be able to show any billing information, though.

Preparing the presentation layer can take a lot of time, as we need to agree with the dashboard or report audience on how the data looks and is presented, which often requires a lot of time and effort from the development teams.

Again, this point is way beyond the purpose of this book, but the presentation layer should present the information in a clear way, which is often aided by using diagrams, plots, or tables to present raw information in a more graphical way that our audience will appreciate.

Dashboards and reports, like many things in the FinOps domain, can be improved iteratively using the **crawl, walk, and run** approach. Don't worry if your first developments are simple or incomplete as long as they bring value. Let us not forget that Rome wasn't built in a day!

Summary

With the process of how to prepare custom dashboards and reports, we conclude this chapter.

During this chapter, we covered how to interpret and understand cloud invoices and billing information in general from cloud providers. We also described the differences between dashboards and reports, introduced some financial basics that can help throughout our FinOps practices, and understood how to track savings and waste and make use of the unit economics concept. In addition to this, we reviewed some built-in assets that we can use as a starting point for FinOps analysis, as well as some guidelines and cloud services we can use to create our own dashboards and reports.

With all this information, we hope that you will be able to tackle future challenges in the *Inform* pillar far more easily.

In the next chapter, we will begin to explore the *Optimize* pillar, in which we will analyze cost optimization initiatives from a much more technical and architectural perspective to make the most out of our compute IaaS and PaaS, database, and storage services.

Part 3:
Optimize – How to Get the Most out of Cloud Resources

This part describes in depth the **Optimize** pillar, which is based on making use of all available features of services offered in the public cloud around cost optimization.

This part has the following chapters:

- *Chapter 6, Implementing IaaS Compute Optimization*
- *Chapter 7, Implementing PaaS and Other Compute Optimization Initiatives*
- *Chapter 8, Implementing Database Optimization*
- *Chapter 9, Implementing Storage Optimization*

6
Implementing IaaS Compute Optimization

Compute, in all of its different flavors, is almost always where we can have the most room for optimization when implementing FinOps practices.

Even when building solutions with cost optimization in mind, some things might be overlooked. To avoid this, across the different sections of this chapter, we are going to introduce a lot of initiatives and ideas to keep in mind, as well as new purchasing and licensing models, such as Reserved Instances and Spot Instances, that need to be in the toolbox of each FinOps practitioner.

The aim of this chapter is to provide FinOps practitioners with a checklist of sorts that they can resort to any time they need to optimize any compute-related resource or solution. Apart from the concepts and ideas, we will provide examples when possible to illustrate the magnitude of savings we can obtain by applying the ideas and architecting techniques we will be covering.

We will begin by describing how to optimize IaaS resources, which will be then followed up with PaaS and other initiatives in the next chapter.

In this chapter, we will cover the following topics:

- Compute optimization key concepts
- IaaS optimization

Compute optimization key concepts

Before getting deep into compute optimization, let's introduce a few concepts and ideas that need to be fully understood to better grasp the ideas behind each initiative presented across the chapter.

To begin with, we are going to introduce a key concept that will be used throughout the optimize foundation, which is **cost optimization quick wins**.

Quick wins

We consider any cost optimization initiative a quick win if it satisfies the following conditions:

- Has no impact, or minimal impact, on running workloads. It does not need to be applied during a maintenance window.

- Does not change a solution design in a major way. Usually, a quick win is a change in a configuration that can be easily applied without redeploying or recreating a resource or the whole solution.

- Can provide instantaneous savings, even if small.

Quick wins are key during the first steps of every FinOps practice, as it takes minimal or no effort to implement them and track their implementation.

By implementing quick wins, we can begin to create trust in FinOps practices by demonstrating the following:

- FinOps practitioners can also aid on the technical side by providing ideas for the technical teams that can be invaluable going forward. These ideas will also help to increase the cloud maturity inside an organization.

- Almost at the beginning of the FinOps journey, we can kickstart it in the best way possible by generating some instant savings that technical teams and stakeholders will be grateful for. This point is key to ensuring that FinOps practices will have enough sponsorship, budget, and support from all parties involved.

- Once they are applied, we can enforce their use by setting up policies that ensure that these conditions are met on newly designed solutions.

Now that you understand this key concept, we will go deeper into different initiatives that may be proposed in this area of work. Let's set the ball rolling by introducing some key concepts about optimization for IaaS workloads.

Introduction to IaaS, PaaS, and serverless

Let's go through some cloud basics. Depending on the level of management for resources, we have different flavors available in cloud services:

- **Infrastructure as a Service (IaaS)**: This is the oldest management model. Using IaaS, the provider makes **virtual machines (VM)** available for clients. Virtualization and the underlying infrastructure are handled by the cloud provider with some added features, such as backup, encryption, and monitoring, while everything else is managed by the client.

- **Platform as a Service (PaaS)**: In PaaS, we go one step forward. The cloud provider also takes responsibility for the operating system and infrastructure management, and we are provided with services such as storage, databases, and web servers directly.

For the sake of this chapter, let's also cover another service model, **serverless computing**, which is a subset of PaaS with specific features. In serverless computing, these are the key takeaways:

- We are charged for services only when the services are used. For example, in a web application, we are charged for execution time after a request; in a database, we are charged when the database is being queried.

- Scaling is also managed from the cloud provider side to ensure that we have enough resources when needed. The client will only set out scaling limits to limit the pricing according to their needs.

- We often have less visibility on the backend side, as it is managed by the vendor. Due to this, troubleshooting can be more challenging.

I want to share a quote from Bruce Zhang, the current director of engineering at Google (https://www.linkedin.com/in/ruofei/), on this topic, which I really like:

"It is like the difference between tap water and drinking fountains.

For developers, both can open the switch to drink water. Serverless computing is like tap water. It flows out as much as you need. And PaaS is like a drinking fountain, it can also provide a lot of water.

Once there is no water in the bucket, you need to purchase a bucket delivery service from the provider. The provider will deliver the water purification package to you.

Their charges are similar. Serverless computing charges according to how much you use. PaaS charges are based on buckets, not usage."

The key differences between these service models are the level of control that we have and how we are charged for them. We can summarize the cost difference in the following table:

Cost driver	IaaS	PaaS	Serverless
Time	Virtual machine uptime (hours)	Application uptime (hours)	Number of requests Function uptime Memory consumption
Rate	Virtual machine class	Environment class (compute size, often vCores and memory size)	N/A
Additional charges	Disk type and size Backup Other attached resources (e.g., Public IPs)	Attached resources Added storage Added features Backup	Attached resources Added storage Added features Backup

Table 6.1 – IaaS, PaaS, and serverless cost drivers

Also, other key concepts are mentioned a lot around cloud professionals such as the following:

- **Containers as a Service** (**CaaS**): In CaaS, the service provider makes a container or a set of containers available for our workloads to run. We don't use this term much as there are a lot of CaaS flavors, from more IaaS-oriented ones, such as OpenShift in cloud offerings, to more PaaS-oriented ones, such as managed Kubernetes or **AWS ECS/Azure Container Instances**, so we think the term is not really representative of a specific offering.

- **Function as a Service** (**FaaS**): This is a type of serverless offering on which we can execute fragments of code in a modular way, which can be executed upon invocation in a lot of different ways (triggered by other events or by HTTP requests as a REST API, for example). Some examples of these services are **Azure Functions**, **AWS Lambda functions**, and **GCP Cloud Functions**.

- **Backend as a Service** (**BaaS**): This is another serverless flavor that offers a lot of backend features for developers, so they can fully work on frontend development. It is used for mobile or web applications and it includes database, hosting, identity, and notifications management, among other features.

Let's use a simple table that involves containers to highlight the differences between IaaS, PaaS, and serverless in regards to who manages what:

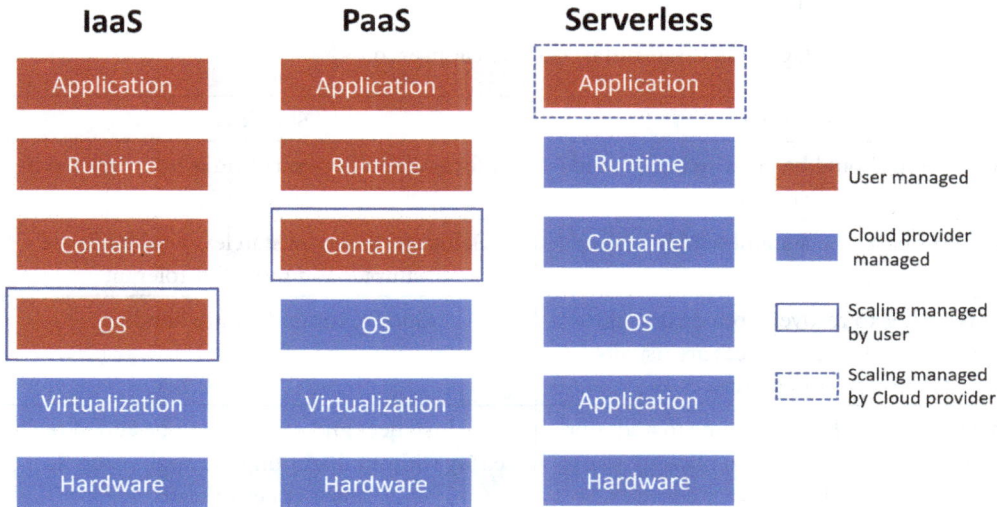

Figure 6.1 – IaaS versus PaaS versus serverless

Now that we've introduced all these strange acronyms that we, as IT people, like to use, let's also have a look at an interesting architectural concept, which is stateless services.

Stateless versus stateful

Let's introduce another key architectural concept that must be incorporated before discussing PaaS optimization, and that is stateless and stateful services.

A service can be considered stateful when it requires persistent storage. This means that data is persisted over time, for example, in a container that hosts some databases on which we need to perform transactions. In these stateful workloads, the following applies:

- Underlying infrastructure process requests take into account past requests
- Data should be made available and shared with any other service that may need it, taking into account the latest requests or updates at any given moment

On the other hand, in a stateless service, no data is persisted after execution. This means the following:

- It does not take into account previous executions, so effectively, it does not need that information.
- It can be executed by any server. This fact makes horizontal scaling really easy.

Stateless services are key for a microservices architecture and also allow for highly elastic and loosely coupled architectures as well. On the other hand, not all workloads are compatible with this paradigm, as it has many implications.

Let's analyze the advantages and disadvantages of each approach:

Stateful	Stateless
Information should be persisted and saved after execution	Information after execution is not persisted in any way
Solutions based on stateful workloads are more complex	Solutions based on stateless workloads are simpler and more fail-tolerant
It is harder to recover from errors and failures, as stateful services are usually long-running services	Good error and failure handling
It's harder and it takes more time to add more instances as data should be replicated across all of them. Scaling is more complex.	More instances and capacity can be added really easily without disrupting the application flow. Much easier to scale.
Usually requires more compute power	Usually takes less compute power
Less network bandwidth is needed and more storage is usually needed	Network bandwidth can be impacted due to a larger amount of data being transferred but usually requires less storage

Table 6.2 – Stateful and stateless comparison

Combining orchestration with stateless workloads allows determining at execution runtime which nodes are most adequate to execute the task at hand, allowing for a more resilient infrastructure.

Stateless workloads are highly adaptable and better suited for modern applications, so solution design should work on adapting applications to this new paradigm when possible. Doing so unlocks a lot of optimization initiatives that we will cover in this chapter, such as scaling and using Spot Instances.

IaaS optimization

IaaS is a purchase model offered by cloud service providers that enables infinite capacity without any kind of limit or restriction on networking, compute, storage, and whatever resource we may need.

As great as these service offerings are, they are also a double-edged sword: if we don't have enough control and visibility on cloud costs at all times, the costs can grow exponentially in no time.

As we often tell our clients and peers, *"The only limit that compute capacity has in the cloud is the size and limit of your credit card"* or, as Uncle Ben in Spider-Man says, *"With great power comes great responsibility."*

> **Example – IaaS cost gets out of hand**
>
> Sometimes, all that it takes for things to get out of control is one misstep.
>
> Imagine that we have given more permissions than needed to a developer in a sandbox environment, or we have failed to restrict what users can and cannot do. The developer would be able to create the highest compute VM to test out some machine learning processes and examples.
>
> In Azure, for example, there are virtual machines that can cost around $138 per hour. Imagine a week with a VM like that running for 24 hours a day; the cost of compute only could reach around $23k. Not good, right?
>
> It is not only important to optimize but also to ensure that we won't trip over the same stone twice. In other words, we need to ensure that bad practices won't be repeated anymore.

Despite the risks that we highlighted in the example, this purchase model provides a lot of value if used properly, as you can use any configuration or specification you need in minutes while forgetting about the underlying storage or virtualization infrastructure. Also, the package includes great features such as automated patching, which will reduce administrative overhead from our cloud operations teams.

With this in mind, let's delve into different initiatives that can be implemented to achieve cost optimization in our IaaS workloads.

Quick win – orphaned resources

It is very common that, when a VM is deleted, the attached resources such as disks or public IPs may remain if we are not careful.

These resources are left in an *orphaned* state, as the resource that they were attached to is no longer available so they can't be used anymore, but we are still charged for them, nevertheless.

One of the first points to go through at the beginning of the FinOps journey is to audit and delete these orphaned resources, as their elimination requires almost no downtime and has no impact. It also can generate some savings that, even if small in volume, will begin to build confidence around the practice.

Some examples of resources are as follows:

- Disks
- Public IPs
- Network interfaces
- Snapshots (if they are not needed anymore)

There are a lot of cloud offerings where orphaned resources may appear and it's always a good practice to clean up after them, especially the ones that imply additional costs.

There are multiple tools we can use to check for the existence of these resources, such as a cloud console in different cloud providers, policies, automation, code review if we are using CI/CD and **infrastructure-as-code (IaC)**, and others. Some examples of these tools are as follows:

- **Azure:**
  - **PowerShell**
  - **Azure CLI**
  - **Azure Resource Graph**
  - **Azure Policy**

- **Amazon Web Services (AWS):**
  - **AWS CLI**
  - **AWS Config**
  - **AWS SDK** to manage AWS resources using the most popular programming languages

- **Google Cloud Platform (GCP):**
  - **Google Cloud Shell**
  - **Google Cloud SDK** to manage GCP resources using the most popular programming languages
  - **Google Cloud Asset Inventory**

Please ensure that you have enough permissions to search for orphaned resources in your cloud environments. It may happen that when searching for these orphaned resources with proper permissions, you are not able to find any, even after double-checking. If that's the case, congratulations! Things are done properly in your organization. Don't relax, though, as we also need to ensure that these resources are not going to appear anytime soon in the future. We can do the following for prevention purposes:

- Ensure that all resources are deleted properly and review the deletion process if that's not the case. Implementing IaC, for example, is a really good method to prevent this from happening.

- Set up policies, automation, or monitoring to get a notification whenever these kinds of resources appear.

- Incorporate the orphaned resources count for all orphaned resources that we have searched for in FinOps dashboards and ensure that the count is always 0.

The cost savings we can obtain by removing and cleaning these resources may be small, but it is one of the easiest tasks to accomplish in the first stages of the practice while more advanced FinOps initiatives are prepared and set up, and it is a very good starting point to create value and confidence for all teams involved.

Having described how to work on orphaned resources, let's move on to another key initiative, which is VM version upgrades.

Virtual machine version upgrades

Every cloud provider needs to regularly upgrade the specifications of their data centers and infrastructure to use the latest technology, and this means using newer generation CPUs, disks, and memory for cloud services offerings. In parallel to this renewal, they also need to keep older versions of infrastructure running to some extent, to cater to clients that are already using them to avoid any kind of disruption.

To be able to accommodate this upgrade process, they usually use *virtual machine versioning or generations*, with newer generations providing more compute power, more efficiency, and better pricing.

Quick win – virtual machine upgrade

When you create a virtual machine, you are presented with all the possible generations that can be used for said VM before its creation. On the other hand, for already existing virtual machines, you are able to upgrade their version to the latest generation if the VM specs, and your workloads, support the move. Newer generation virtual machines offer faster CPU chipsets, as well as more bandwidth and overall efficiency while having a slightly lower price tag. What's not to like?

Honestly, there is no reason to justify *not* upgrading all virtual machines to the latest generations whenever possible. However, the following considerations must be taken into account before doing so:

- Not all generations are available for all regions. Please ensure that the latest version is available in your region.

- The VM upgrade should be done during a maintenance window, as it requires the VM to be shut down.

- Please ensure that all the requirements (accelerated networking, maximum number of NICs, and so on) of the workload are also available on the newer version virtual machines, to ensure that the upgrade is possible.

- If there are Reserved Instances in place or Saving Plans, make sure that changing VM generation does not affect current reservations usage in any way (for example, Azure Reserved Instances are purchased in vCores for a VM family in a specific version).

Azure example – previous versions versus current

In this example, we can illustrate the differences between old VM versions and new ones (D8 v3 versus D8 v5):

Instance	Region	On-Demand rate	Configuration	Price/month
Standard D8 v3	West europe	0.4800	8 vCPU 32 GB	$350.40
Standard D8 v5	West europe	0.4600	8 vCPU 32 GB	$335.80

Table 6.3 – Azure VM versions pricing comparison – prices obtained using the Azure Pricing Calculator in April 2023

As we can see, the price differences on the same VM are around 5%, with newer families providing more performance and efficiency as well.

If we analyze the differences in performance level, v5 virtual machines use the latest generation Intel Xenon processors, which result in higher performance and scalability, as well as better network bandwidth.

AWS example – previous versions versus current

In this example, we can illustrate the differences between old VM versions and new ones (M3 versus M5):

Instance	Region	On-Demand rate	Configuration	Price/month
m3.2xlarge	eu-west-1	0.585	8 vCPU 32 GB	$311.94
m5.2xlarge	eu-west-1	0.428	8 vCPU 32 GB	$271.61

Table 6.4 – AWS EC2 versions pricing comparison – prices obtained using AWS Pricing Calculator in April 2023

As we can see, the price differences on the same VM are around 26%, with newer families providing more performance and efficiency as well.

Apart from the cost differences, by using new generations, we can also benefit from higher performance and lower power consumption, as M5 virtual machines use the latest generation Intel Xenon processors, as well as increased network performance. In addition to this, in AWS, newer generation VMs include EBS-optimized feature, which enables a dedicated bandwidth for AWS EBS disks to be used.

GCP example – previous versions versus current

In this example, we can illustrate the differences between older VM versions and newer ones (N2 versus N2D):

Instance	Region	On-Demand rate	Configuration	Price/month
n2-standard-8	europe-west-4	0.427319	8 vCPU 32 GB	$311.94
n2d-standard-8	europe-west-4	0.37208	8 vCPU 32 GB	$271.61

Table 6.5 – GCP Compute Engine versions pricing comparison – prices obtained using Google Cloud Pricing Calculator in April 2023

As we can see, the price differences on the same VM are around 13%, with new families providing more performance and efficiency as well.

N2D virtual machines result in higher CPU and network performance, as they use newer AMD EPYC processors.

But there are other options, apart from upgrading to the latest generation, to make the most of your VM workloads, and one of them is leveraging the use of AMD and ARM processors.

Upgrading to AMD/ARM processors

An additional way to perform a VM upgrade is to use AMD or ARM processors, a different CPU architecture that comes with great performance, high efficiency, and high scalability while being more cost-effective than the standard x86 CPU counterpart.

Mainly, we are going to cover two different processors apart from the x86 standard, which are ARM and AMD x86:

- For ARM processors, each cloud provider offers their own take on this CPU architecture:

  - Ampere in Microsoft Azure

  - AWS Gravitron in AWS

  - Tau in GCP

- Regarding AMD x86 architectures, we are referring to AMD EPYC chipsets in its different generations, which are really efficient processors with great performance, while keeping the same x86 architecture. They also include the security feature AMD Infinity Guard to increase the defenses against malicious attacks.

These CPUs are even offered in some PaaS offerings such as AWS RDS. However, using these processors sometimes requires a change in application architecture, as not all workloads support its use.

> **Side note**
>
> Usually, ARM architectures are widely used in mobile devices or tablets, which are less demanding in performance terms, and in smaller computers such as Raspberry Pi. With the rise of highly efficient and powerful processors such as Apple M1/M2, ARM architecture usage is being used more and more and its popularity keeps rising, as efficiency and sustainability are a great concern for everyone, as well as performance, of course.
>
> Microsoft has even created a version of Windows 10/11 that supports this processor architecture to be competitive with other Linux distributions in this market sector (`https://learn.microsoft.com/en-us/windows/arm/overview`).

Before considering its use, please test it thoroughly and ensure that your workload fully supports this CPU architecture. If that's the case, you can leverage the many benefits of this architecture while generating some additional savings!

Azure example – different CPU chips offered

In this example, we will compare the pricing of different CPU architectures for the same VM size on a Linux VM in Azure:

Virtual Machine	Hours/month	Region	Processors	vCPU	Memory	Price/month
D4 v5	730	West Europe	x86	4	16 GiB	$167.90
D4ad v5	730	West Europe	AMD	4	16 GiB	$151.84
D4pls v5	730	West Europe	Ampere ARM	4	16 GiB	$113.15

Table 6.6 – Azure ARM/AMD pricing comparison – prices obtained
using the Azure pricing calculator in April 2023

The price differences are notable, with ARM being around 32% cheaper than the x86 counterpart with similar specs.

If we compare AMD with x86, the pricing is 10% cheaper.

AWS example – different CPU chips offered

In this example, we will compare the pricing of different CPU architectures for the same VM size on a Linux VM in AWS:

Virtual Machine	Hours/month	Region	Processors	vCPU	Memory	Price/month
m5.xlarge	730	Ireland	x86	4	16 GiB	$156.22
m5a.xlarge	730	Ireland	AMD	4	16 GiB	$140.16
m6g.xlarge	730	Ireland	Graviton	4	16GiB	$125.56

Table 6.7 – AWS EC2 ARM/AMD pricing comparison – prices
obtained using AWS Pricing Calculator in April 2023

As shown in the table, the pricing for the same VM size in ARM/Gravitron compared to standard x86 architecture is around 20% cheaper.

If we compare AMD with x86, the pricing of AMD is 10% cheaper.

GCP example – different CPU chips offered

In this example, we will compare the pricing of different CPU architectures for the same VM size on a Linux VM in GCP:

Virtual Machine	Hours/month	Region	Processors	vCPU	Memory	Price/month
n2-standard-4	730	europe-west4	x86	4	16 GiB	$156.10
t2d-standard-4	730	europe-west4	AMD EPYC	4	16 GiB	$135.81
t2a-standard-4	730	europe-west4	Tau ARM	4	16 GiB	$123.72

Table 6.8 – GCP AMD/ARM pricing comparison – prices obtained
using Google Cloud Pricing Calculator in April 2023

As shown in the table, using t2a/ARM over n2 virtual machines results in around 20% savings for this VM size. Using t2d/AMD VMs, on the other hand, results in 13% cost savings.

For newer applications, it is definitely worth it to analyze and consider its usage in new solutions, where the effort to make a workload compatible may pay off in the long run.

Virtual machine rightsizing

Optimal usage of resources is essential to build a good FinOps strategy. We should size our resources properly and in accordance with requirements, both technical and from the business side.

This effectively means that it makes no sense to use a high-compute VM with 36 cores for a development environment, for example, the same way that it makes no sense to use a lower-tier VM with 2 cores to run a mission-critical workload that requires a lot of compute power to run properly.

One of the first points of this long journey is to begin by analyzing long-running workloads and checking for resource usage, mainly CPU, memory, and disk throughput/IOPS usage.

Cloud providers offer visualization and monitoring services for our infrastructure resources, which will allow us to deep dive into a set of highly detailed metrics in a granular way. Having these services is a great driver for better decision-making in operations related to cost and performance optimization.

There are multiple ways to access this information, which can be done in an interactive way using the administration console for each cloud provider, or in a programmatic way by accessing Azure, AWS, and GCP REST APIs.

These are the services offered to do so in each public cloud:

- **Azure**: Azure Monitor
- **AWS**: Amazon CloudWatch
- **GCP**: GCP Compute Engine metrics

> **Important note**
>
> Keep in mind that, during this rightsizing process, we may find overstressed virtual machines that we need to scale up as well. This may have an impact on costs, but it may solve potential issues in the long run.
>
> Remember that FinOps is not about getting savings but optimizing cloud resources. *Rightsizing is not only downsizing.*

Having these metrics at hand may not be an easy task. In most public clouds, it is easy to check out the metrics of one resource, but it may not be as easy to have a complete list of the average use of all resources across our cloud environments.

We recommend increasing visibility on resource usage by leveraging dashboards and reports, where all metrics are aggregated, so we can have updated information at hand at all times, to be able to react in a swifter way should there be inconsistent or non-ideal metric data.

After all this information has been gathered, we can take the following actions:

- **Upgrade or scale up**: If we have virtual machines that are showing constant errors or failures and we check that they are fully using all the compute resources they have available, we can choose to scale them up to get better and more consistent performance. Keep in mind that we can scale up on the same family, change the family for another one that offers a better match for the use case, or even begin using horizontal scaling if the resource supports this scenario.

- **Downsize or scale down**: If metrics are showing consistent usage under 25%, we may have a candidate for downsizing. We can downsize in a couple of ways: by scaling down in the same family or by choosing another family with better pricing and a slightly smaller compute capacity.

- **Terminate**: Sometimes we may be even able to fully terminate or delete the instance if its use is not justified. For example, it may not make sense to use a five-piece VM cluster in a sandbox environment, or there may be machines that are no longer used. Of course, terminating virtual machines yields bigger cost savings.

Before terminating a virtual machine, please ensure that the technical teams are fully aligned and support the idea. We don't want to be blamed for your one-week production downtime!

For upgrading or downsizing workloads, we should have periodic checks to ensure that metrics don't change over time, as rightsizing is an iterative exercise that never ends.

Virtual machine family standardization

VM family homogenization is key to simplifying the governance of virtual machines for big organizations. Also, having a standardized set of families allows for greater savings when using Reserved Instances, as we will cover later in this chapter.

This initiative also eases the financial side, as cloud bills get simpler to understand with not that much heterogeneity in sizes and families used.

This concept consists of the following:

- Choosing specific families, VM SKUs, and regions to be used per environment

- Once the standard has been set, it is important to enforce its use using policies to guarantee that only families and sizes outside the general rule are used when there is enough justification to do so

Did you know? Region cost differences

It is important to note that there are huge differences in VM costs between regions. For example, this is the pricing difference in AWS between the US and Europe, considering the same VM size and family:

Virtual Machine	Hours/month	Region	vCPU	Memory	Price/month
m5.xlarge	730	Ireland	4	16 GiB	$156.22
m5.xlarge	730	N.Virginia	4	16 GiB	$140.16

Table 6.9 – AWS EC2 pricing difference between regions – prices obtained using AWS Pricing Calculator in April 2023

A whopping 10% difference – who would have known? Now, we can think twice before dismissing the use of North American data centers in cloud providers for our development environments, right?

By using VM family standardization, we could, for example, set out a new company policy to use the cheapest regions (which are the regions hosted in the US for Azure, AWS, and GCP) for development environments if latency is not an issue.

Before considering changing current virtual machines for standardized sizes and families, please keep in mind the following considerations:

- Changes to virtual machines should be applied during agreed maintenance windows to avoid any downtime in running workloads. Such changes in a VM usually require a restart.

- Before proceeding with the changes, please ensure that all the features that are needed for running workloads are also ensured in the new family and size, to avoid any issues.

With these concepts in mind, let's delve into a great initiative for IaaS optimization, which is to shut down services during off-hours, or, as we call it, power scheduling.

Virtual machine power scheduling

Cloud costs on virtual machines depend on two variables or cost drivers: the time that the services are running and the rate that we are charged for them. The rate can also include multiple variables to calculate pricing. In virtual machines, for example, we can have a variable for storage capacity in GB, another for compute size, and so on.

This simple formula that we presented effectively means that, for every minute that a VM is stopped instead of running, we are generating savings, should we compare with the cost of having it running.

Due to this fact, shutting down virtual machines, especially in non-productive environments, is one of the main sources of IaaS cost savings in the cloud.

> **Important note**
>
> It's essential to note that, while virtual machines are shut down, we are not charged for compute costs until they are powered up again.
>
> However, we are still charged for disks and other attached resources such as public IPs and storage disks. This is the case in Azure, AWS, and GCP.

To illustrate what we can achieve by using this simple concept, we have calculated the total hours of use of the most commonly used schedules for virtual machines.

Let's do a simple calculation:

1. A year has 365 days

2. A year is equal to *365 * 24 = 8,760* hours

3. If we divide those hours per 12 months, we obtain 730 hours in an average month

Having a VM on for a whole month has a cost that is easy to calculate, as we can get that information easily from any of the calculators that the major cloud providers offer, but how much can we save should we follow power scheduling strategies?

Schedule (hours x days)	Total hours of use	Savings
24 x 7	730	N/A
8 x 5	180	75%
12 x 5	270	63%
12 x 5 + 48 (weekends)	318	56%

Table 6.10 – Savings depending on hours of use

We don't often have that information in the pricing calculators that cloud providers offer, do we?

We know that going to the technical teams and asking them to shut down unused virtual machines is going to take some work and effort, but with such big advantages, there is honestly no reason not to do so.

Keep in mind that, for preproduction and development environments, there will always be some workloads that cannot be shut down, such as some Active Directory services, for example. It is perfectly fine to leave them running if that's the case, as the purpose of FinOps practices is to also better understand the limits for cost optimization.

> **Important note**
>
> This concept of shutting down resources that are not used off-hours can even be taken to an extreme if we decide that, for some workloads such as batch processes, we can use IaC and deploy the complete environment in minutes, use it while it's needed, and then destroy it afterward instead of having the environment deployed at all times.
>
> This concept is often referred to as **ephemeral environments**, and we fully support its use for cost optimization.
>
> Instead of having a development environment, we can use an older backup from production, for example, and create an environment from scratch for testing purposes. After the tests have ended, we can delete everything. Is there anything more cost-effective than that?
>
> The cost savings in this case can be huge should we use this strategy.

Virtual machine scaling

One of the biggest benefits of the cloud is the possibility to have unlimited compute on tap. On the other hand, we need to find the sweet spot for our solutions to run smoothly while limiting the cost as much as possible.

To do so, we have multiple features that we can leverage, and one of them is scaling. There are many ways to scale our workloads and, in this section, we will describe its uses and benefits. To begin with, let's analyze the different types of scaling that we have available and how they work.

Virtual machines' horizontal and vertical scaling

There are two different flavors of VM scaling:

- **Vertical scaling**: Also known as rightsizing or scaling up/down; it consists of upgrading or downgrading the compute specs of the virtual machine.

- **Horizontal scaling**: Also known as scaling in/out; it consists of adding more virtual machines to a pool that executes the same processes or runs the same workloads, distributing the work among nodes. The idea behind horizontal scaling is to optimize by running tasks in parallel:

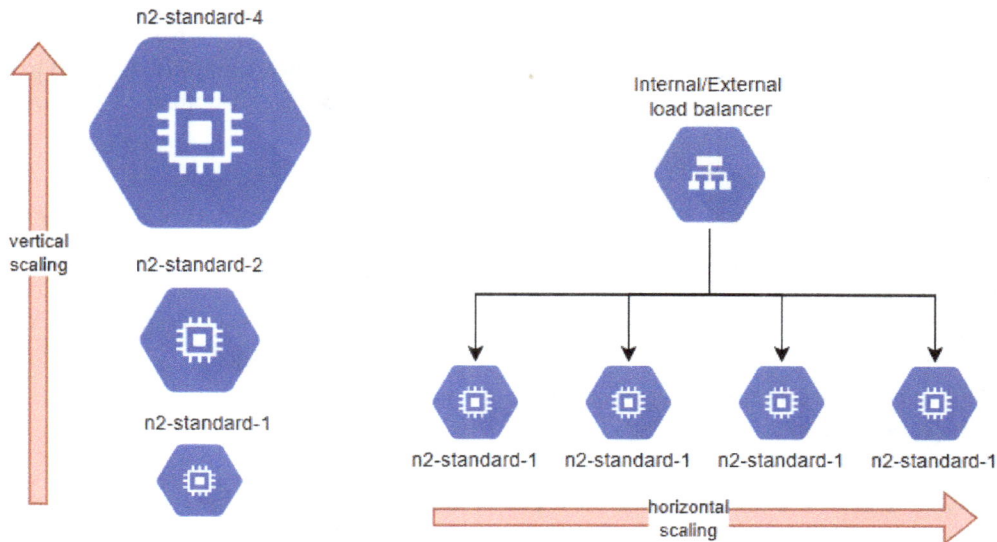

Figure 6.2 – Google Cloud vertical versus horizontal scaling on Compute Engine/virtual machine

To leverage all the cloud has to offer, we need to make use of both of these scaling mechanisms and to do so, we need to better understand when to choose one or the other, and how to get the most out of it.

Vertical scaling is often better for traditional or legacy solutions, where one server does the heavy lifting in stateful architectures for web applications, databases, and similar workloads. We need to rightsize the servers first and only scale up if we know in advance that there is going to be a period of high demand (for example, when the month is closing or at the end of a fiscal year).

To apply vertical scaling, we need to restart our virtual machine, so make sure that this is planned accordingly during a maintenance window. After the period of high demand is over, we can go back to the initial VM size.

Horizontal scaling, on the other hand, is a great feature that is offered by cloud providers but it is not used that often, in our experience. It is a great way to reduce waste, as it enables the use of additional resources just when they are needed. After the period of high demand has passed, we can go back again to the minimum number of virtual machines that we need.

However, it is important to note that not all workloads support this kind of scaling. This is ideal for the following:

- Stateless services or applications
- Batch services or processes
- Containers

When working with horizontal scaling, we can also decide when to scale in or out based on standard metrics such as the percentage of CPU consumption or memory. Some cloud offerings even have *autoscaling* mechanisms to automate this process, which can lead to really cost-optimized workloads.

> **Example**
>
> The perfect example to use for this is a web application that handles payments every morning.
>
> If we know in advance the time at which we expect more demand from users, we can use horizontal scaling to add multiple nodes to our payment application. We can even set it up using automation, so one hour before the big spikes in demand, we can have everything ready, and then after the spike has passed, reduce the nodes to the minimum again.
>
> A VM using vertical scaling is never able to adapt as fast as a set of virtual machines working using horizontal scaling.

Now, there is another key question to ponder as we also introduced the possibility of rightsizing in the picture. When should we choose to rightsize and when to scale?

It depends; we will try to summarize the advantages and disadvantages of each option so you can make the best choice depending on each use case with this table:

	Advantages	Disadvantages
Vertical scaling	Easier to execute, as it applies to only one instance	It can only scale up to some point, which depends on the family and the instance type
	Does not depend on application or workload design	High availability is not guaranteed, as scaling up or down requires a restart
Horizontal scaling	Allows for highly available workloads	Has a strong dependency on application design
	It can scale way higher than rightsizing	If rules are not designed and tested properly, the cost can get out of hand pretty easily
	Monitoring and automation of scaling rules are managed by cloud providers, simplifying the process for IT ops teams, which are only responsible for autoscaling rules design	It is more complex to maintain and manage than vertical scaling

Table 6.11 – Horizontal versus vertical scaling

With the different mechanisms that we can use to scale clearly described, let's move on to a special type of VM that can scale to some extent without changing its size: burstable virtual machines.

Burstable VMs

A burstable VM is a special kind of VM that can scale up above its weight in CPU and memory for limited periods of time without changing its tier.

It uses a system based on credits. While the VM CPU usage is under a baseline, credits are accumulated. If the VM is using a percentage of CPU above the baseline, credits are consumed.

These VMs also have the possibility to scale for more than 100% of the allocated CPU, given that you have enough credits to pay. The limit for scaling up usually depends on the VM family and size.

The number of credits generated per hour depends on the VM size, the same as the baseline and the maximum CPU that a VM can scale up to.

This diagram shows how credits are used and generated:

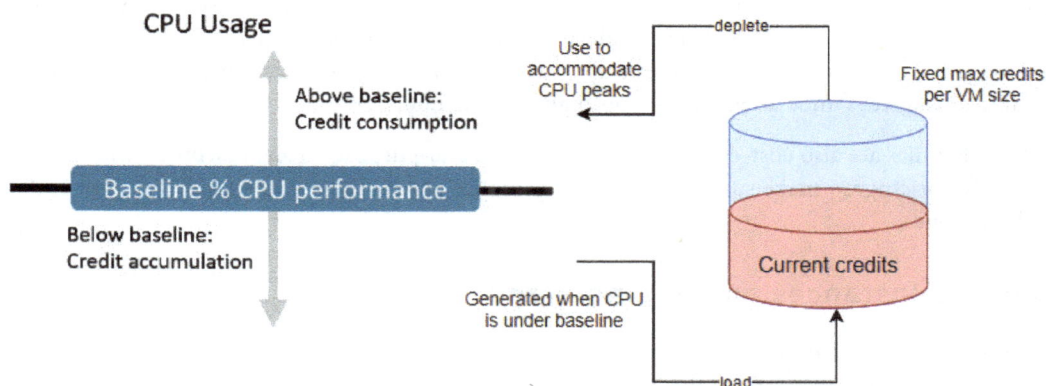

Figure 6.3 – Burstable virtual machines

To calculate how many credits are generated or consumed per hour, the following formula can be used:

Credits generated/consumed per hour = (baseline CPU performance – CPU utilization) / 100

Example – AWS bursting standard versus unlimited

In AWS, the virtual machines that offer bursting capabilities are T4g, T3a, T3, and T2 Instances.

There are two different settings for these instances:

- In *Standard* mode, bursting is limited by the credits that are generated by your virtual machine. This means that, at the cost level, we are always charged for the same amount (base price of the VM).

- In *Unlimited* mode, bursting is not limited by the current credits that are accumulated. If the VM CPU demands are higher than the credits, the instances will burst anyway, and the instance is able to pay pending credits after the burst. If the additional consumed credits are not compensated in a 24-hour time frame, we will be charged a flat additional rate per vCPU-hour.

Keep in mind as well that in GCP, even though bursting is possible in shared core virtual machines, a credit system is not used. This effectively results in limited bursting capabilities compared to AWS and Azure.

Burstable virtual machines are ideal for development environments and other use cases where the usage patterns are not constant but have sudden spikes. These instances will be able to accommodate these usage spikes without the need to scale up to a higher size VM size, which results in way higher costs if the instance is underused during non-high demand hours.

Burstable families are also cost-effective compared to general-purpose virtual machines, with really competitive pricing, so make sure that these instances are used if they are a good fit for your IaaS cloud solutions.

Reserved Instances and Saving Plans

Once the basic initiatives for cost optimization are being implemented, and maturity in FinOps improves, it's time to tackle one of the most important, yet complex, initiatives, which is the use of **Reserved Instances (RIs)** or **Saving Plans (SPs)**.

It's a purchase model that some cloud providers offer in their services portfolio. Basically, you commit (in Azure, GCP, and AWS, it is for one or three years) to pay for a number of virtual machines from a specific family in a specific region or to a monetary amount of money to be spent in virtual machines. In exchange for this long-term commitment, you get big discounts on the services compared to pay-as-you-go retail prices.

We need to have a stable perimeter on which big changes are not expected in the future before even considering purchasing RIs.

On the other hand, it is so easy to benefit from an RI, as we only need to purchase them and wait for eligible virtual machines to be allocated to that reservation, which will diminish our charges considerably.

Let's introduce a key concept that is essential to understanding RIs, which is the **break-even point**. The break-even point is the point from which we will begin to benefit from the savings generated by using RIs compared to on-demand. Consider that RIs are often paid upfront (even though this concept applies to upfront or no-upfront-payment models) and the benefit may not be fully visible from the beginning, so this is essential to justify its purchase.

This break-even point can be calculated using the following formula:

*Break-even point = (1- RI/SP Discount Rate) * RI/SP Term (result in months)*

Let's use an AWS example to calculate this break-even point to illustrate how it works and what it means. As an example, here's the pricing of a Linux general-purpose VM in the different purchase models:

Virtual Machine	Region	Term	Payment Option	RI monthly fees	RI effective hourly rate	Savings over On-Demand	On-Demand rate
m5.xlarge	Ireland	1yr	No Upfront	$98.55	$0.14	37%	$0.21

Table 6.12 – On-demand versus RI rates

Which, taking the data from the table and following the formula, results in the following break-even point

*m5.xlarge break-even point = (1 – 0.37) * 12 months = 7.56 months*

Let's see what this break-even point looks like in a plot:

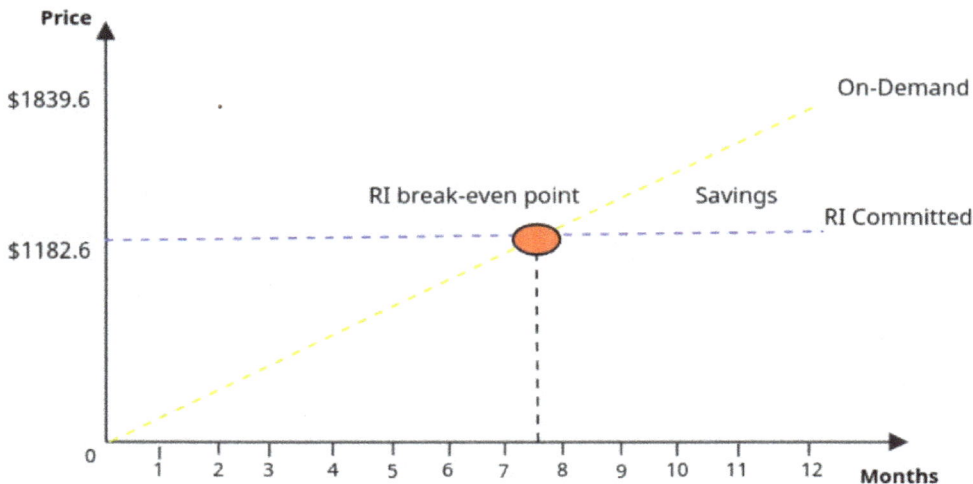

Figure 6.4 – Break-even point for m5.xlarge example VM

For a 1-year reservation, we will begin generating savings for this specific VM after 7.5 months of use, when using the no-upfront-payment option.

From this moment onward, we will be generating savings if we compare it with the price that we would be paying for the same VM with pay-as-you-go.

> **Important note**
>
> We don't recommend at all using RIs and SPs without having the rest of the initiatives in place. FinOps practices need to be mature enough to avoid making big mistakes.
>
> It is of the utmost importance to have a rightsized infrastructure perimeter as much as possible and to use power scheduling when possible. If this is not the case, the use of RIs can be more detrimental than beneficial, as it essentially forces you to commit to a set of virtual machines for a long period of time. Also, it is recommended to avoid RIs in greenfields with not a lot of infrastructure or for lift-and-shift migration scenarios. If this commitment is in place, you won't be able to reduce or scale down resources without wasting money.
>
> We have seen a lot of situations where technical teams in search of generating savings (we don't like this, as we have stated before) purchase badly planned RIs that are impossible to cancel.

One of the first points we need to decide upon when considering purchasing Reservations is the scope on which we should apply them. There are mainly two options:

- **Specific scope (fixed)**: They only apply to one account (AWS), subscription/resource group (Azure), or project (GCP). Reservations will only cover virtual machines in a specified scope.

- **Broad scope (flexible)**: They apply to multiple management containers, either accounts (AWS), subscriptions (Azure), or projects (GCP). This flexible scope reduces waste by ensuring that reservations will be fully used at all times by maximizing the scope to include more eligible virtual machines.

In the following table, we describe the advantages and disadvantages of each approach:

	Pros	Cons
Broad Scope	<ul><li>Reservation use is maximized while reducing the risk of underutilization (more VMs can benefit from it)</li><li>Easier to track reservation usage at the business unit or project level</li></ul>	<ul><li>More difficult to make the purchasing decision with more stakeholders involved</li><li>Less autonomy to manage reservations due to higher permissions needed</li><li>Post-billing processing is required for cost allocation</li><li>Difficulties in allocating reservations to specific resources</li></ul>

Specific Scope	• Billing simplicity due to clear reservation allocation • Better autonomy to decide reservation parameters and configuration	• Risk of reservation underuse, as fewer resources can benefit from them • Administrative overhead and management complexity

Table 6.13 – Broad versus specific scope for RIs

After having introduced the concept, let's differentiate between different Reservation flavors that are offered by cloud providers.

How do RIs and SPs work?

In RIs, we usually commit to a number of virtual machines from a specific family in a specific region, for a specific term of 1 or 3 years.

There is, however, a special type of RI called Saving Plans (at least in Azure and AWS), on which the commitment is monetary instead of several machines of a specific family. Basically, you commit to pay a monthly rate for one or three years, and you can benefit from discounts on compute costs.

Once Reservations or Saving Plans are in place, virtual machines that match the conditions of the Reservation will benefit from it. If there are no virtual machines matching the conditions stated, you will get billed anyway.

The discounts will be lower than the discounts from RIs but, in exchange, we get more flexibility, and it is easier to ensure that they won't be wasted.

Azure example – pricing differences between on-demand and RI/SP

As an example, here's the pricing of a Linux general-purpose 4 vCore VM in the different purchase models:

	Virtual Machine	Hours/month	Region	Price/month 1 year	Price/month 3 years	1 year savings	3 years savings	Total Commitment 1 year	Total Commitment 3 years
On Demand	D4 v5	730	West Europe	$327.04	$327.04	N/A	N/A	N/A	N/A
Saving Plans	D4 v5	730	West Europe	$263.07	$173.66	19.56%	46.90%	$3,156.84	$6,251.76
Reserved instances	D4 v5	730	West Europe	$192.92	$124.28	41.01%	62.00%	$2,315.04	$4,474.08

Table 6.14 – Azure RI/SP/on-demand comparison – prices obtained
using the Azure pricing calculator in April 2023

As you can see in the table, the price difference is huge, but the difference in commitment is also vast. SPs in Azure offer less discount but better flexibility, as you are not tied in any way to a specific family or VM version.

Keep in mind that SPs in Azure don't cover licensing costs; these are billed separately.

AWS example – pricing differences between on-demand and RI/SP

As an example, here's the pricing of a Linux general-purpose VM in the different purchase models:

Purchase model	Type	Term	Payment Option	Virtual Machine	Hours/month	Region	Price/month	Savings
On Demand		N/A					$156.22	N/A
Reserved Instances	Standard	1yr	No Upfront	m5.xlarge	730	Ireland	$98.55	36.92%
			Partial Upfront				$93.55	40.12%
			All Upfront				$91.83	41.22%
		3yr	No Upfront				$67.16	57.01%
			Partial Upfront				$62.64	59.90%
			All Upfront				$58.75	62.39%
	Convertible	1yr	No Upfront				$126.29	19.16%
			Partial Upfront				$119.94	23.22%
			All Upfront				$117.53	24.77%
		3yr	No Upfront				$89.60	42.64%
			Partial Upfront				$83.22	46.73%
			All Upfront				$81.03	48.13%

Table 6.15 – AWS EC2 RI/on-demand comparison – prices obtained
using AWS Pricing Calculator in April 2023

Purchase model	Type	Term	Payment Option	Virtual Machine	Hours/month	Region	Price/month	Savings
On Demand		N/A					$156.22	N/A
Savings Plans	EC2 Instance	1yr	No Upfront	m5.xlarge	730	Ireland	$98.55	36.92%
			Partial Upfront				$93.44	40.19%
			All Upfront				$91.98	41.12%
		3yr	No Upfront				$67.16	57.01%
			Partial Upfront				$62.78	59.81%
			All Upfront				$58.40	62.62%
	Compute	1yr	No Upfront				$126.29	19.16%
			Partial Upfront				$120.46	22.89%
			All Upfront				$117.53	24.77%
		3yr	No Upfront				$89.60	42.64%
			Partial Upfront				$82.49	47.20%
			All Upfront				$81.03	48.13%

Table 6.16 – AWS EC2 RI/on-demand comparison – prices obtained
using AWS Pricing Calculator in April 2023

As we can see in both tables, the discounts are really attractive. As the term grows larger, the discounts get more interesting, as is expected due to larger monetary commitment.

All the payment options and RI and SP types in AWS add more flexibility to this purchase model, which is always welcome.

GCP example – pricing differences between on-demand and Committed-Use Discounts

As an example, here's the pricing of a Linux general-purpose 4 vCores VM in the different purchase models:

Purchase model	Term	Virtual Machine	Hours/month	Region	Price/month	Savings	Total commitment
On demand	N/A	n2-standard-4	730	europe-west4	$156.10	N/A	N/A
Sustained Usage discounts	N/A	n2-standard-4	730	europe-west4	$129.91	16.78%	N/A
Resource-based CUD	1 year	n2-standard-4	730	europe-west4	$98.34	37.00%	$1,180.08
Resource-based CUD	3 years	n2-standard-4	730	europe-west4	$70.25	55.00%	$2,529.00
Flexible CUD	1 year	n2-standard-4	730	europe-west4	$112.39	28.00%	$1,348.70
Flexible CUD	3 years	n2-standard-4	730	europe-west4	$84.29	46.00%	$3,034.58

Table 6.17 – GCP CUD/SUD/on-demand pricing comparison – prices obtained using Google Cloud Pricing Calculator in April 2023

In Google Cloud, monetary commitment reservations are called flexible **Committed Use Discounts (CUDs)**, and if we commit to a number of virtual machines, we will use resource-based CUDs.

There is also another discount we can benefit from, which is **Sustained Use Discounts (SUDs)**. If a VM is eligible for SUDs (depending on the kind and family of the virtual machine) and it is used for more than 25% of the total billing month, you will get the discount (as long as it does not benefit from other discounts).

By definition, SPs are more flexible as it is way easier to find virtual machines that match the conditions, which are mainly region and compute type.

How to plan around RIs and SPs

Apart from the scope, there are other settings that can be configured when purchasing RIs, such as the following:

- **Reservation type**: Some cloud providers also offer different types of reservations depending on exchange and modification terms. This is allowed on AWS, where convertible (exchangeable) reservations provide less discounts but higher flexibility.

- **Payment type**: You can choose between **All Upfront** and **Monthly** payments. In AWS, there is a third option available, which is **Partial Upfront**. Also, in AWS, you get further discounts if upfront payments are used.

- **Term**: Duration of the reservation. This is the period of time for which we will benefit from the discount.

> **Important note**
>
> Don't overlook the importance of the term that is chosen!
>
> Three years is a long time and a lot of monetary commitment. If the life cycle of a specific workload or application is not clear, it's better to use 1-year reservations and get less discount than to commit to 3-year reservations and waste money.
>
> Even though you can exchange reservations, each cloud provider imposes terms and limits to do so.
>
> For example, in Azure, before the launch of SPs, you could exchange reservations for up to 50,000 dollars/total commitment annually per agreement, and now that SPs are in place, exchanges won't be allowed after January 2024.
>
> In AWS, you cannot exchange a standard RI, you can only change some parameters such as Availability Zone, scope, or size.
>
> Please plan your reservations carefully with this idea in mind.

Example – AWS RI types

Let's use AWS as an example, as it is the cloud provider that offers more options when purchasing reservations.

For RIs (commitment based on resources), these are the different options that are available:

	Standard RI	Convertible RI
Services	EC2 instances, Amazon RDS, Amazon ElastiCache, ElasticSearch Service, Amazon Redshift, Amazon DynamoDB and AWS elemental media convert	EC2 instances
Term	1 or 3 years	
Purchase Option	All Upfront, Partial Upfront or Not Upfront	
What are reservations?	You commit to a consistent usage (nº of VMs/hour)	
Max discount	72%	66%
Reservation parameter and flexibility	Select scope (regional or zonal) and region or AZ. Scope and AZ can be changed at any time, without doing an exchange Reservation splits and merges are possible too. You can split a current reservation into parts and assign them to different AZs.	
	Select family, size, quantity, OS, tenancy. Only size can be changed, and only in Linux	Select family, size5, quantity, OS, tenancy.
Exchange a reservation	Exchange is not allowed by default. If any change is required, a support case must be opened, and it's not always authorized	Family, size, quantity, OS and tenancy can be converted. For region and master payer modifications, a support case must be opened, and it's not always authorized

Table 6.18 – AWS RI settings and services covered in April 2023

For SPs, this table summarizes the options as well:

	EC2 SP	Compute SP
Services	EC2 instances	EC2 instances, Fargate and Lambda
Term	1 or 3 years	
Purchase Option	All Upfront, Partial Upfront or Not Upfront	
What are reservations?	You commit to a consistent usage (€/hour)	
Max discount	72%	66%
Reservation parameter and flexibility	Saving plan automatically applies to any instance size, tenancy or OS within an instance family in a particular region	Saving plan automatically applies to any instance family, OS, tenancy, Fargate or Lambda in any region
Exchange a reservation	Exchange is not allowed by default. If any change is required (instance family, region, master payer), a support case must be opened, and it's not always authorized	Move the saving plan to a different master payer account is not allowed by default. A support case must be opened, and it's not always authorized

Table 6.19 – AWS SPs settings and services covered in April 2023

All these options allow us to tailor our reservation usage to the workloads and use cases of each one, greatly enabling cost optimization.

Reservation amortization

Amortization is the process of dividing the cost of a resource that has been upfront based on usage, so we can get daily or hourly rates.

If, due to financial processes in your organization, you need to perform chargeback or showback on cloud costs, amortization is important.

In Azure and AWS, you can get detailed amortized costs from Azure Cost Management and AWS Cost and Usage Reports, respectively.

If we combine the information from cost reports for virtual machines for both on-demand usage (hours in on-demand at a certain rate) and usage under Reservations (hours that the VM benefits from reservations at an amortized cost rate), we can effectively perform efficient chargeback or showback.

Example – Azure pricing differences between on-demand and RI/SP

As an example, here's the comparison of upfront costs against amortized costs for a D4v5 Linux VM in Azure:

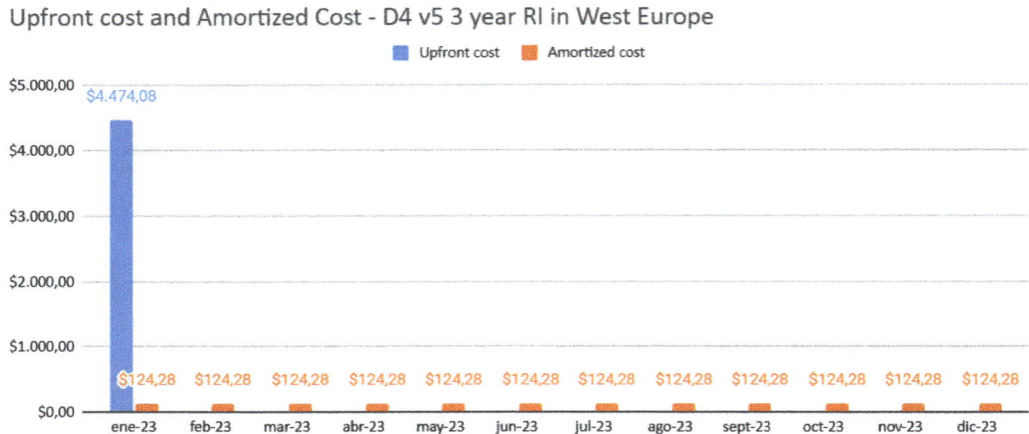

Figure 6.5 – Azure RI amortization – prices obtained using the Azure pricing calculator in April 2023

From this information, we can also deduct an hourly fee for this virtual machine:

Hourly cost = $124.28 / 730 = $0.17 per hour (RI rate)

Here, we get the pricing for on-demand pricing:

Hourly cost = $0.23 per hour (on-demand rate)

With this information at hand, we can analyze Cost Management reports for each VM (or even get the information using automation in a programmatic way) to track the total number of hours in each payment model per month and calculate the total price to be used for chargeback or showback purposes.

For example, if a specific RI is not used, and the scope is limited to a business unit, we can report the total cost in both on-demand and RIs, to demonstrate how much waste (we are effectively paying almost twice for a resource) we are creating by not making full use of paid reservations, to the business unit via chargeback or showback.

Tracking reservation usage

If reservations are not fully used, we are effectively wasting money. To ensure that reservation usage is maximized, let's define some KPIs that can aid in this process, and then track them in each public cloud to ensure optimal use of RIs:

- **RI utilization**: This KPI reflects, in a percentage, how much time a VM has benefited from a reservation out of all the hours it has been paid for. This KPI should be always 100%.

- **Reservation coverage**: This KPI reflects, out of all the hours of virtual machines used throughout your environments, the percentage of Reserved Instances used. This value highly depends on your reservation strategy (what environments to reserve, on which conditions, and to which extent), and it is really useful to track how Reservations are used across an organization.

Reserved Capacity for software licenses

In Azure and GCP, it is also possible to purchase Reserved Capacity for Linux SUSE distributions and Red Hat licenses.

As an example, this table reflects the savings that we can get should we use this purchase model for these software licenses in GCP:

Product	Number of vCPUs	1-year CUD savings	3-year CUD savings
SLES	1-2	77%	79%
SLES	3-4	54%	59%
SLES	5+	45%	50%
SLES for SAP	Any	59%	63%
RHEL	Any	20%	20%
RHEL for SAP	Any	20%	20%

Table 6.20 – CUD discounts for software licenses

With the possibilities and specifics of the Reserved Capacity purchase model already covered, let's move on to some general recommendations to get the most out of them.

General recommendations

To close this topic, we would like to give out some suggestions based on our experience to get the most out of RIs and SPs:

- It's a good idea to use reservations in stable environments such as production. We recommend provisioning a percentage of the virtual machines in RI/SPs (such as 50% to begin with) and analyzing the RI/SP usage along the way, only adding more RI/SPs if we are sure that none of them will be wasted. There is nothing worse than an RI that is under 100% utilization from a financial perspective.

- If the life cycle of environments is not clear, we don't recommend choosing 3-year commitments. It is always best to get lower savings but to be careful than to purchase 3-year instances and end up wasting them because an application has been suddenly retired out of nowhere.

- We always prefer to use SPs (monetary commitment) over RIs for their flexibility.

- For non-productive environments, we always recommend the following:

  A. First, rightsize.

  B. Then, use power scheduling to shut down workloads during off hours. It is always best to prioritize power scheduling to RIs when possible, as there is no commitment attached to it

  C. Lastly, use Reservations *only* for the remaining virtual machines or workloads that cannot be shut down for any reason. On those, we also recommend beginning with a small RI/SP percentage, as non-productive environments are more variable and everchanging than productive ones.

We hope all this information has been useful to get a better grasp on how to optimize the use of RIs. With this said, let's move to another alternative purchase model, which is Spot VMs.

Spot VMs

Spot VMs is a newer purchase model that is available on most of the major public cloud providers. This purchase model allows us to make use of unused or spare compute capacity from the cloud providers at a much better cost compared to retail.

The lower cost, however, comes with some considerations:

- You need to specify the price you are willing to pay for a virtual machine

- If the retail price at any moment surpasses your established price, you will get evicted from the virtual machine

This implies that if you set up a really low cost, you will have more probability of getting interrupted, whereas by setting a higher price (closer to retail), you will get interrupted less often.

These virtual machines are ideal for clustering scenarios, where you can have a stable set of virtual machines running and use this purchase model to add additional capacity to accommodate periods of high demand, for example.

You can use RIs for workloads that are always on, and on-demand to accommodate for workloads that are shut down when their usage fluctuates over time. On top of RIs and on-demand virtual machines, you can use additional capacity to accommodate the highest peaks in demand for workloads, as depicted in the following figure:

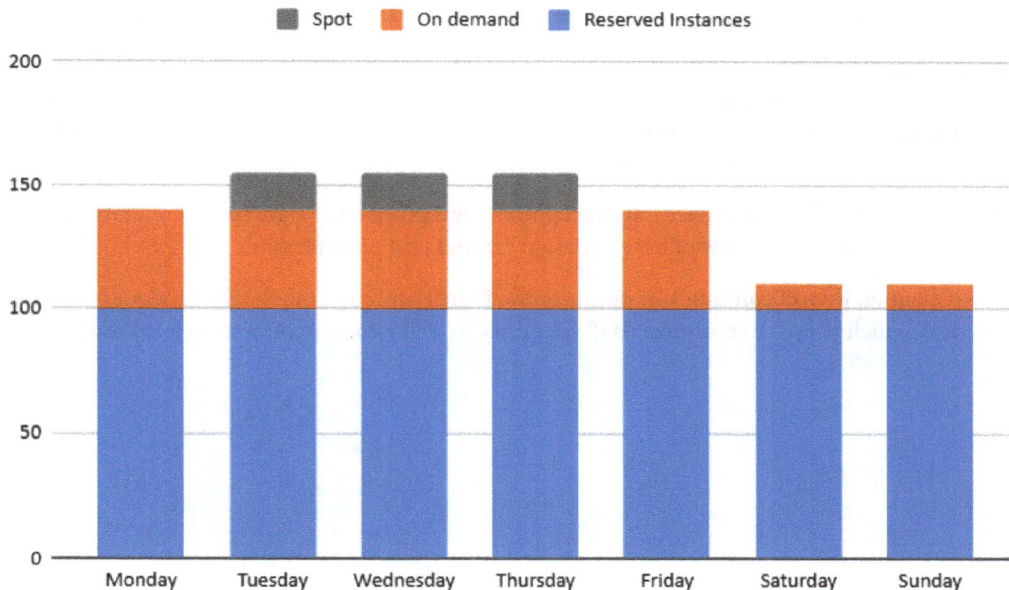

Figure 6.6 – On-demand/RI/Spot ideal use

However, not all workloads are compatible with this purchase model. It is not recommended to use these virtual machines for workloads that require storing or persisting data (stateful).

It also requires additional planning on the technical teams, as the price must be set in an iterative way until we find the sweet spot for each use case between cost optimization and stability, while the operational teams keep a careful eye on monitoring for errors and issues due to this provisioning model.

Here are some use cases for Spot Instances:

- Batch processes
- Stateless containers or applications
- CI/CD self-hosted agents

- Big data and analytics workers
- Web services in microservices architectures

With this last initiative, we put an end to our review of IaaS initiatives.

Summary

It is time to close this lengthy chapter. We hope it has been useful for you to gain a better view of the different initiatives that can be used to optimize IaaS compute service offerings in the cloud.

During the chapter, we have reviewed initiatives for IaaS services such as rightsizing, vertical and horizontal scaling, and power scheduling to have a broad palette of tools to apply FinOps in cloud services. We have also described new purchase models such as Reserved Instances and Spot VMs, which are great yet complex offerings to manage, and may be new for cloud newcomers.

Also, having covered a lot of architectural concepts and best practices, we are confident that you have ended up with a higher-level view of how to design around cost optimization.

In the next chapter, we will continue our optimization journey by covering the remainder of compute optimization, which is how to optimize PaaS solutions, as well as other compute-related initiatives.

7

Implementing PaaS and Other Compute Optimization Initiatives

The use of PaaS and SaaS resources is one of the key milestones in the journey to the cloud. If we modernize our solutions to be cloud native, we are effectively reducing the administrative overhead that is needed for our cloud operations, as well as enabling much more agile and fast development and solutioning.

But with the wide offering of services available, it is not difficult to lose sight of which initiatives we should focus on first. In this chapter, we will give an overview of different PaaS initiatives, as well as covering some other key topics, such as licensing optimization and data transfer costs.

In this chapter, we will cover the following topics:

- PaaS optimization
- Data transfer costs optimization
- Licensing optimization
- Cloud provider agreements and resource allocation

PaaS optimization

Using PaaS instead of IaaS services is a form of optimization by itself, as the level of management and administrative overhead that services require is much lower. However, some IaaS concepts still apply in PaaS services as well, such as rightsizing.

In this section, we will also cover newer PaaS flavors, such as **Serverless** which can be a great way to optimize costs while improving solution design to adapt to newer architectural paradigms, such as stateless services.

This chapter is an introductory overview of some ideas that we will develop further in the next two chapters, which are dedicated to databases and storage optimization. Both of them include a lot of initiatives for PaaS services as well.

But for now, let's get into some ideas to get the most out of this management model.

PaaS rightsizing and workload consolidation

PaaS services, just as with IaaS services, need to be rightsized to better adapt to the workloads that are running on them.

They are, after all, running under a compute size that can be selected and scaled up and down, which has a big impact on costs, and almost always provide metrics to track resource consumption such as CPU or memory usage percentages.

Also, the same as with databases, applications can run on the same compute to save costs, especially on smaller applications.

When analyzing PaaS resource usage, please follow these guidelines:

1. Ensure that PaaS resources are making enough use of the compute that is provisioned. We could begin by identifying resources with less than 25% usage, and then apply more restrictive thresholds as the practice evolves, such as 50%.

2. If resources are not making enough use of compute, we can do either of the following:

    *   Try to group applications on the same project to use the same compute, based on dependencies and resource usage analysis
    *   Rightsize the compute used to a smaller size

The best option to choose depends on the project, usage, and dependencies.

Example – Azure App Service and App Service plans

As an example of rightsizing and consolidation of PaaS workloads, we want to highlight the case of one client, where after a proper assessment, we found out that a Service Plan was provisioned per application on all environments.

We won't present here the exact case but a similar one. Let's say that, for example, we have three smaller applications and we choose *Standard S1*, which is the smallest size available for an app service.

This service has a price tag of $69.35 a month in West Europe (pricing as of April 2023).

Let's consider the alternatives. The first one is to provision three service plans:

Application	Service Plan	Region	Hours of use	Tier	Price/mo	CPU Pct usage
app1	sp1	West Europe	730	App Service Plan S1	69,35$	30%
app2	sp2	West Europe	730	App Service Plan S1	69,35$	20%
app3	sp3	West Europe	730	App Service Plan S1	69,35$	10%

Table 7.1 – Azure PaaS rightsizing (1/2). Prices obtained using Azure pricing calculator in April 2023

The second option is for all smaller applications to share a service plan, allowing much better usage of compute:

Application	Service Plan	Region	Hours of use	Tier	Price/mo	CPU Pct usage
app1	sp1	West Europe	730	App Service Plan S1		
app2	sp1	West Europe	730	App Service Plan S1	69,35$	60%
app3	sp1	West Europe	730	App Service Plan S1		

Table 7.2 – Azure PaaS rightsizing (2/2). Prices obtained using Azure pricing calculator in April 2023

The savings are immense in this scenario, as we can reduce the total spending by 66%. This initiative is especially fruitful in development or sandbox environments, where resources are not used all the time throughout office hours. Instead, their use is concentrated at specific times (e.g., before a new release in production).

Serverless versus provisioned compute

PaaS' popularity is only increasing, due to all its benefits and the features that are offered. But some services can take it one step further and offer PaaS services on which no underlying compute is needed.

In those Serverless services, the pricing model usually depends on the time of use or execution of those services, a fully pay-as-you-go cost model on which you only pay for the services when they are used. There are still servers running the workloads, but they are abstracted away from the development process, simplifying it in a big way, while the cloud provider manages provisioning, scaling, and maintenance.

These services have many advantages:

- Lower costs
- Higher scalability
- Simplified code
- Time-to-market reduction

Apart from the benefit of its simplicity, let's analyze how, by leveraging these services, we can save on cloud costs for some workloads:

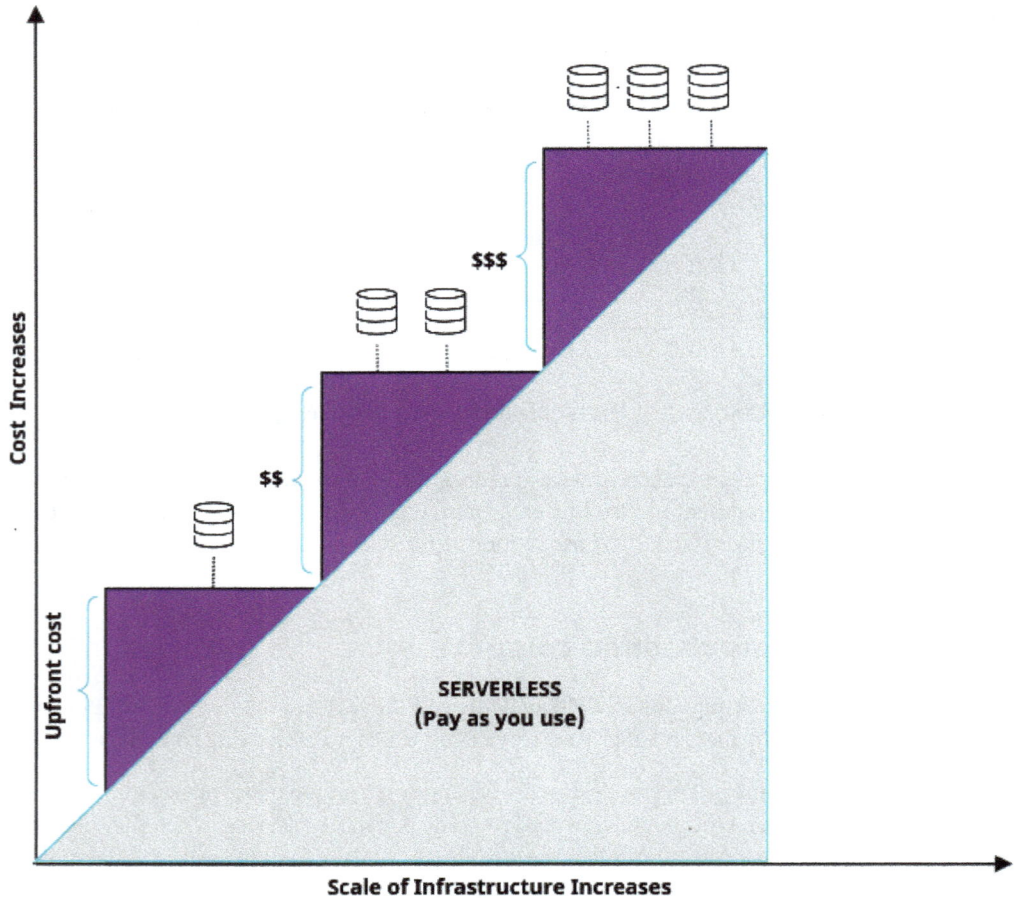

Figure 7.1 – Serverless versus pay as you go

As we can see from the figure, instead of paying in *chunks* based on the size of our infrastructure, we can benefit from a much more flexible cost model on which we are only charged when services are used.

Let's give some examples of which services we can use in PaaS and their Serverless counterparts, and the differences between them regarding costs.

Azure Functions versus Azure App Service

In **Azure App Service** (which is a PaaS offering), you need to provision the underlying compute (App Service plan) that is offered in different tiers (vCores, RAM, and storage). Once you have a Service Plan, you can allocate app services on it. A Service Plan cannot be shut down or turned off; it can only be scaled vertically or horizontally once it's been deployed.

In **Azure Functions** (which is a FaaS offering), on the other hand, you can use a special type of Service Plan on which you are only charged when the Function is executed. You also have the option to use Functions that make use of regular service plans with provisioned compute (non-serverless) to have some prewarmed instances, so the activation time for Functions is reduced to a minimum (no cold start).

The underlying engine is the same, but the way we are charged for them differs in a big way.

AWS Elastic Beanstalk versus AWS Lambda

In **AWS Elastic Beanstalk** (which is a PaaS offering), when you provision an application, the underlying infrastructure that supports it is generated (AWS EC2 virtual machine, AWS S3 storage, and other services, such as load balancers). Elastic Beanstalk has no attached cost by itself when using it, but we are charged for all the infrastructure resources that support the application.

On the other hand, when you use **AWS Lambda** (which is a FaaS offering), you are not provisioning any infrastructure by default. You will be charged per execution time, storage, and memory allocated for your Function You also have the option to use the **Provisioned Concurrency** feature, which lets you have some prewarmed instances ready that will start without any delays if a request comes (this option comes at an additional cost). If Provisioned Concurrency is not activated, Functions will have an activation time (cold start)

GCP App Engine Standard versus GCP Cloud Functions

In **Google Cloud Platform** (**GCP**) **App Engine Standard** (which is a PaaS offering), you select the instance class you want to be deployed, similar to Azure App Service. Your code will run in a sandbox in a container hosted in GCP. In this service, you will be charged for the characteristics of the container you have chosen, which are basically memory, CPU limit, and scaling capabilities. You will also be charged for additional features, such as storage and outgoing network traffic. Once the App Engine instance is created, you will be charged per hour while the resource is deployed, and you can disable it on demand if needed to stop incurring charges.

In **GCP Cloud Functions** (which is a FaaS offering), you also select an instance tier depending on how much CPU speed and memory is needed for your workload, but there is no underlying infrastructure. You will only be charged per execution time when your Function is executing your code, as well as some additional charges, such as network egress and number of concurrent instances for the Function.

The benefits of Serverless

It should already be clear that both PaaS and Serverless have ideal use cases.

In our opinion, Serverless is well suited for smaller applications, such as the following:

- REST APIs
- Small worker batch processes
- Microservices architectures
- Event-driven architectures

Apart from these use cases, it is also ideal for use in development or sandbox environments, where resources are not used all the time, to fully benefit from the billing model of these services.

Example – Azure SQL Serverless

As an example of how to apply this idea, Azure SQL offers a Serverless option on which the hourly rate is around double compared to the provisioned compute option, but you are only charged when your databases are used.

Imagine a development environment where a database is used only 1 or 2 days a week (around 200 hours a month).

Let's now calculate the cost of Serverless against the non-serverless model using similar compute characteristics:

Service	vCores	Type	Hours/mo	Region	Price/hour	Price/mo
Azure SQL Provisioned Compute Gen5 GP	2	Single database	730	West Europe	$0,53	$386,90
Azure SQL Serverless	0,5(min) 2 (max)	Single database	100	West Europe	$0,57	$57,00

Table 7.3 – Azure SQL provisioned compute versus Serverless

Even with the higher price of Azure SQL Serverless per hour, we could get a lot of savings (around 70% in this example) just by changing to the Serverless option in Azure SQL in some workloads.

This is a really specific example, but it illustrates how useful this approach can be in some situations.

Now that we've described different ideas of how to leverage Serverless service offerings, let's now move on to the last section on PaaS optimization, where we will look at managed Kubernetes optimization.

Managed Kubernetes cluster optimization

A **Kubernetes** cluster is formed of a set of nodes or virtual machines that run containerized applications.

Let's take a look at the official definition from the Kubernetes creators (`https://kubernetes.io`):

Kubernetes, also known as K8s, is an open-source system for automating deployment, scaling, and management of containerized applications.

These clusters are ideal for scenarios such as the following:

- Microservices-based applications
- Applications that are using CI/CD combined with DevOps and infrastructure as code
- Modernization of applications by migration to containerized workloads

Kubernetes is a really powerful engine to run containerized workloads both in the cloud and on-premises. However, one of its drawbacks is the management complexity and how difficult it is to hire resources with knowledge on the matter.

To simplify its management, cloud providers and software vendors offer a **managed version of Kubernetes**, on which a lot of features are managed by the provider. We have, for example, the following:

- **Azure Kubernetes Service (AKS)**
- **AWS Elastic Kubernetes Service (EKS)**
- **Google Kubernetes Engine (GKE)**

The main differences between the unmanaged version (Kubernetes installed and managed by you) and their managed counterpart are as follows:

- In managed Kubernetes, some parts of Kubernetes, such as the control plane, are managed on the provider side, making it easier to deploy services and perform maintenance. You can deploy a container that runs on cloud virtual machines.
- It makes it simpler to scale your node pools.
- Identity and access management is also included in the package, so you can integrate with the user directory of your choice.
- Networking is also provided, with easier integration with load balancers, networks, and other cloud services.
- Kubernetes core upgrades are also managed from the cloud provider side.
- Backup is offered out of the box.
- Managed Kubernetes clusters are continuous integration/continuous deployment ready.

We are not going to get deeper into Kubernetes as we could dedicate an entire book to the topic (many great authors have done that, in fact). But we are going to try and provide some ideas for cost optimization in managed Kubernetes clusters for Azure, AWS, and Google Cloud.

Keep in mind that most of the virtual machine optimizations we have previously covered also apply here, such as rightsizing, power scheduling, and use of Spot VMs, which we have covered thoroughly in *Chapter 6*, of this book.

Azure Kubernetes Service

These are key strategies to try out when running workloads in AKS clusters:

- One of the first points to work on is **rightsizing your cluster**. Our recommendation is to use as a starting point one of the pre-set cluster configurations that have been made available by Microsoft, such as Standard, Dev/Test, Cost Optimized, or Batch. Make sure that you are using the right virtual machines and storage, and that they are correctly sized for the workloads they are going to run.

- You can also apply **power scheduling** strategies on AKS clusters by using the cluster start/stop feature on non-production environments when clusters are not used.

- As we covered in the *Spot virtual machines* section in *Chapter 6*, including **Spot virtual machines** can be very cost-effective for some workloads. You can add secondary node pools to be used as additional capacity to primary node pools running in normal virtual machines.

- Make use of the **Cluster Autoscaler** and **Horizontal/Vertical Pod Autoscaling** features to scale either pods or nodes up and down automatically.

- Try **ARM** and **AMD** architectures, as we covered previously in the chapter. If your applications support them, they can lead to a huge improvement in both performance and pricing.

Important note: AKS start/stop

This feature is only available on clusters that use Virtual Machine Scale Sets.

A cluster that is powered off for more than 12 months won't be recoverable, as its state will be lost after that time. It may seem like a long time, but in our experience, time flies when there's a lot of work on your desk.

Also, keep in mind that you won't be able to scale or upgrade your cluster when it is powered off. Keep in mind that if you shut down the cluster during off hours all the time, you won't be able to perform other important maintenance operations that need to happen during this time.

A cluster can take a lot of time to be fully operational after it has been stopped. Please test this feature in development environments to ensure that your cluster supports this initiative before going forward.

AWS EKS

These are key strategies to try out when running workloads in EKS clusters:

- One of the first points to work on is **rightsizing your cluster**. Our recommendation is to use general-purpose virtual machines as a starting point and thoroughly analyze your clusters using monitoring tools such as `kube-resource-report` (`https://github.com/hjacobs/kube-resource-report`) to ensure that your containers are not wasting any resources. Also, make sure that you have chosen the right instance size, keeping in mind that for smaller workloads, you can use the most cost-effective EC2 instances available.

- **Power scheduling** can also work. We can set up automation to scale in or out depending on the time of day (for example, use `kube-downscaler`: `https://github.com/hjacobs/kube-downscaler`).

- **Autoscaling** is also possible here, by using the Cluster Autoscaler tool in combination with Horizontal Pod autoscaler, to try to align EC2 billable hours with demand as much as we can.

- Consider using **AWS Fargate** for a Serverless flavor on EKS, as it may be really cost-effective and simpler for some use cases.

Google Kubernetes Engine

These are key strategies to try out when running workloads in GKE clusters:

- Make use of **GKE Autopilot** mode to optimize the usage of compute while reducing administrative overhead, letting GCP manage resources, nodes, autoscaling, and pod placement. It leads to much simpler billing as well, as you only pay for CPU, memory, and storage.

- **Autoscaling** is also possible in GKE, by using Cluster Autoscaler, **Horizontal Pod Autoscaler (HPA)**, and **Vertical Pod Autoscaler (VPA)**, combined with node auto-provisioning to create new node pools automatically based on resource usage information (CPU, memory, requests, and so on)

- Consider using **Spot virtual machines** for Kubernetes node pools as added capacity, if your pods are fault-tolerant and support the Spot eviction process.

- **Rightsize** by testing and choosing the right virtual machine type for each job. Consider also choosing cost-efficient machine types such as Tau and AMD offerings, or even E2 virtual machines for cheaper pricing in less demanding workloads.

- Use scheduled automatic downscaling rules to implement **Power Scheduling**, so your cluster can be downsized off hours.

Let's now move on to another topic that is not often brought to the spotlight and is still a mystery for a lot of cloud professionals, which is data transfer costs.

Data transfer costs optimization

One of the more unknown and most hidden cost drivers in the cloud is **data transfer costs**. It is not easy to understand how they work, and it can be very time-consuming to keep them at bay and have some degree of control over them.

In this section, we will try to give some general insights on how to tackle this and avoid unwanted costs. Keep in mind that when working in the cloud, you are charged for egress and ingress traffic in and out of cloud services:

- **Data Transfer Ingress/Inbound**: This is basically all the information that we load or move to cloud services, such as uploading a file to an AWS S3 bucket or an Azure Storage account. This traffic is often free, as cloud providers want their clients to upload as much information as possible to their cloud offerings

- **Data Transfer Egress/Outbound**: This type of data transfer is the one we use when we download or transfer information from the cloud to other services or other cloud regions. It generally gets higher as the distance on which the data travels grows. This cost is mostly harder to keep at bay, and it can have a huge impact on the cloud bill if we don't limit it properly. To illustrate this with an example, if we download a file from the same AWS S3 bucket or Azure Storage Account to our computer, we will be charged some egress costs.

Let's describe in detail how this essentially works for Azure, AWS, and GCP, and how we can keep costs at bay in each of these services.

Azure – Data transfer costs

In Azure, as per the Microsoft documentation, these are the different types of traffic that cloud workloads generate and whether they are free of charge or not:

Type of traffic	Free of charge?
Ingress – Data transferred in	Yes
Data transfer between Availability Zones	No
Data transfer within an Availability Zone	Yes
Data transfer from Azure to Azure CDN/Front Door	Yes
Data transfer between regions within geographies	No
Data transfer between continents	No
Internet egress from a Region to any destination	No (100 GB free/month)

Table 7.4 – Data transfer charges in Azure

Also, one of the key points to understand is that on storage services such as disks and storage accounts, on some of the tiers, you are also charged per transaction. We will cover this topic in depth in the next chapter.

But for the topic at hand, we can provide a set of general recommendations:

- In Azure, *100 GB per month of Internet egress is free*. This means that if your workloads are not generating that much traffic, your data transfer costs may be contained.

- For development or non-productive (without any high-availability needs) workloads, avoid using multiple regions or Availability Zones when a lot of data is transferred to avoid high data transfer charges.

- If possible, choose the cheaper pricing data transfer regions for your workloads (in Azure, South America is the one with the highest cost, with the US and Europe being the lowest).

- Avoid routing traffic through the internet when possible. Instead of the public internet, you can use the Microsoft network by using private endpoints/links, peerings, site-to-site VPNs, and ExpressRoute.

- Leverage Azure CDN services to avoid repeatedly sending the same data over and over when applications support its use, as data transfer from workloads or resources hosted in Azure to Azure CDN is free.

- Provide some visibility on the costs of data transfer using Azure Cost Management. There is nothing worse than acting blindly! If the cost information is available on dashboards and in reports, you will be able to optimize and reduce costs as much as possible, targeting the right resources on which to reduce data transfer.

- Compress data if feasible, to reduce data transfer packet size as much as possible.

Now, let's analyze how AWS charges for data transfer egress and ingress.

AWS – Data transfer costs

In AWS, data transfer charges are similar to Azure.

Let's use the following table to analyze which types of traffic are free of charge and which aren't:

Type of traffic	Free of charge?
Ingress – Data transfer in	Yes
Data transfer between Availability Zones	No (Yes for multi-Availability-Zone resources)
Data transfer within an Availability Zone	Yes
Data transfer to CloudFront	No
Data transfer between regions within geographies	No
Data transfer between continents	No
Internet egress from a Region to any destination	No

Table 7.5 – Data transfer in AWS

Here are some general recommendations to minimize data transfer costs on AWS:

- For development or non-productive (without any high-availability needs) workloads, avoid using multiple regions or Availability Zones when a lot of data is transferred to avoid high data transfer charges.

- For services in production, make use of multi-Availability-Zone features when possible, instead of providing replication in other manual ways, to avoid cross-region or cross-Availability-Zone data transfer costs.

- If possible, choose the cheaper pricing data transfer regions for your workloads (in AWS, the highest-priced region is South America, with the US/Europe being the lowest priced).

- Avoid routing traffic through the internet when possible. Using VPC gateways to allow communication between services is a great way to generate some cost savings for this traffic. Also, VPC interfaces are available for some services. Please consider using these mechanisms, as well as AWS Direct Connect, so internet traffic can be minimized.

- Leverage AWS CloudFront services to avoid repeatedly sending the same data over and over when applications support its use, as data transfer from workloads or resources hosted in AWS to AWS CloudFront is free.

- Provide some visibility on the costs of data transfer using AWS Cost Explorer and Cost and Usage reports. There is nothing worse than acting blindly! If the information is available on dashboards and in reports, you will be able to optimize and reduce costs as much as possible, targeting the right resources on which to reduce data transfer.

- Compress data if feasible, to reduce the data transfer packet size as much as possible.

Lastly, let's close with GCP's take on data transfer costs.

GCP – Data transfer costs

In GCP, the data transfer costs are really similar to what we covered for Azure and AWS, but with some particularities:

Type of traffic	Free of charge?
Ingress – Data transfer in	Yes (No for resources that process traffic: load balancers, Cloud NAT, and protocol forwarding services)
Data transfer between Availability Zones	No
Data transfer within Google Cloud Zone	Yes
Data transfer from Google Compute/ Kubernetes Engine to Media CDN and Cloud CDN	No
Data transfer between regions within geographies	No
Data transfer between continents	No
Internet egress from a Region to any destination	No
Internet egress to selected Google services (for example, Gmail, YouTube, and Drive)	Yes

Table 7.6 – Data transfer in GCP

The following are recommendations for managing data transfer costs in Google Cloud:

- For development or non-productive (without any high-availability needs) workloads, avoid using multiple regions or Google Cloud Zones when a lot of data is transferred to avoid high data transfer charges. Keep data within the same Google Cloud Zone to minimize costs.

- If possible, choose the cheaper data transfer regions for your workloads (in GCP, Indonesia or Oceania is the one with the highest cost, with the US being the one with the lowest cost).

- Avoid routing traffic through the internet when possible. You can use Dedicated Interconnect or Partner Interconnect and VPC Network Peering instead to minimize the cost.

- Avoid long travels through the internet between the users consuming the services and the services themselves to avoid higher data transfer costs. This will result in additional latency as well, which is never desirable, especially for user facing applications.

- Leverage the GCP Cloud CDN and Media CDN services to avoid repeatedly sending the same data over and over when applications support its use, as data transfer (your workloads) from Google Compute Engine and GKE to GCP Cloud CDN and GCP Media CDN is free of charge.

- Avoid routing traffic through the internet when possible. Instead, it is better to use internal IP addresses to get better rates on egress traffic.

- Provide some visibility on the costs of data transfer using GCP VPC Flow Logs. There is nothing worse than acting blindly! If the cost information is available on dashboards and in reports, you will be able to optimize and reduce costs as much as possible, targeting the right resources on which to reduce data transfer.

- Compress data when feasible, to reduce the data transfer packet size as much as possible.

Now, we will proceed to another interesting topic that can yield cost optimization with little effort from technical teams, which is licensing optimization.

Licensing optimization

Licensing model cost optimization is one of the first things to work on when carrying out a FinOps exercise. We need to take a step back and try to look at the licensing needs for solutions from a higher-level perspective. The idea is, instead of managing the licenses at a solution scope, we manage them at the enterprise level instead.

This allows for much better savings due to licensing volumes if all licenses are purchased under the same conditions and contracts.

This topic is huge, and we could go on about all kinds of software vendors and how to optimize them. But as that goes beyond the purpose of the book, we are just going to cover one key concept here, which is to make use of the **bring-your-own-license model** in the cloud.

Bring-your-own-license model

Bring-your-own-license model, as the name suggests, it is about purchasing separate licenses directly to the vendor instead of paying them through the cloud provider. For this concept to work, we need licensing model for this to be supported from the different cloud providers on their services and offerings. In this section, we will see examples of applying this model and go through all the services that support such configurations.

Please keep in mind that the cost savings that will be attained come with a different, non-monetary cost. This licensing exercise requires planning, often in advance, of current and future license needs for workloads, as well as centralized management. This needs to be understood before going deeper into this topic.

Azure – Hybrid Benefit

Hybrid Benefit is a licensing benefit offered by Microsoft and associated with the Software Assurance licensing, which allows you to use your Enterprise Agreement licenses (traditionally, only used on-premises) on the Azure cloud.

The services that support this are the following:

Offering type	Type of service	Option
Virtual machine	IaaS	Windows Server
Virtual machine	IaaS	Red Hat Enterprise Linux
Virtual machine	IaaS	SUSE Enterprise Linux
Database	IaaS	SQL Server IaaS
Database	PaaS	Azure SQL Database Single database/Elastic Pool (Only in vCore provisioned compute)
Database	PaaS	Azure SQL Managed Instance

Table 7.7 – Hybrid Benefit supported services

For Microsoft licenses, there is a table you can use to compare on-premises licenses, and their correspondent cloud licenses that can be used in exchange by making use of BYOL licensing model:

EA license (Software Assurance required)	Azure	Equivalence	Other benefits
Windows Server Standard Windows Server Datacenter	Azure virtual machine	2 processor EA license = 2 Azure instances up to 8 vCPU or 1 Azure instance up to 16 vCPU	Windows Server Standard: 180 days coexistence on prem-azure / Windows Server Datacenter: dual usage on prem-azure
SQL Server Standard SQL Server Enterprise	SQL Server on Azure VM (minimum eligible: 4vCPU)	1 EA core license STD = 1 vCPU (STD) 1 EA core license ENT = 1 vCPU (ENT) 4 EA STD licenses = 1 vCPU ENT	180 days coexistence on prem-azure
SQL Server Standard SQL Server Enterprise	Azure SQL Single Database / Elastic pool (only in vCPU-Provisioned compute) Azure SQL Managed Instance	1 EA core STD license = 1 vCore GP 1 EA core STD license = 1 vCore HYP 4 EA core STD license = 1 vCore BC 1 EA core ENT license = 4 vCores GP 1 EA core ENT licenses = 4 vCore HYP 1 EA core ENT license = 1 vCore BC	180 days coexistence on prem-azure

Table 7.8 – Azure Hybrid Benefit equivalences

Another key feature of Azure Hybrid Benefit is **dual usage**, which allows you to use a license for a server hosted on-premises and the same license in a cloud server at the same time. This is of particular benefit in migration scenarios, where sometimes you need a grace period for licenses while services are migrated to the cloud.

Microsoft has created a simple calculator to help you with this process, to estimate the savings by using this initiative, which can be accessed here: https://azure.microsoft.com/en-us/pricing/hybrid-benefit/#overview.

Example – Azure Hybrid Benefit

As an example, we have prepared a cost comparison between two Ubuntu virtual machines, one with Windows Server installed without Hybrid Benefit and one with Windows Server with Hybrid Benefit:

Virtual Machine	Region	Hours of use	Operating System	Hybrid Benefit	OS Cost	Compute Cost	Total cost	License Type
D4 v5	West Europe	730	Ubuntu	N/A	$0,00	$167,90	$167,90	N/A
D4 v5	West Europe	730	Windows Server 2022	No	$132,32	$167,90	$300,22	Windows Server Datacenter provided by Azure
D4 v5	West Europe	730	Windows Server 2022	Yes	$4,00	$167,90	$171,90	Windows Server Standard + Software assurance per processor

Table 7.9 – Azure Hybrid Benefit example. Prices obtained using
the Azure Hybrid Benefit calculator on April 2023

> **Note**
>
> OS pricing varies from organization to organization depending on how many licenses we purchase, as we get better discounts at higher volumes, we have used a ballpark price for a Windows Server license based on our experience.

Keep in mind that using the equivalence table, you can use Windows Server Standard licenses acquired for on-premises in the cloud for Windows Server Datacenter virtual machines. This is where most of the savings come from.

In this example, using Hybrid Benefit for this virtual machine can provide up to *42% savings* on the total virtual machine price. The savings can be even higher if we consider multiple virtual machines, as one on-premises two-processor license entitles you to use Windows Server Datacenter in two virtual machines for up to eight vCores.

AWS bring your own license

In AWS, it is also possible to make use of the bring-your-own-license model for some service offerings, such as the following:

Offering type	Management level	Option
Virtual machine	IaaS	AWS EC2 Windows Server (Dedicated Hosts)
		Amazon Workspaces for Windows Desktops
Virtual machine	IaaS	AWS EC2 with
		Red Hat Enterprise Linux
Database	IaaS	AWS EC2 SQL Server
Database	PaaS	AWS RDS Oracle

Table 7.10 – AWS services that support bring your own license

Also, there is a key service offering that is provided by AWS: **AWS License Manager**.

AWS License Manager lets you centrally track all licenses used throughout cloud environments, allowing for much easier management and simplifying audit processes. It also helps to avoid licensing overage, as you can track allocation per resource (vCores, physical cores, or other allocation methods) for each license as well.

It supports a lot of vendors, such as Microsoft, IBM, Oracle, and SAP. Using this feature allows for the use of bring your own license, which is key to attaining some savings as you can use your licenses instead of paying a premium for licenses to be provided by AWS.

It is, however, difficult to estimate cost savings by using this initiative on AWS, as pricing per license from vendors varies greatly depending on licensing volumes.

Google Cloud bring your own license

In Google Cloud, making use of already owned licenses when using Google Compute Engine/Virtual Machines is also supported.

Keep in mind that for licenses that require dedicated hardware, sole-tenant nodes are needed to ensure that licensing will not malfunction.

There is a Google-developed and open source tool as well, available to aid with license management in GCP, called **GCP License Tracker** (`https://github.com/GoogleCloudPlatform/gce-license-tracker`).

It can be used to track license placement on running nodes, which can be really helpful for auditing purposes.

Cloud provider agreements and resource allocation

Choosing the right contract and account, project, and subscription structure under which you will create your cloud resources is one of the first key points that should be considered for cost optimization purposes.

Even though the concepts described in this section won't result in direct savings being generated, it will allow for better cost visibility and easier management overall.

We are going to cover the differences between a **Cloud Solution Provider** (**CSP**) and Enterprise Agreement In Azure, which are licensing or contract models to make use of Azure services, while for AWS and GCP, we will describe strategies to organize your workloads and allow for better cost allocation and visibility.

Azure – Enterprise Agreement versus CSP

In Azure, there are mainly two different models:

- **Enterprise Agreement**, where you sign a direct contract with Microsoft that lasts three years, with which you are able to add software licenses and buy cloud services
- A **CSP** program, on which the services are provided through a Microsoft partner or reseller, instead of directly through Microsoft

The key takeaway is that Enterprise Agreement is a fixed contract that lasts three years, as opposed to a CSP program, where you can use the services for as little as one month if necessary, without any further commitment.

Let's analyze the differences between them:

Model	Enterprise Agreement	Cloud Solution Provider
Offering	Commitment-based cloud services and software licensing	Transactional (based on services used) cloud services and cloud licenses
Term	Three years with yearly payments	One year (auto-renew) with monthly payments
Purchase coverage	Organization-wide	Multiple CSPs could coexist in one organization
Sales model	Directly through Microsoft or indirectly through a Partner	Indirectly through a partner
Minimum usage	500+ users or devices	N/A
Support	Direct support with separate purchase, such as Premier Support	Through partner
Discounts	Volume discounts and software assurance pricing Fixed prices during the contract period	Negotiated discounts with partner
Admin overhead	Agreement renewals or true-up	N/A

Table 7.11 – Enterprise Agreement versus CSP

From the table, it is clear that CSP is better suited for smaller companies, as it has no requirements and requires no commitment. The payment is monthly so you can adapt and change volumes or eliminate services if needed. The discounts are lower, though, as you need to negotiate with the partner, and the support may not be as straightforward, having a man in the middle, when issues arise or support tickets are needed.

However, for bigger organizations, it is much better to opt for Enterprise Agreement, where the discounts can be higher and you can largely benefit from volume discounts. Also, even though it requires additional administrative overhead, it allows for central license management and Reserved Instances/Saving Plans management, which enables bigger savings in the long run.

From a pricing perspective, Enterprise Agreement allows for bigger discounts and protects prices from inflation rise throughout the contract period. In our experience in different organizations, we have seen huge price differences over a really short time due to inflation (e.g., the recent Ukraine/Russia conflict raised cloud prices through the roof, as compared to a few years before), which should not be dismissed at all.

Once we have our contract in place, we can use different management containers to organize our resources based on our needs. Our recommendation in this regard is to follow the Cloud Adoption Framework from Microsoft to the letter as a starting point and then adapt it to your needs (`https://learn.microsoft.com/en-us/azure/cloud-adoption-framework/ready/landing-zone/design-area/resource-org-management-groups`).

AWS organizations, billing accounts, and OUs

Let's describe how we can organize our workloads in AWS.

To begin with, we assume that a company will be comprised of multiple business units or domains. We can segregate our resources in this manner by making use of AWS **Organizational Units (OUs)**, and when we have those OUs in place, create accounts for different environments, as depicted in the following figure:

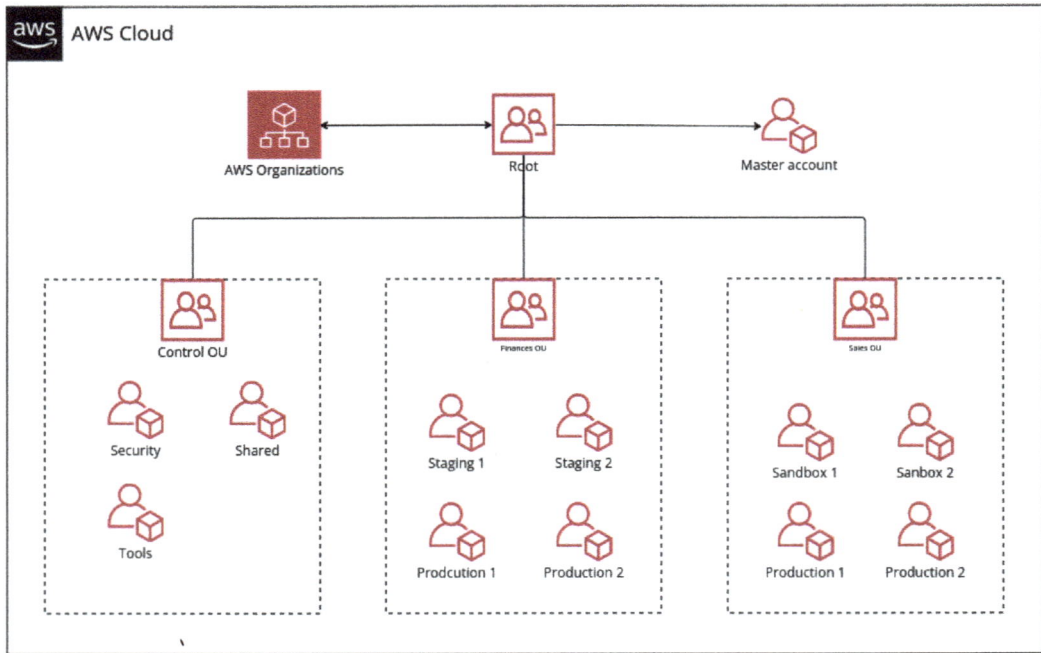

Figure 7.2 – AWS OU and account structure

This account centralization allows for easier management and more overall control over resources created in each one. But it also allows for easier cost allocation, as the costs can be divided per account and OU or aggregated at the highest level.

Regarding billing, we need to have a master account that will be responsible for the billing of all linked accounts assigned to the master account.

With this business unit distribution established, it is a good practice to create linked accounts in each OU for different environments. By doing this, we will be adding another layer of segregation for cost allocation purposes.

Having this environment separation will allow for the better tracking of initiatives that are usually applied to non-production environments, such as Power Scheduling, and the ones that are applied to productive ones, such as Reserved Instances and Saving Plans.

Having this structure also allows us to apply policies and enable easier IAM in AWS. On Reserved Instances and Saving Plans, should we purchase them at the organization level, multiple environments can benefit from the discounts, which is really good to add flexibility and ensure that Reserved Instances won't be wasted. If needed, we can even configure that Reserved Instances should only be applied to specific accounts (production ones, for example).

This straightforward way of structuring AWS accounts and OUs provides a lot of benefits from a governance point of view and for cost allocation and visibility purposes, which is essential for the fluid implementation of the FinOps initiatives discussed in this chapter and others.

GCP organization, folders, projects and resources

GCP also enables us to logically organize our resources in GCP Resource Manager. We can establish a hierarchical structure with different levels:

- Organization
- Folders
- Projects
- Resources

The use of folders, in the same way we described for AWS, is really beneficial for cost allocation and simpler overall management. It is the perfect layer to be used for Business Units or different entities inside an organization.

Once we have some folders in place, we can begin creating projects on which resources will be hosted. Projects used can depend on region, environment, department to better fit the logical segregation of each organization, as shown in the following figure:

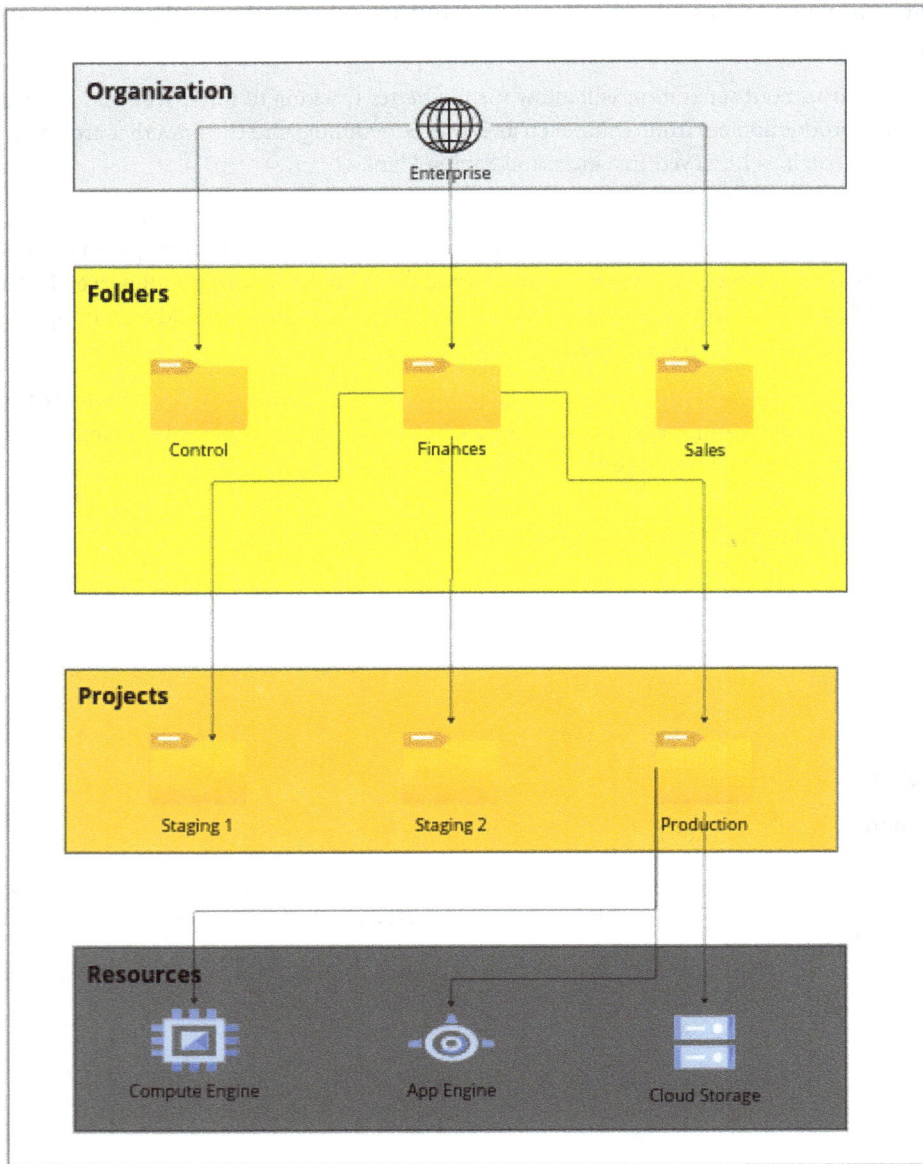

Figure 7.3 – GCP management structure

Now, for billing purposes, the best practice is to have one global billing account that will be shared across all projects for an organization. We can see what this looks like in the following figure:

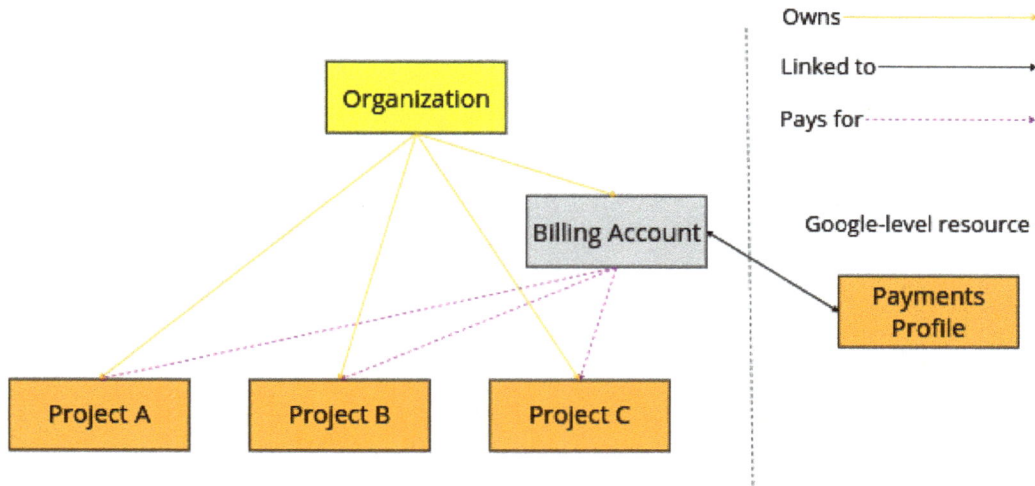

Figure 7.4 – GCP Billing account and projects

Regarding committed use discounts, they are only assigned to one project by default, but you can activate a feature in the billing account for those discounts to be shared between projects linked to that account, allowing for more flexibility and avoiding waste as well.

Summary

With all the ideas described in this chapter, we close our exploration of compute optimization. We discussed the initiatives, concepts, and tools that every FinOps practitioner should know about, as well as deciding whether to apply them or not depending on the use case and the context.

We also explained some key topics, such as data transfer costs and resource management, which, even though they don't result in direct cost optimization, can be a great help in your cost optimization journey.

In the next chapter, we will delve into database optimization, keeping the same mindset we've utilized during the chapters dedicated to the compute domain.

8
Implementing Database Optimization

A database is, in essence, a platform where data can be organized, stored, and accessed electronically.

There is no denying that databases are one of the most used ways of storing information for enterprise applications. With the rise of big data, data science, and machine learning, their popularity has skyrocketed, making them a great asset to build reports, store information, and build all kinds of projects around them.

Such popularity is a double-edged blade because if databases are not used in an optimal way, the cost can increase exponentially, especially if their key cost drivers are not taken into account from the solution design phase.

With the rise of the public cloud, you can spin a database for your data in minutes, allowing a lot of agility for projects and IT teams, but it is also as easy to fall into common mistakes if important details are overlooked.

There are also a lot of different types of databases, each with strengths and weaknesses and with specific use cases. For FinOps purposes, it is crucial to start from the beginning, which means choosing the right type of database for your workload. If this part of the process is not done in a proper way, inefficiencies are going to appear in the long run, when they will be much harder to correct.

To avoid these situations, we will try to cover all that's needed to make the right decisions from the solution design phase. Non-optimized databases equal storage or compute overage, which always results in additional costs that no one wants and are difficult to justify.

In this chapter, we will look at concepts and ideas that can be used to keep your databases in an optimal state, along with strategies to align the use of each environment with the right SKUs and sizing. To top it off, we will also give some hints on how to manage licensing and scaling for databases, and when to go for PaaS solutions over IaaS offerings, as well as covering new serverless database offerings such as AWS Aurora and Azure SQL.

In this chapter, we are going to cover the following main topics:

- Which database is correct for my workload?
- IaaS database optimization
- PaaS database optimization
- Reserved capacity
- Licensing optimization

This is one of the key questions to ask ourselves, and it is not one that is easily answered. As a starting point, let's analyze all the different options we have for databases, to better understand the possible choices that we need to carefully consider.

Relational versus non-relational/NoSQL databases

There are mainly two types of databases, and choosing one over the other depends on how data is formatted.

On one hand, we have **structured data**, based on rows and columns, as we have in Excel or CSV files, in which data always has a pattern or predefined format, thus it is often easier to read or search. On the other, we have **unstructured data**, which is data that does not follow a predefined data model and is not limited to a set of rows or columns, such as images, videos, and business documents such as contracts.

Organizations use structured data on a day-to-day basis, to keep track of business or financial processes or human resources, for example. But there are also a lot of use cases for unstructured data as well, such as social media, IoT, and invoice management, to begin with. For all these use cases, unstructured data is needed, and huge volumes of data can be generated due to this fact.

Different types of data require different types of databases if we want to achieve optimization, not only on a cost level but also to get the best performance we can. If we focus on cost optimization, deciding on one type or another can have a huge impact on the costs of storing data, as well as performance, so we will introduce all the key differences between them in a nutshell and when we should choose which one.

Relational databases

A relational database consists of tables that are formed from columns and rows. Tables can have relationships by using links that join tables together.

Let's use an example to illustrate how they work:

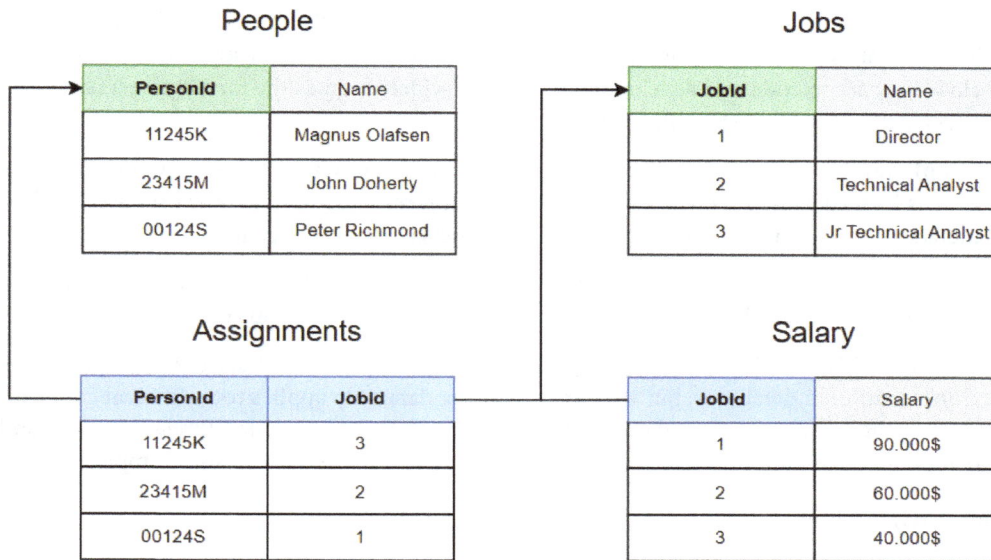

Figure 8.1 – Relational database example

In this example, there are four different tables: one for people, one for jobs, another for job assignments, and the last one for job salaries. In `people`, we store different people's information (identifier and name), and the rest of the tables describe the jobs that have been assigned to each person and the salary that each job implies.

The `PersonId` and `JobId` (Jobs) columns are **Primary keys (PKs)** and the `PersonId` and `JobId` (Salary) columns are **Foreign keys (FKs)**. A PK is a unique row in a table, while an FK references another PK from a different table. PKs contain values that make each table record unique (in `People`, `PersonId` for example), while FKs reference primary keys from other tables.

Data in a relational database is always structured by definition, as each row in a table has different fields or columns with a specific format. With the links between tables, we can enrich our data, making it more complete and more complex, which allows us to make complex queries that span different tables, such as a list of all persons with their jobs and salaries.

Non-relational or NoSQL databases

NoSQL databases or non-relational databases are unstructured databases that store data in documents rather than relational tables, using a much more flexible approach in which we are not limited to a specific data model or format.

There are four data structures that are used in NoSQL databases:

- **Document databases**: These store data in text files, such as JSON, BSON, or XML documents. These databases are really popular because of their flexibility in data structures and for how close they are to code and development. They are widely used as the format to get data from REST APIs, such as AWS S3 REST APIs.

- **Graph databases**: This NoSQL data structure focuses on relationships between elements. Data is formed by nodes interconnected by links. It is really useful for some use cases, such as metadata and advanced analytics and real-time transactional applications.

- **Key-value stores**: These are as simple as it gets. Data is stored in keys that have values assigned to them. It's the model used in Azure Key Vault to store secrets, for example.

- **Column-wide or column-oriented databases**: These databases still use the concept of rows from traditional databases, but writing or reading data only applies to individual columns, making them much faster, and great for use cases such as data analytics, time series data, and databases to store logs. This data model is used in Azure Log Analytics, for example.

As a summary, this is how each of these data structures can look:

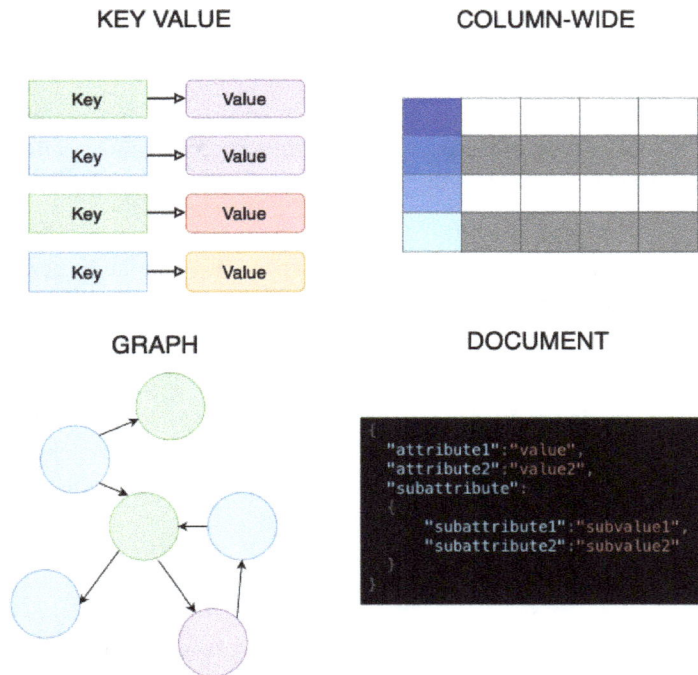

Figure 8.2 – NoSQL data structures

In this example, we are going to use a NoSQL document database in JSON to represent the data depicted in the example that we provided in the previous section:

```
{
        "name":"Magnus Olufsen",
        "PersonId":"11245K",
        "Job":"Jr Technical Analyst",
        "Salary":"40,000",
}
```

In this fragment of JSON, we can see that information is presented in different lines defining each field and its value and separated by commas. It is a simple yet powerful way to represent data that offers a lot of flexibility.

Now that the reasoning behind relational and non-relational databases is more or less clear, let's move on to the key question: which one should you choose?

Which one should you choose?

With the different examples in mind for both relational and NoSQL databases, let's ask ourselves some key questions.

Let's say we have a really simple use case, for a really simple application that stores the hierarchical structure of our company, so we can keep it updated and available for all employees:

- *Does it make sense to have a full-blown relational database for such a simple use case?* Maybe not.
- *Would it be possible to use a relational database for this use case?* Yes, for sure, the application will work nevertheless.
- *Which one will be more cost-efficient?* A NoSQL graph database will probably be much more scalable and cost-efficient. We could reduce its size to the minimum and grow all we need. On the other hand, if we have a relational database, scaling will be much harder and we could only scale down to a point.

So, yes, choosing one or the other has a big impact on our **Total Cost Of Ownership (TCO)**, among other things. Because of this fact, it's vital to understand why this impact on cost occurs and make the best choice for your database to be cost-effective.

Also, the paradigm of the database has implications for how databases can scale. Relational databases are more difficult to scale, especially if we try to use horizontal scaling, while it is usually a breeze to scale out in NoSQL databases. On the other hand, relational databases are the way to go in applications where data integrity, compliance, and security are key points to consider.

It is impossible to cover all the different products and use cases, and it goes far beyond the purpose of this book, but we will try to summarize, using a table, the advantages and disadvantages of each one:

Type of database	Pros	Cons
Relational	<ul><li>Centralized</li><li>Maturity</li><li>Data integrity and normalization</li><li>Data model simplicity and ease of use (SQL-type languages)</li></ul>	<ul><li>Requires more maintenance</li><li>Costs</li><li>Requires more storage</li><li>Limited scalability</li><li>Performance is limited when data volumes or demand are high</li><li>Low flexibility</li></ul>
Non-relational	<ul><li>Decentralized</li><li>Data ingestion from multiple locations</li><li>Can store huge volumes of data</li><li>Performance and scalability</li><li>High flexibility</li></ul>	<ul><li>Lack of maturity</li><li>Lack of standardization</li><li>More complex to use and to ensure data integrity</li><li>Requires more expertise and specific tools</li></ul>

Table 8.1 – Relational versus non-relational databases

With relational and non-relational databases in mind, let's move on to the next topic that it is vital to understand when designing solutions that use databases, and that's choosing the right database management system.

Which database management system?

A **database management system** (**DBMS** from now on) handles all the operations in the database, such as providing user access to stored data, as well as all the required operations on the database, such as backup, queries, permissions management, and so on.

A **database engine** is the underlying interface, part of the DBMS, that is used to **Create**, **Read**, **Update**, and **Delete** information from a database, which is often referred as **CRUD**.

Choosing the best DBMS and database engine is a key point in solution design, as it has a lot of implications – one of them being cost.

Example – SQL Server versus Oracle pricing for AWS RDS

Let's use a simple example for clarification. In this example, we will analyze the price difference between the exact same AWS service offering – in this case, AWS RDS, where the only difference is the database engine:

Service	Region	Configuration	Type of license	Price/month
Amazon RDS Custom for Oracle	eu-west-1	db.r5.xlarge Single-AZ gp2 SSD 100 GB storage	Enterprise	464.57 $
Amazon RDS Custom for SQL Server	eu-west-1	db.r5.xlarge Single-AZ gp2 SSD 100 GB storage	Enterprise	1312.54 $

Table 8.2 – Prices obtained using AWS Pricing Calculator in April 2023

I know you must be thinking, it is not the same! For sure, it isn't; there are a lot of considerations to ponder:

- **Public Cloud**: SQL Server has a higher price on Amazon Web Services than on Azure. Compared to a similar enterprise license, it can cost five times more. Of course, it makes sense for Microsoft to charge a higher price for their licenses that their cloud competitors' PaaS databases offering. This has a huge impact on cost. If we compare pricing in Azure, the numbers vary a lot.

- **Licensing**: I have chosen the Enterprise license for both database engines, which is a good choice for big organizations' databases. For clarification, the features offered are different in each one, though, but it is a good way to compare similarly tiered licenses.

- **Price**: Oracle pricing, when support is included and only considering the software price, is considerably higher than the SQL Server price offering similar features. However, in this example, as you can see, the SQL Server cost is significantly higher, due to the cloud of choice.

- **Operating System**: If this is considered for migration from on-premises to the cloud, and your organization has historically used Linux, there is a big chance that you are not using SQL Server on Linux, as it has only been available since 2017. Technological debt plays its part here, as some companies may be somewhat forced to choose one over the other if they already have operations in place on a specific DBMS.

It should be clear now the impact that this decision has on so many levels.

The idea of this example is not to determine which one is better but to highlight how big the differences are even on the same services with the same configuration, and how big the impact on costs is when choosing one database engine over another.

Now that the implications of choosing one DBMS over another are more or less clear, let's move forward and analyze the main advantages of the most well-known DBMS.

SQL Server

Microsoft SQL Server is offered in different editions (Developer, Express, Standard, Web, and Enterprise) with different features for different use cases. As a DBMS though, and compared to other alternatives, these are its main advantages and disadvantages in our view:

Pros	Cons
• Data security • Market presence • Optimized data storage • Standardization • Shows you an integrated view of your business • Support for use on Linux since SQL Server 2017 • It works very well with other Microsoft products	• High price • Licensing: license changes are very difficult if you don't stay on Azure • Restricted compatibility • Resource-intensive scaling • Complex functionality that requires specific skills and configuration/maintenance • Supports only structured data

Table 8.3 – SQL Server pros and cons

Oracle

Oracle, like SQL Server, is an industry standard that offers multiple editions (Enterprise, Standard, and Express) that can be used to adapt to different needs. These are the pros and cons of this DBMS:

Pros	Cons
• Data security • Market presence • High compatibility • High performance • Support of all major cloud providers • Shows you an integrated view of your business	• High price • Resource-intensive scaling • Complex functionality that requires specific skills and configuration/maintenance • Supports only structured data

Table 8.4 – Oracle pros and cons

PostgreSQL

PostgreSQL is an open source DBMS that is widely used for a lot of applications and use cases. These are the main strengths and weaknesses compared to other DBMSs:

Pros	Cons
• Data security • Open source license • Support of all major cloud providers • Supports structured and unstructured data • Highly expandable • It is compatible with various platforms and major languages (Python, Java, Perl, PHP, C, C++, etc.)	• In terms of performance metrics (complex queries), it is slower than other DBMSs • Not easy to install for beginners • Lack of documentation: doesn't have the best documentation compared to other database engines

Table 8.5 – PostgreSQL pros and cons

MySQL

MySQL is another open source DBMS that is widely used in both on-premises and cloud solutions. It offers the following features compared to the other DBMSs:

Pros	Cons
• Data security • Open source license • High compatibility • Support of all major cloud providers • Easy to install • Large community support	• In terms of performance metrics (complex queries), it is slower than other DBMSs • Scalability • Lack of stability • Supports only structured data

Table 8.6 – SQL Server pros and cons

MongoDB

MongoDB is an open source industry standard in NoSQL/non-relational databases that is document-oriented. These are its main features:

Pros	Cons
• Data security	• Memory limitation
• Open source license	• Limited data size (documents)
• Performance	• Limited nesting
• Supports structured and unstructured data, but the best features relate to unstructured data	
• Easy to install	
• Flexibility – dynamic schematic architecture	
• Technical support available from MongoDB for enterprises	

Table 8.7 – MongoDB pros and cons

It is no easy task to decide on one or the other, but we will provide some general recommendations to keep in mind when doing so:

- Open source alternatives may be a good choice if the budget is limited or for proof of concept or test scenarios.
- If, due to technological debt or any other requirements, the chosen DBMS must be SQL Server, the cost-effective option is to host it on Azure. Hosting it in non-Microsoft public clouds will probably be a costly exercise.
- If the workload requires low latency and great response times in structured data scenarios, Oracle may be a good choice.
- If the schema is unclear or unknown, MongoDB and PostgreSQL may be good choices, if their scalability limitations are not an obstacle.
- Lightweight DBMSs such as PostgreSQL and MySQL may be a good choice for containerized scenarios.

With all this information, let's tackle the last key point to choose the most efficient database for your solutions, which is to choose how to host it. Which one is the most adequate – IaaS, PaaS, or serverless?

IaaS versus PaaS versus serverless

When choosing which database is better for your workload, there is always a major decision to make: should you go for IaaS or for PaaS? If you are going to opt for the PaaS option, when should you use serverless offerings?

It is never an easy question, and there is no unique answer, so let's analyze this question from different perspectives.

Let's revisit the differences, which we already covered in *Chapter 7*, but specifically for databases:

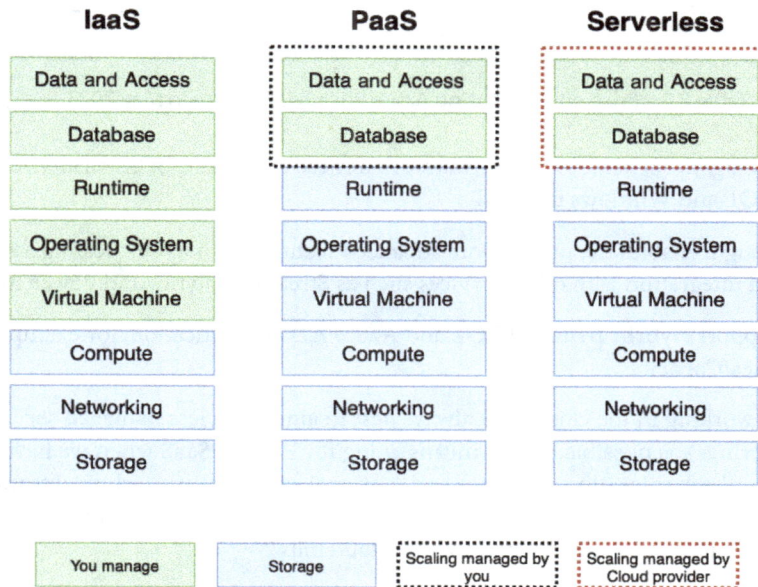

Figure 8.3 – IaaS versus PaaS versus serverless

In general terms, having your database in IaaS means that you need to take care of the virtual machine (specs, disk, network), operating system, and the database engine software installation. In high availability settings, it also means that you are in charge of creating a cluster and maintaining it, which can sometimes be troublesome if there is a lack of experience with databases and clusters among teams, as it requires additional administrative overhead.

In PaaS, you just deploy the service and the database will be available in a minute. However, if you need your database to work in a private network, it requires additional setup and the complexity of the solution rises, as it requires additional planning such as DNS management. In PaaS, you also need to manage and take care of scaling for your databases, while with serverless options these settings are managed by the cloud provider.

To begin the discussion, let's analyze the cost of a similar database in Azure, AWS, and GCP hosted both on IaaS and PaaS versions.

Example: Azure Virtual Machine with SQL Server versus Azure SQL

Let's use a really simple example for clarification. In this example, we will analyze the price difference between the exact same database in both IaaS and PaaS offerings with similar configurations.

Service	Region	Storage	Redundancy	Hours of use	Price
Azure SQL Provisioned Compute General Purpose (Gen 5) 4 vCores	West Europe	128GB	LRS	730	$803,59
Azure Virtual Machine D4 v5 with SQL Server Standard	West Europe	128GB (E10 SSD Disk)	LRS	730	$603,82

Table 8.8 – Price obtained using Azure Pricing Calculator in May 2023

The cost of PaaS is slightly higher (25% more), but we don't need to take care of installation, configuration, and backup, or SQL and Windows updates.

Having PaaS, though, enables the use of more advanced features such as autoscaling, creating read-only replicas, and integration with other services such as Stream Analytics and Power BI.

Both versions support **Hybrid Benefit**/BYOL and Azure AD authentication, for example, as well as Reserved Instances/Capacity.

Generally, when working in the cloud, it is always best to aim for as less managed services (such as PaaS or SaaS offerings) as possible, which means going for PaaS or SaaS when we have the chance instead of IaaS, for both migration scenarios and new projects. The time technical teams spend on cloud operations and administrative overhead are also resources that cost a lot of money, even though they may be more intangible compared to monthly cloud bills.

This does not mean that IaaS is not recommended – sometimes, it can be easier if teams are more accustomed to working with IaaS databases and we have well-oiled cloud operations running backup and monitoring in highly available setups that are already built. If this is not the case, you should always choose PaaS services.

On the other hand, there are other key differences that should be taken into consideration:

- We should aim for as few managed services as possible. This means that we should prioritize PaaS over IaaS.

- Reserved Instances and Saving Plans for PaaS resource offerings are limited to only a few services. If you have running operations with reservations in place for virtual machines, it may make sense to host databases on virtual machines as well, while you extend reservations for really stable workloads with a long lifecycle.

- If your cloud operations are focused on IaaS and you have clear backup policies and operations that handle monitoring and maintenance in IaaS already, it may make sense to host databases in IaaS, as you don't need to build operations from the ground up but reuse them for databases.

- There are some PaaS database offerings that have serverless/consumption models on which you are only charged when the database is used. This makes sense in a lot of scenarios, as we will explain later in this chapter.

- In PaaS, you manage scaling for services, either manually or automatically. If the patterns of usage are unclear for your workload, it is always best to choose serverless offerings. If patterns are clear and regular, you can tailor and test them, so they fit your needs.

- Some databases are only available in the IaaS model, with a traditional install. BYOL is also limited for a lot of databases with the IaaS model. You must carefully plan your needs and then select the right product and license for cost optimization.

Now let's deep dive into how to optimize databases running on IaaS models, and the different initiatives that can be applied to those.

IaaS database optimization

When working with databases installed on virtual machines in the IaaS management model, most of the compute initiatives that we covered in the previous chapter apply here as well, such as rightsizing the virtual machine specs.

However, there are concepts and ideas that we want to specifically highlight to optimize databases hosted on IaaS. As we already discussed, FinOps is far from one-size-fits-all.

Rational database use

This concept is a really simple one that we may forget sometimes. Databases are not black holes where we can put all the data we want. We must use databases wisely, and only store the data that is essential in them, as not doing this may have implications on performance and cost, to begin with.

Let's not forget that we have evolved from on-premises traditional IT, where this concept was not as important as it is right now. For every GB of unneeded data in our databases, we may be paying top dollar, which everyone will see at the end of each month when the cloud bill comes, whereas with traditional on-premises hosted databases, this was not as important due to spare capacity.

There are a number of strategies you could follow to reduce your databases:

- Colder data that is not needed for day-to-day work can be exported to datalake services such as GCP BigQuery, AWS Athena, or Azure Data Lake Storage Gen2, where storage is much more cost-efficient. In these services, apart from having the possibility to query data using a powerful query engine, it is also possible to create dashboards and reports based on the data.

- With the same idea in mind, if data does not need to be consulted with query engines, we can just store it in cold tiers in storage services such as AWS S3 or Azure Blob storage. We can set up different tiers for different data temperatures in these services, as well as setting up lifecycle policies for data to be moved between tiers automatically based on conditions that we can configure.

- The same as virtual machines have a lifecycle, which we may use to plan for Reserved Instances and other strategies, data itself should have a lifespan. We should ensure that a data retention and data deletion policy is defined and reasonable, as well as aligned with cloud operations. For sure, the definition of these key policies is also subject to compliance and legal considerations, and should also be discussed with different business areas in some cases. We need to ensure that there are processes to clean and delete old data past its lifespan, as well as to move it to colder tiers of storage if it needs to be retained for a long period of time.

- We must also consider, for some use cases, making use of cache services such as Azure Cache for Redis. Services such as these can reduce the load of our databases while reducing the costs of queries that are repeated a lot over time and requested from different clients. Let's not forget that in the cloud we are also charged for network egress and ingress in some cases, so reducing it is always a good idea. It will also reduce the load of the database, which is also a great benefit.

With this concept in mind, though abstract, let's focus on backup storage used for databases.

Backup storage optimization

As we will cover in the next chapter, keeping storage optimized can be a daunting task.

When covering database optimization, there is one important initiative to take into consideration related to backups, and it is to always try to store the backups in object storage such as Azure Blob, AWS S3, and GCP Cloud Storage instead of storing them directly on disks.

Considering that backing up a database is a task that needs to be done periodically and, given data retention regulations for most companies, the backups need to be retained for years sometimes, backup storage size (and cost) may not be a small thing at all.

These storage services offer different tiers based on data temperature, which allow for backups to be stored in colder tiers, on which the storage capacity is cheap but the price per transaction goes up. As these backups are the kind of data that is not touched every day (we usually access the backups just when they are going to be used), they are an example of a really good use case for these cold tiers, allowing for huge cost savings.

Traditionally, I have seen a lot of backup files stored on data disks when disk capacity was available and not scarce, but as times change, we need to adapt as well and change our ways to leverage cloud service offerings.

Apart from where these backups are stored, we also need to keep in mind that backups can take up a lot of storage, so another thing that must be carefully analyzed is the current **Recovery Point Objective (RPO)** and **Recovery Time Objective (RTO)** needs for our solution.

We may be storing many more backups than RPO and backup retention policies require. Also, we need to consider the RTO if we store our backups in the coldest storage tiers, such as AWS S3 Glacier, Azure Blob Archive, or GCP, as in these storage tiers data cannot be retrieved instantly and it can take some time to have our data available, usually in the order of hours.

We will provide some examples on Azure, AWS, and GCP as to how much of an impact this optimization initiative can have on costs.

Azure example: 1 TB of backups – Object/Blob Storage versus Block Storage/ Managed Disks

We are going to compare 1 TB stored on a data disk attached to a virtual machine with the same amount of capacity but stored in an Azure Blob Storage , which is Azure Blob, in all the different tiers that Azure offers for each service:

Type of Storage	Capacity/GB	Redundancy	Region	Price per GB	Tier	Price
Azure Managed Disk	1024	LRS	West europe	$ 0,1452	Premium SSD	$ 148,68
Azure Managed Disk	1024	LRS	West europe	$ 0,0750	Standard SSD	$ 76,80
Azure Managed Disk	1024	LRS	West europe	$ 0,0400	Standard HDD	$ 40,96
Azure Blob Storage	1024	LRS	West europe	$ 0,0181	Hot	$ 18,53
Azure Blob Storage	1024	LRS	West europe	$ 0,0092	Cold	$ 9,46
Azure Blob Storage	1024	LRS	West europe	$ 0,0017	Archive	$ 1,71

Table 8.9 – Prices obtained using Azure Pricing Calculator in May 2023

The results speak for themselves. The price per GB on Standard SSD is more than 8 times the price of Azure Blob Storage on the cold tier, and we could obtain 88% savings should we choose to store that TB in cold storage.

AWS example: 1 TB of backups – Object/S3 Storage versus Block Storage/EBS

We are going to compare 1 TB stored on a data disk attached to a virtual machine with the same amount of capacity but stored on AWS S3, which is the object storage offering from Amazon, in the different tiers that are suited for backup use cases:

Type of Storage	Capacity/GB	Redundancy	Region	Price per GB	Tier	Price
Amazon EBS	1024	Single-AZ	Ireland (eu-west-1)	$ 0,1100	gp2 - General purpose SSD v2	$ 112,64
Amazon EBS	1024	Single-AZ	Ireland (eu-west-1)	$ 0,0880	gp3 - General purpose SSD v3 3000 IOPS	$ 90,11
Amazon EBS	1024	Single-AZ	Ireland (eu-west-1)	$ 0,0021	sc1 - Cold HDD	$ 2,10
Amazon S3	1024	Single-AZ	Ireland (eu-west-1)	$ 0,0230	S3 Standard	$ 23,55
Amazon S3	1024	Single-AZ	Ireland (eu-west-1)	$ 0,0125	S3 Standard-Infrequent access	$ 12,80
Amazon S3	1024	Single-AZ	Ireland (eu-west-1)	$ 0,0040	S3 Standard-Glacier Instant retrieval	$ 4,10
Amazon S3	1024	Single-AZ	Ireland (eu-west-1)	$ 0,0036	S3 Standard-Glacier Flexible retrieval	$ 3,71
Amazon S3	1024	Single-AZ	Ireland (eu-west-1)	$ 0,0010	S3 Standard-Glacier Deep Archive	$ 1,03

Table 8.10 – Prices obtained using AWS Pricing Calculator in May 2023

The results are quite interesting here compared to Azure. In this case, on AWS, there is a really cool offering called **Cold HDD Volumes**, which is really well suited for storing backups, so from the cost perspective, either object or blob storage can work.

In this case, comparing a normal disk with cold tier storage, the standard data disk storage (gp3) price per GB is 7 times the price of the same storage in S3 Standard-Infrequent Access.

Glacier Deep Archive is still king from a pricing perspective, but keep in mind that the retrieval time can take up to 12 hours, as per the Amazon documentation.

GCP example: 1 TB of backups – Object Storage versus Block Storage/Disk volumes

We are going to compare 1 TB stored on a data disk attached to a virtual machine with the same amount of capacity but stored in GCP Cloud Storage, which is the object storage offering from Google, in the different tiers that are suited for backup use cases:

Type of Storage	Capacity/GB	Redundancy	Region	Price per GB	Tier	Price
GCP Persistent Disk	1024	Zonal	Netherlands (europe-west-4)	$ 0,3489	pd-extreme 3000 IOPS	$ 357,31
GCP Persistent Disk	1024	Zonal	Netherlands (europe-west-4)	$ 0,1870	pd-ssd	$ 191,49
GCP Persistent Disk	1024	Zonal	Netherlands (europe-west-4)	$ 0,0440	pd-standard	$ 45,06
GCP Cloud Storage	1024		Netherlands (europe-west-4)	$ 0,0181	Standard Storage	$ 20,48
GCP Cloud Storage	1024		Netherlands (europe-west-4)	$ 0,0092	Coldline Storage	$ 4,10
GCP Cloud Storage	1024		Netherlands (europe-west-4)	$ 0,0017	Archive	$ 1,23

Table 8.11 – Prices obtained using GCP Pricing Calculator in May 2023

The results are similar to the ones we obtained for this exercise on Azure, with Coldline storage being 10 times cheaper than pd-standard storage, allowing for savings of up to 90% if data is moved from disks to Coldline storage.

> **Important note**
>
> In these examples, we are solely comparing the costs for capacity, not considering the costs for transactions, as we expect them to be really low for backup files (there should not be a lot of reads or writes apart from the daily backup schedule).
>
> Write transactions and reads have a higher cost on colder tiers.

Let's keep optimizing with some clustering ideas for database scenarios.

Shared Disks for database clusters

A cluster in IT is a group of servers that act together as one, usually through a load balancer, to provide high availability and fault tolerance for workloads, while allowing for scalability. In IaaS databases, you need a cluster if you want your database to be fault-tolerant, which is often a requirement for production environments.

There are different ways to provide high availability for database instances. In this section, we are going to illustrate a way to get some savings for database clusters that may be ideal for some workloads.

Traditionally, when designing clusters, there are different possible architectures that can be chosen for your database clusters in the cloud. Mainly, we are going to focus on the following, which are the most usual in cloud scenarios:

- **Shared Disk clusters**: In a Shared Disk cluster, all the connected nodes share the same disk devices. Every node has its own memory and CPU.

- **Shared-Nothing clusters**: In Shared-Nothing clusters, every node has its own independent disk, and data is either partitioned (distributed) or replicated between nodes.

We can see the differences in each architecture with the following diagram:

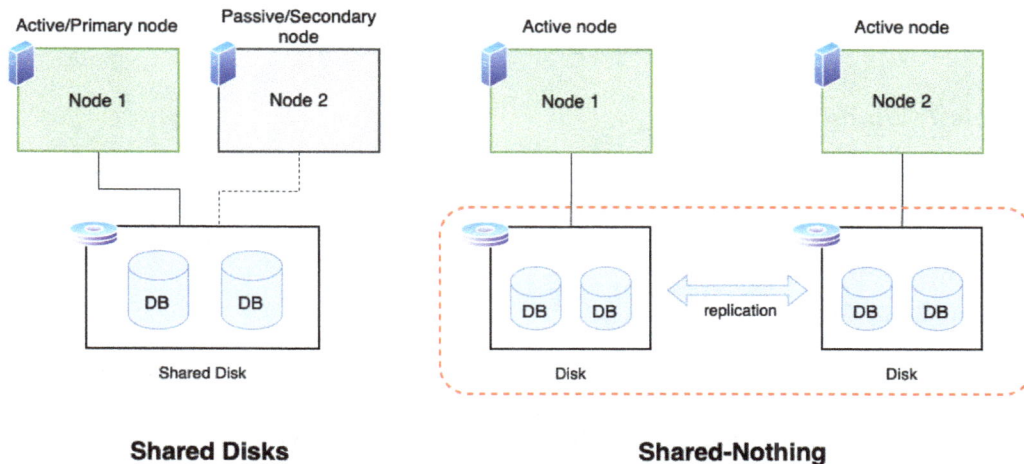

Figure 8.4 – Shared Disk versus Shared-Nothing architecture examples

To sum it up, these are the main advantages and disadvantages of choosing one over the other:

Type of cluster	Pros	Cons
Shared Disk	• Considerably lower cost, as there is only one disk shared between nodes • Simpler setup with no replication or partitioning that ensures data consistency	• Limited scaling • Restricted to one region, as Shared Disks cannot be used across regions
Shared-Nothing	• Reduces the possibility of failures, as each node is self-reliant. It also simplifies upgrades and maintenance on nodes • Better scaling • Allows for multi-region clusters, which adds up to higher available and fault tolerant solutions	• Has a strong dependency on cross-communication between the nodes performance (network throughput) • Higher cost, as requires to have a copy of all the databases in each disk

Table 8.12 – Shared Disk versus Shared-Nothing architecture comparison

Example 1: Windows clusters

As an example, for SQL Server hosted on Azure Virtual Machines, you can create a SQL Server cluster in two different ways:

- **Failover Cluster Instances (FCIs)**
- **Availability Groups**

There are many key differences between the two: in the way that the failover takes place, how long it takes to fail over, and the way that the nodes respond to requests, among others. But the key difference is that Failover Clusters can be set up with Shared Disk architecture, while an Always On Availability Group uses Shared-Nothing architecture.

Let's use a table to highlight the differences between the two:

Feature	Failover Cluster Instances	Availability Group
Shared storage	Yes, it uses a **Cluster Shared Volume** (**CSV**)	Not required – each node has its own disk
Load balancing	Active-Passive	Active-Active (secondaries are readable)
Failover level	SQL Server instance	SQL Server database
Failover time	Minutes, based on load	Seconds
Multi-regional	Same region	Can span across multiple regions if load balancers support it

Table 8.13 – FCI versus Availability Group comparison

Example 2: MySQL Linux Clusters

In MySQL, the most used clustering configurations, depending on how storage is used, are the following:

- **Synchronous Replication using Distributed Replication Block Device (DRBD)**: Apart from cluster nodes, there is a cluster manager node that listens for each node's heartbeats and, in the event of a failure in the primary node, promotes a secondary node to primary. Data is replicated from primary to secondary nodes.

- **Shared Storage Cluster**: In this setup, replication is not needed, as storage is shared across nodes. We have two servers that share the same storage, one of them being the active node while the other remains as the passive or secondary node.

- **Network Database (NDB) cluster**: This uses NDB cluster technology in a Shared-Nothing architecture. It is designed to not have any single point of failure, as hosts coexist with one or more manager servers.

The following table highlights the key features of each one:

Feature	Sync replication with DRBD	Shared Storage cluster	NBD
Shared Storage	No, as synchronous replication is in place instead	Yes	No, as synchronous replication is in place instead
Load balancing	Active-Passive	Active-Passive or Active-Active with limitations	Active-Active
Failover time	Depends on transactions	Depends on transactions	Seconds
Multi regional	Can span across multiple regions if load balancers support it	Same region	Can span across multiple regions if load balancers support it

Table 8.14 – MySQL Linux cluster types

With these examples of databases clustered for both Windows and Linux/MySQL in mind, let's move on to how to set up Shared Disk architectures in each public cloud.

Shared Disk options in Azure, AWS, and GCP

These Shared Disk options were not available until recently. At the moment, these are the services offered in the clouds that we cover in this book:

- **Shared Disks in Azure**: It uses SCSI Persistent Reservations, which enables shared block storage to be accessed from multiple VMs. Shared Disks cannot be accessed directly by nodes without a cluster manager such as Windows Server Failover Cluster or Linux Pacemaker, as it is not supported.

- **Multi-Attach enabled Amazon EBS volumes**: Limited to io1/io2 EBS volume types, it supports multiple instances to use storage within an Availability Zone. It also requires a cluster manager to access data and ensure data consistency.

- **Multi-Write Persistent disks in GCP**: Currently in preview and supported for SSD-type persistent disks, it also provides shared storage for virtual machines through an access coordinator.

As always, take this advice with a grain of salt. We are not saying that Shared Disks are the way to go for all clusters and replication is never to be used; we just want you to understand that Shared Disks are a really cost-effective way to create clusters in IaaS solutions, if this setup fits your solution requirements.

Shrinking relational databases

To shrink a database is to reorganize or compact the information, freeing up some disk space in the process. This is not an operation that should be done on a regular basis, as it takes a lot of resources from the machine. Due to this fact, it is recommended to perform the shrinking during the maintenance window.

Most DBAs agree that shrinking a database regularly is not a good practice, but it has its use cases, such as when tables are dropped, or big DELETE queries are run. In those cases, doing a database shrink can be beneficial, because the freed-up space can be used to store more data, or that space can be reclaimed for the disk at the operating system level.

One idea that comes to mind is that with databases that are not backed up regularly but are heavily used, such as development databases, the transaction log can grow until it is cleared when a backup is done. In these specific cases, shrinking can do marvels and free up a lot of disk space.

From that point on, you can even reduce the disk size if needed on the virtual machine, or at least you will have a clear understanding of how much space is left on the virtual machine as it won't be obscured anymore.

Shrinking a database is supported on some DBMSs, such as SQL Server, Oracle, PostGreSQL (it is called VacuumM), and IBM DB2.

Database grouping in SQL Server

This concept is related to database allocation in instances.

Let's think for a moment. Imagine we create a new database for an application, db-A, which we host in instance I-A. After some time, we have another database for a new application, db-B, which we can host in one of the following:

- A new instance I-B, hosted in one of these:

  - A different server/cluster, newly created for this workload

  - The same server/cluster as I-A

- The instance that we created for the application, I-A

This diagram summarizes the scenario we propose:

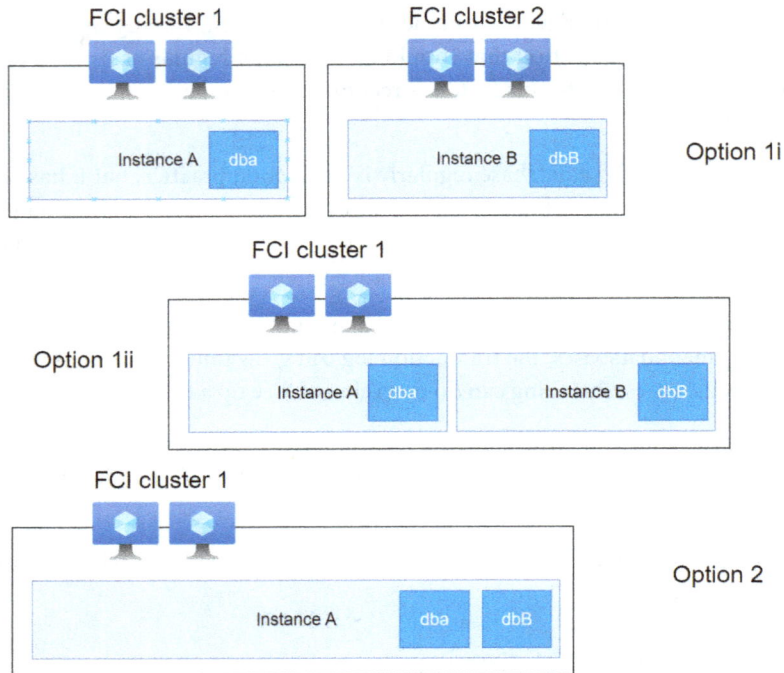

Figure 8.5 – Database grouping options

With the current context in mind, let's analyze the options we have:

- Should we choose `option 1i`, we will deploy and prepare new infrastructure for this workload. If we keep doing this, we will probably end up with a lot of servers dedicated to SQL, which will probably end up with non-optimal resource consumption (low CPU usage if we don't make full use of the VMs) and will be harder to manage and operate.

- If we choose `option 1ii`, we can reuse the same infrastructure as `I-A` and even divide the CPU, memory, and other resources between those instances.

- If, finally, we opt for `option 2`, the outcome will be similar to that of **1ii**, but the segregation of permissions and users between databases will get more complex and harder to manage as well.

Imagine now that the infrastructure hosting `I-A` is a Windows Failover Cluster, or an Always On Availability Group, in order for the databases to be highly available in case of contingency or hardware failure, for example, in a production environment.

It's clear that these architectural concepts have an impact on costs when we speak about cloud operations, for which more servers imply more money spent and higher cloud bills. The TCO for an environment in which option 1i is chosen is bigger from the start and will keep growing. It will also be harder to control in time and will require a lot of effort to set right.

With options 1ii and 2, the costs are contained, and infrastructure resources are leveraged in a better way, but we need to think about the criteria for which instances we should use for certain databases. We can provide some examples of possible groupings, such as environments, regions, and business units, but it's up to you, in the end, to decide which one best fits your organization.

To sum this idea up, it's always good practice to try to group databases using a logical criterion to make better use of infrastructure resources instead of creating new infrastructure for new workloads.

PaaS database optimization

As we discussed previously in this chapter, PaaS databases offer a lot of features. The sets of features that are included can also be cost drivers, which can raise the cost if not looked at and planned for carefully.

In this section, we will provide some strategies that can be used in PaaS databases for cost optimization, as well as zooming in on some concepts that were covered in the previous chapter, such as scaling, but with the specific aim of databases.

Compute optimization and rightsizing

For database optimization, it is key to have a clear definition of the workloads, which should include the type of data to be processed (structured or unstructured), the solution (migration scenarios, for example), as well as user concurrency, high availability, and other key factors.

In this section, we are going to review different metrics that we can use to determine whether our databases are correctly sized, which begins by applying common sense when we develop cloud solutions. We should be aware of how solution decisions today may affect TCO and, therefore, the huge impact that they may also have on costs.

In order to rightsize PaaS resources, as discussed in the previous chapter, the starting point is to review whether the compute resources allocated to our PaaS services have enough use or, on the other hand, they are underused and could be potentially replaced for smaller-sized virtual machines or compute tiers so we can make some savings.

To do this exercise, we need to focus on the following metrics:

- **CPU and RAM consumption**: PaaS resources also allow checking the CPU and RAM consumption of the underlying compute, so this should be the starting point for any rightsizing exercise. When reviewing these metrics, ensure that both average and maximum values are considered. If our workloads have consistent usage patterns and usage never goes above 50% in memory and CPU percentages, we may have a candidate for rightsizing. Also, in some PaaS services, we may have other measurement units such as DTU (Azure SQL) and RU/s (Azure Cosmos DB), among others. We can also use them to evaluate how much of the resources allocated to the database is actually in use.

- **Disk Space**: When working with PaaS resources, we usually decide upon creation how much storage we need for our workloads. From that moment on, we will be charged for that storage regardless of whether it is actually used or not. It is always good practice to check whether storage is used in its entirety, as if that's not the case, we could generate some savings by reducing the storage associated with PaaS resources. We also consider disk space a form of rightsizing. Apart from liberating free space, we must ensure that our database is used rationally, with the same principles that we covered in the *Rational database use* section earlier in this chapter.

- **Network Traffic**: We also need to consider the bandwidth of our network as part of the analysis, as we may have provisioned throughput that costs money that we are not fully using, or even the other way around – that our database is almost exhausted by a lot of concurrent connections. By having more information on how everything is going, we create value.

- **Database connections**: This is another important metric related to network usage. We need to evaluate whether the use of our databases is the one that was expected and adjust sizing accordingly. Most PaaS databases allow us to limit concurrent sessions to databases, so we can ensure that performance and the user experience won't be degraded. On the other hand, it is important to test out the limits so we understand the usage patterns for our databases supporting different workloads.

- **IOPS:** IOPS is defined as input/output operations per second. This parameter is related to disk bandwidth, and it limits the number of operations that we can do at the same time. There are some workloads that have a requirement on the number of IOPS needed, especially in high compute/performance scenarios. Please ensure that the IOPS associated with PaaS databases and usage are aligned for cost optimization purposes.

- **Multizone/Replication**: To conclude this exercise, we need to analyze the replication and redundancy options that we are using for our databases. As we already covered in this chapter and the previous one, this setting may have a high impact on costs, not only on capacity, but on data transfer, backups, and other services. These settings should be also considered as part of the rightsizing exercise, in which we need to determine whether redundancy is really needed if it is already activated.

All these settings must be carefully reviewed, so we can fully ensure that the current configuration is aligned with the use case and the needs of our workloads – no more and no less. We also recommend having this information at hand for iterative rightsizing reviews, as these metrics' behavior may change over time.

With these metrics at hand, we can follow the same principles that we described in the previous chapter, so we are able to decide how to act based on our findings.

Database grouping

When we deploy databases in a managed service offering, we host them in logical containers, which are often called instances.

The same concepts that we covered in the *IaaS database optimization* section apply here from a grouping perspective.

Especially in PaaS, this concept may be confusing, as we have seen a lot of organizations that create one new instance per application, which can result, the same as in an IaaS counterpart, in unused compute and wasted resources and costs.

This initiative can generate a lot of savings if done right. If there are a lot of separate instances that are not used enough (we can check monitoring to see the actual usage of compute resources), it is always a good plan to create new logical instances based on criteria we can decide upon (business unit, region, environment) and host the databases there, which will also simplify management. We can consider that database instances will be a shared service used by different workloads if needed.

These new instances will host multiple databases, maximizing CPU and memory usage and generating some savings in the process. We can consider this point a rightsizing of sorts because, in the end, that's what we are doing after reorganizing the logical grouping of databases.

Azure example: Azure SQL for databases and Azure SQL

In Azure, Azure SQL has different mechanisms to ease database grouping.

If we want to deploy Azure SQL, we need to create an Azure SQL Server first. An Azure SQL Server is only a logical resource used to group databases. Once the Azure SQL Server is in place, you can create databases. You are able to choose the compute type and storage options for each database, then better adapt the database resources to the use case of each workload. This effectively means that you pay separate charges per database, instead of paying for the server.

But there is also an additional option that can be used to ease the database grouping, which is **Elastic Pools**. Elastic Pools allow you to use the same compute and storage for multiple databases, and you are only charged for the cost of the Elastic Pool compute. The compute allocated is bigger than in a single database, as it is meant to include multiple databases at once.

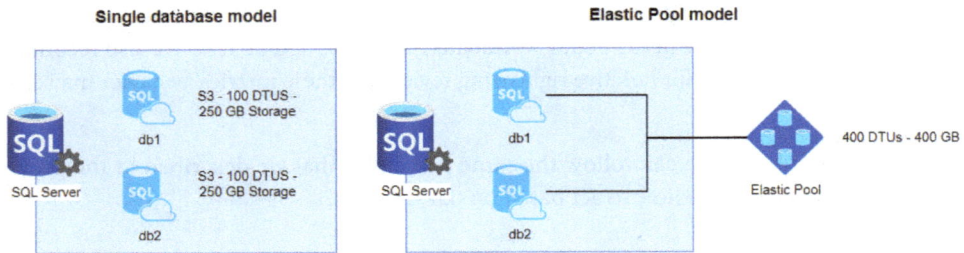

Figure 8.6 – Single database versus Elastic Pool

In addition, it allows for scaling mechanisms based on compute resource consumption, to grow when demand is higher and decrease when usage is low. Using simple automation and scripting, a lot of savings can be achieved using this method if usage patterns are known in advance.

On the other hand, when we need an Azure SQL database for PostGreSQL or MySQL, a flexible server is offered as a database grouping mechanism, and for automation, allowing to start and stop the cluster during off hours to generate additional savings, for example.

Regarding Azure SQL Managed Instances, they work similarly to IaaS infrastructure, for which you pay for the whole instance compute size, as you do with virtual machines, with the particularity that you cannot shut it down during off hours as opposed to virtual machines.

AWS example: AWS RDS

In AWS RDS, before we can create databases, we need to create an AWS DB instance. When we create an instance, we select the compute that's adequate for our workload. At this point, we need to decide which kind of compute we need (General Purpose or Memory Optimized, for example), as well as how many vCPUs and how much memory we need.

The DB instance also has storage capacity attached to it, where we can also select which kind of storage is needed for our workload (General Purpose or Provisioned IOPS, for example). This will be the storage that our databases will use.

Once the instance has been created, we can proceed to create databases in it.

GCP example: Cloud SQL

In GCP Cloud SQL, databases are also hosted in instances.

When we create an instance, we select the compute type we need, the memory, and the vCPUs, as well as the storage type and the rest of the settings that can be configured, such as encryption, data protection settings, and region and zonal availability.

With all PaaS database grouping covered, let's move on to the next topic: database scaling.

Database scaling

In the last chapter, we covered the possibilities that we have to scale virtual machines and the advantages and disadvantages of vertical and horizontal scaling. As this chapter is dedicated to databases, we want to specifically deep dive into scaling for databases, as there are some key considerations that must be taken into account for cost-optimization purposes.

When our database needs scaling, it is mainly due to one of these reasons:

- Applications or users are making too many requests that overload our compute resources (mainly CPU/RAM). Exhausting computer resources may lead to higher response times or even failure in the databases.

- Our database has run out of disk space, and it is not able to store more data or information, or even register more transactions.

- It is also possible that the network adapter may be overwhelmed by too much traffic, which can create a bottleneck, impacting database performance as well.

Let's analyze when to use vertical and horizontal scaling and the different possibilities that these mechanisms open up for our databases.

Vertical scaling

This is essentially the fastest way to solve issues, but sometimes it is not the best road to choose. Vertical scaling is often used when applications are not elastic enough (tightly coupled instead of loosely coupled) and the only way to solve issues is to increase the compute or storage resources associated with our database. We also consider vertical scaling as changing parameters such as disk tier or type, instance size, or network throughput.

Changes such as these in the configuration must be implemented during off-hours, and during maintenance windows, to avoid any downtime in our workloads throughout the process.

If you are going to use vertical scaling for databases, please also consider the following:

- Database licensing often depends on how many processors/vCores our database has provisioned, so please make sure that licenses stay fully compliant throughout the whole process.

- In some cases, scaling may affect reserved capacity. Please also ensure that no Reserved Instance will be underutilized as a result of scaling.

Vertical scaling can be applied only to some extent, as all cloud providers have hard limits you cannot surpass for virtual machine size. Also, please consider rational use, as it may not be the best solution to have, for example, a 128 vCore database, when we can divide that database into smaller ones that we can manage more easily.

Scaling can usually be performed from any cloud console, such as the Azure portal or AWS Management Console, or programmatically using APIs, the **Command-Line Interface (CLI)**, or SDKs for most popular programming languages.

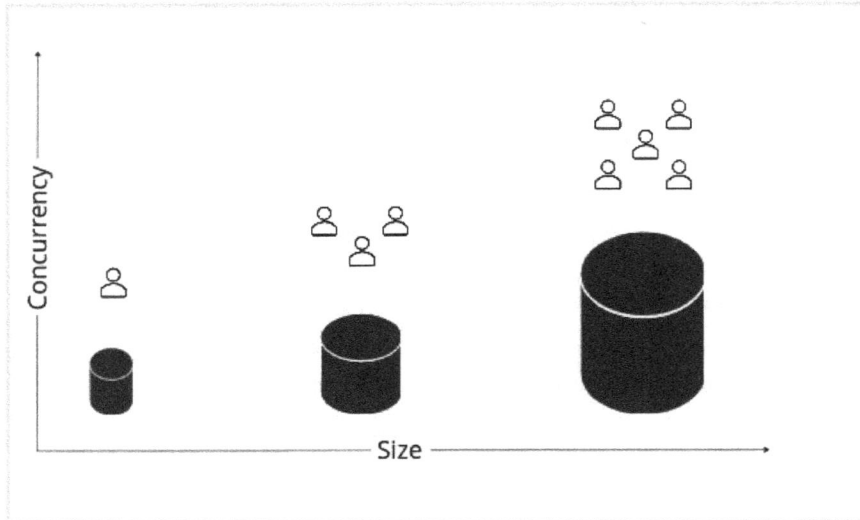

Figure 8.7 – Example of vertical scaling versus user demand

Horizontal scaling

In horizontal scaling, we are essentially adding more nodes that can act as read or write replicas for our databases. Through the use of load balancing, we can distribute the requests across different databases so, for example, all the writes go to the master node while reads are distributed between read-only replicas. Doing this will alleviate the load of the master database, which will improve data consistency and performance overall while ensuring that the database is consistently replicated from the master node to secondary read-only nodes.

Read replicas are also a really good option to allow for data analytics on our workloads. Instead of getting the information directly from the source while increasing the database load, which may not be ideal, we can prepare a copy in parallel where we can read all we want without disrupting the main function of our database.

For NoSQL databases, horizontal scaling is widely used, as data is often distributed between all the nodes instead of having this master-slave setup.

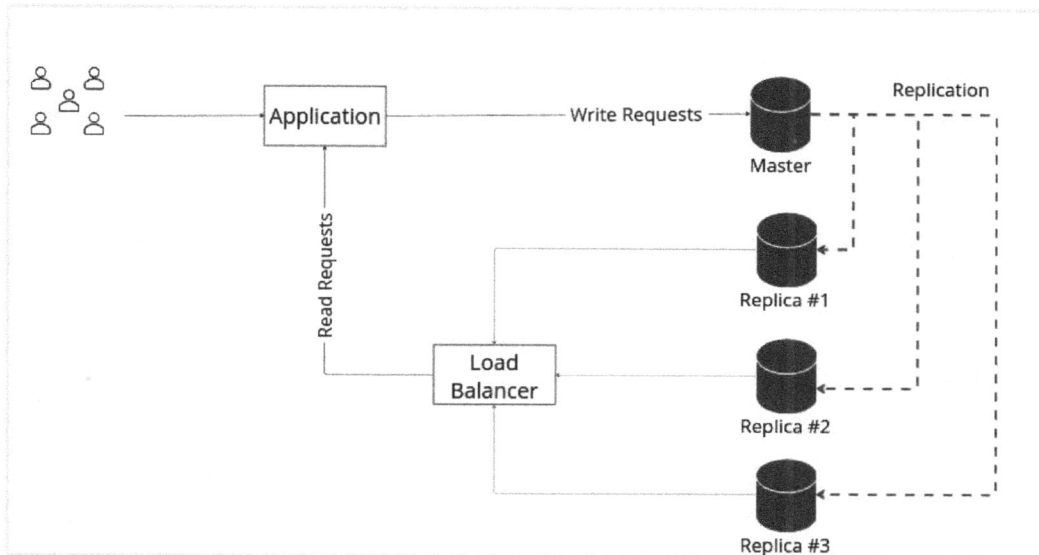

Figure 8.8 – Example of horizontal scaling

Autoscaling

Now that it is understood how scaling can help us design better solutions, let's focus on the extra mile, which is to do it in an automated manner. Most cloud providers offer autoscaling rules for PaaS servers, where you can set up a number of rules for your workloads to scale horizontally or vertically depending on key metrics such as % CPU or memory used.

With these rules, we can establish that, when our database reaches 80% CPU usage, horizontal scaling will be used to provision another read-only copy to balance requests between them. Also, after the usage peak is passed, we can set up a rule to decrease the available nodes to the minimum based on % CPU rules.

This is truly the way to effectively use cloud services, as it is a model that resembles pay-as-you-go, in which **we only pay for what we use when we are using it**.

Using PaaS services in this way should be our focus for cost optimization. However, autoscaling rules often require intense and careful testing so we can fully guarantee that our workloads are fully compatible with this scaling model.

When considering setting up autoscaling rules, keep in mind that you must also update them dynamically in an iterative manner, as the workload's patterns of usage, as well as the rules baselines, may change over time. Make sure to carefully review the metrics associated with autoscaling rules using services such as AWS CloudWatch or Azure Monitor, so the technical teams can be on top of what's happening and understand how everything is working.

To be able to consider these initiatives, make sure to go over the rightsizing exercise first, as that should be the starting point before autoscaling mechanisms are set up. Also, it is good practice to set up some alerts when potential thresholds that may be used are reached if we are considering setting up autoscaling rules, so we can imagine how our databases would react to load using autoscaling before applying it.

Autoscaling rules can also be used as a power scheduling method, allowing us to reduce the tier of our databases during off hours. We may not be able to fully shut down the databases, as we would with their IaaS counterparts, but at least it will allow for some cost optimization, especially in non-productive environments.

Serverless versus Provisioned Compute

Following the same principles that we already covered in *Chapter 7*, we can benefit from using serverless database services for some use cases.

With a serverless database, maintenance is considerably simplified, so developers can focus on their use cases to build workloads and applications. Serverless databases are ideal for unpredictable workloads in rapidly changing environments.

The payment model of these databases is pay-as-you-go or consumption-based billing, on which you are only charged when the databases are used. We could define the billing mode as **database-as-a-service**.

Serverless databases are available for both relational databases and non-relational/no-SQL databases. We will go through the different offerings of Azure, AWS, and Google that are proprietary to each public cloud.

> **Important note**
>
> In this book, we are only going to cover these proprietary solutions, but there are amazing offerings in the serverless world, such as **CockroachDB** (SQL), **PlanetScaleDB** (MySQL), and **Fauna DB**.
>
> Make sure to also check these serverless options when deciding where to host your data.

These are different offerings in Azure, AWS, and Google Cloud that provide serverless options to be considered when serverless fits the solution requirements.

Azure SQL Serverless

In Azure SQL, we have a Serverless option in Provisioned Compute SKU. At the time of writing this book, it is available in General Purpose and in preview in Hyperscale and only in Standard-series hardware (Gen5).

It has two main features:

- An automatic autoscaling mechanism that automatically allocates vCores ranging from the minimum to the maximum vCores selected.

- An auto-pause feature, which allows the database to be paused after an auto-pause delay amount that the user can set. After the database is paused, it can take around 1 minute to be warmed up and respond to requests.

Here are some points from the cost perspective:

- When the database is shut down or paused, you only pay for storage costs. Storage costs are the same as in the non-serverless Azure SQL Provisioned Compute tier.

- When the database is in use, you are charged for the current vCores and memory allocated to the database. If the vCores are below the minimum, you will be charged for the minimum vCores nevertheless.

- The price for one hour of a vCore is approximately double if we compare it with the Azure SQL Provisioned Compute tier.

This database tier is ideal for the following scenarios:

- Development environments or environments with large periods of inactivity

- For testing purposes, to analyze how many vCores a database may require in the first stages of development, so it can be correctly sized later in the project. It can also be useful when user demand is unclear for an application.

For these use cases, using this serverless model can be very cost-effective compared to Provisioned Compute. Also, the good thing about Azure SQL serverless is that you can change from serverless to Provisioned Compute in minutes and just a few clicks, as no migrations are necessary from one to the other.

Cosmos DB Serverless

Cosmos DB is a PaaS offering from Microsoft that offers both relational and non-relational databases. To do so, it offers different APIs for relational and non-relational databases: NoSQL, MongoDB, PostgreSQL, Cassandra, Gremlin, and Table. They are summarized in the following table, as well as which ones also support serverless:

API	Database type	Data structure	Serverless
NoSQL	Non-relational	Document	Supported
MongoDB	Non-relational	Document (BSON)	Supported
PostgreSQL	Relational	N/A	Not supported
Apache Cassandra	Non-relational	Column-wide	Supported
Apache Gremlin	Non-relational	Graph	Supported
Table	Non-relational	Key-Value	Supported

Table 8.15 – Cosmos DB API offerings

Cosmos DB works with abstract throughput units that are called **Request Units per second (RU/s)**. When working with non-serverless Cosmos DB, you select a number of RU/s for your databases, and therefore the cost of the databases depends on how many RU/s you have provisioned over time, while storage costs are charged separately.

> **Note: Azure Cosmos DB free tier**
>
> You can enable the free tier on an Azure subscription for Cosmos DB, which offers 1,000 RU/s and 25 GB on a Cosmos DB account for free.
>
> This is ideal for testing out and learning about the service.

There are two additional options apart from traditional Standard Provisioned throughput:

- **Autoscale provisioned output:** It enables an autoscaling mechanism on the database that grows or reduces the RU/s allocated based on usage. You select a maximum throughput and the service automatically manages autoscaling for you.

- **Serverless:** With this offering, you can use your Cosmos DB account in a consumption-based model, on which you are only charged for the Request Units consumed by your database operations and the storage consumed by your data.

There are currently some limitations to this model:

- It is not supported by multi-region Cosmos DB.

- Serverless containers can store 50 GB as the maximum of data and indexes. There is, though, another option that is currently in preview to expand this limit to 1 TB (`https://learn.microsoft.com/en-us/azure/cosmos-db/serverless-1tb`).

- 5,000 RU/s is the maximum throughput for a database in serverless

With the same ideas as Azure SQL serverless, this model is ideal when the throughput needed is unclear, or when workloads are running with intermittent or unknown usage patterns. For those scenarios, using serverless may be a really cost-effective solution compared to the Provisioned Compute model.

For throughput costs, in Provisioned Compute you are charged for RU/s per second, while in Serverless you are charged for the total usage of RUs throughout the whole billing month.

> **Important note: Cosmos DB changing from Serverless to Provisioned Compute**
>
> Keep in mind that it is not currently possible to seamlessly change from Cosmos DB Serverless to Provisioned Compute or the other way around.
>
> If such a change is needed, you need to make use of data migration tools such as **Data Migration Tool** and **Change Feed and Restore**.

When working in prototyping or development environments, make sure to use both Azure Cosmos DB Provisioned Compute free tier and Serverless offerings, to generate some savings using these database offerings from Azure.

AWS Aurora Serverless v2

AWS Aurora is a managed database offering from Amazon, part of the AWS RDS family, and is fully compatible with Postgre and MySQL relational databases.

It offers both a Provisioned Compute and Serverless model, with Serverless having two different serverless generations (v1 and v2). In Provisioned Compute, it works as AWS RDS where you select the instance classes on which your database will live, while in Serverless instances, AWS will manage the underlying compute.

AWS Aurora Serverless works in a similar way to Cosmos DB but with **Aurora capacity units** (**ACUs**). One ACU corresponds to 2 GB of RAM allocated to a database instance. When an AWS Aurora instance is created, you must specify the ACU range that you want (minimum/maximum).

The most recent v2 offers added features to v1, such as multi-region/multi-AZ support and the possibility of using Read-Replicas. This is the list of features of AWS Aurora:

- **API for Data**: Allows you to operate the database using HTTPS APIs for inserts, updates, deletes, and so on.

- **Autoscaling**: You don't need to provision in advance, as autoscaling happens underneath without user interaction.

- **Sleep**: A pause mechanism that puts AWS Aurora to sleep after a period of time. After a sleep period, it will have a cold start of approximately 1 minute.

- **Multi-AZ support (v2)**: Having multi-AZs allows for highly available setups that won't be affected in the event of a failure in a specific AZ.

- **Read Replicas (v2)**: You can access data from read replicas to offload the main database (reporting, for example).

AWS can be a very cost-effective option to use for MySQL and PostgreSQL databases for unpredictable or spiky workloads.

> **Important note: AWS Aurora changing from Serverless to Provisioned Compute**
>
> Keep in mind that it is not currently possible to seamlessly change from Aurora Serverless to Provisioned Compute or the other way around.
>
> You need to create a new cluster and use a snapshot or a logical backup and restore to do so.

AWS DynamoDB

AWS DynamoDB is a PaaS offering in Amazon Web Services for non-relational/NoSQL key-value databases. DynamoDB is based on the principles of Dynamo, a storage system developed by Amazon between 2004 and 2007.

In AWS DynamoDB, your data is distributed on different servers managed by Amazon, and data can be accessed using HTTP APIs or the AWS SDK/CLI. It offers two different modes:

- **Provisioned**: In which you select the capacity, which essentially consists of how many **Write Request Units (WCUs)** and **Read Request Units (RCUs)** per second are needed to read 4 KB of data for reads and 1 KB for writes of data

- **On-demand**: With this option, you only pay for what you use

It is a really good solution to integrate with other serverless offerings such as AWS Lambda, to have fully serverless workloads for which minimal maintenance and management is needed.

DynamoDB Provisioned offers two table classes:

- **Standard**: The default and recommended for most workloads
- **Standard-Infrequent Access**: This can be used for tables where storage is the dominant cost (not accessed regularly)

Each table class offers a price for data storage as well as read and write requests, in a similar way to AWS S3 Object Storage.

Having this temperature option, although basic, can generate a lot of savings in some workloads. Combining these features with a Serverless consumption-based pricing model, this service can be a really cost-effective offering in some use cases.

> **Note: AWS DynamoDB free tier**
>
> You can also make use of the AWS DynamoDB Provisioned free tier, which provides 25 GB of storage, along with 25 provisioned WCUs and RCUs, which is enough to process 200 million requests a month, for free.
>
> This is ideal for testing out and learning about the service.

GCP Firestore

GCP Firestore is a NoSQL PaaS document database offering in Google that is serverless by definition, while being highly scalable, and enables interesting features such as offline data access and a query engine to run transactions against document data.

The pricing model of Firestore consists of how many documents reads, writes, and deletes we run, as well as how much storage capacity data is used. There is a free quota of transactions per day, and you pay for anything over that quota.

It has some limitations that need to be considered, though:

- The document size limit is 1 MB
- Document write operation frequency is limited to 1 per second
- There's no possibility to aggregate queries – as Firestore performance is vital, the possibility is not even offered

Firestore also offers the benefits of **Atomic, Consistent, Isolation, and Durability (ACID)** transactions, which are often more present in relational databases and not that common in NoSQL counterparts.

Regarding offline data access, it enables the database to be used regardless of network latency or internet connectivity. When devices are back online, changes will be synced to the cloud instantly.

The same as other Serverless databases, this service can be paired with other PaaS offerings such as GCP Functions and GCP Cloud Run, to end up with simple serverless solutions that require not that much management while being cost-effective.

Backup storage and redundancy

One of the key points of PaaS resources is that some features are built into the service offerings, which eases the management and the operations that are needed on cloud resources, which is almost always a good thing.

Backup is one of the features that is included out of the box in PaaS resources such as databases.

These backups for PaaS resources are usually stored in object storage that is managed by the cloud provider. But, in some cases, such as AWS RDS, you may want to export snapshots or database backups to object storage such as AWS S3. If these export-to-object-storage features are used, make sure to leverage the benefits of this kind of storage, such as data temperature tiers, and avoid redundancy if it is not needed. We will review the different types of storage thoroughly in the next chapter.

In Azure specifically, even though the storage where your backups are stored is managed by the cloud provider, you can select the redundancy that you want when you create a database.

> **Azure: Azure SQL for …. databases and Azure SQL backup storage redundancy**
>
> In Azure SQL, in general, when you create a database using the portal, **Geo-redundant Storage (GRS)** is selected by default for backup storage. This may not be ideal for some scenarios from a cost perspective.
>
> Our recommendation is to limit its use to production environments, as that's where backup is a requirement most often.
>
> In Azure Database for MySQL and PostgreSQL, backup storage redundancy is not enabled by default.

Reserved capacity

As we covered in the previous chapter, Reserved Instances, even with their attached trade-offs, are a great way to get some savings in stable environments.

For IaaS solutions, please consider the options that we already covered in the previous chapter as well, as all of those also apply to databases.

In PaaS database services, however, Reserved Instance offerings are limited. In this section, we will go through the different database services that support this purchasing model.

Azure

These are the PaaS database services that support Reserved Instances and the discounts that you get by using them:

Service	Region	1 year discount	3 year discount
Azure SQL**	West Europe	16-21%	25-36%
Azure SQL Managed Instance**	West Europe	16-21%	25-34%
Azure SQL Database for MySQL*	West Europe	35-40%	60%
Azure SQL Database for PostGreSQL*	West Europe	32-40%	60%
Azure SQL Database for MariaDB	West Europe	35-37%	53-57%
Azure Cosmos DB PostgreSQL	West Europe	32%	53%
Azure Cosmos DB rest of NoSQL APIs	West Europe	15%	25%
Synapse Analytics Gen2	West Europe	37%	65%

Table 8.16 – Azure Reserved Capacity for PaaS databases

Reserved Instances are only supported for Das v3/v4 virtual machines in the General Purpose tier

**Hybrid Benefit not activated (license included in price)*

The prices and discounts shown are from May 2023, at the time of writing this book, taking into account only the price of the databases (no storage and no backup).

AWS

In AWS, the following services are available to be used with Reserved Capacity:

Service	Region	1 year discount	3 year discount
Amazon RDS for SQL Server Enterprise* **	Ireland	28-32% ****	53-70% ***
Amazon RDS for Oracle Enterprise* **	Ireland	28-38% ****	53-70% ***
Amazon RDS for MySQL* **	Ireland	23-34%	45-57%***
Amazon RDS for PostgreSQL* **	Ireland	23-34%	45-57%***
Amazon RDS for MariaDB * **	Ireland	23-34%	45-57%***
Amazon DynamoDB (Capacity Reservation Not IA)*	Ireland	54%	56%%
Amazon Aurora MySQL – Compatible **	Ireland	20-34%	51-61%***
Amazon Aurora PostgreSQL – Compatible **	Ireland	22-34%	51-61%***
Amazon Redshift	Ireland	21-30%	56%
Amazon ElasticCache *****	Ireland	29-32%	47-50%

Table 8.17 – AWS Reserved Capacity for PaaS databases

* *Single AZ*

** *Payment option No Upfront*

*** *Not available payment option No Upfront for this term, use Partial Upfront to calculate*

**** *Bring your own license (BYOL)*

***** *Node Type Standard*

Google

In Google Cloud, committed use discounts for PaaS databases apply on GCP Cloud SQL services for any of the database engines that are offered: MySQL, PostgreSQL, or SQL Server:

Service	CUD 1yr	CUD 3yr
GCP Cloud SQL for any database engine	25%	52%
GCP BigQuery Enterprise/Enterprise Plus Capacity Compute	20%	40%

Table 8.18 – GCP Reserved Capacity for PaaS databases

Licensing optimization

It may seem obvious, but here is the key question: are you using all the features that the current licensing for your databases offers? Would it be enough if, instead of having Enterprise, the most expensive, we downgraded to Standard? Could you use lower-tier licensing for non-production environments?

I know, these questions are not easy to answer, but we will try to address them one by one and provide examples to illustrate the train of thought we are indicating.

Bring your own license (BYOL)

As we already covered in the previous chapter, the use of the BYOL licensing model can help us attain great cost savings in exchange for long-term commitment to having certain products licensed. This is especially the case for databases as well as DBMSs such as SQL Server or Oracle with higher price tags that we want to reduce as much as possible.

The purpose of this section is to illustrate how we can use this model for databases and which ones are supported, as it varies from cloud provider to cloud provider.

Azure – BYOL for IaaS and PaaS databases

In Azure, Microsoft only allows the use of SQL Server licenses on Standard and Enterprise editions in the BYOL licensing model.

Using Microsoft documentation on the topic (`https://azure.microsoft.com/en-us/pricing/hybrid-benefit/#why-azure-hybrid-benefit`), the savings could be up to 85% in this case.

If using SQL Server on IaaS, make sure that you register SQL Server with a SQL Server IaaS extension, to fully benefit from SQL Server capabilities in the cloud such as integrated monitoring, backup, key vault management, patching, and so on.

One key takeaway on Azure is that using the SQL Server IaaS extension allows for seamlessly changing between pay-as-you-go and BYOL licensing models, as well as upgrading the SQL Server edition on the go.

This is a really useful feature from the cost optimization perspective because, for example, you can use a Developer SQL Server license while the project is being built, and then upgrade to Enterprise or Standard some weeks before go-live for final testing. We will cover development scenarios later in this chapter.

For PaaS databases, the use of Hybrid Benefit is fully supported in the Azure SQL Provisioned Compute model (not DTU) and Azure SQL Managed Instances.

Don't forget to make use of the Hybrid Benefit calculator provided by Microsoft to help you plan and estimate how many licenses you may need.

AWS –BYOL in databases hosted in EC2

In AWS, BYOL is supported through the use of AWS License Manager, which we covered in the previous chapter.

In regards to AWS RDS PaaS databases, BYOL is solely supported on Oracle databases.

GCP – BYOL in databases hosted on virtual machine instances

In GCP, BYOL is also supported for Oracle, SQL Server, and the main DBMS.

Just keep in mind that, if using physical core processor licenses, sole-tenant nodes (`https://cloud.google.com/compute/docs/nodes/sole-tenant-nodes`) should be used to avoid having hardware spread across different data centers, which may not work well with these kinds of licensing models.

Development scenarios

Apart from the features that we already covered, there are additional ways to test out databases for development or non-production scenarios that can be very cost-effective if they match your workload requirements.

Azure Dev/Test subscriptions

In Azure, you can create special Dev/test subscriptions that offer cheaper prices. The use of these subscriptions is limited to Visual Studio subscribers.

Having these special subscriptions allows you to use the following services at a much lower rate:

- Windows Virtual Machines
- Cloud Services, SQL Database
- SQL Managed Instance
- HDInsight
- App Service (Basic, Standard, Premium v2, Premium v3)
- Logic Apps

For example, in Windows Virtual Machines and SQL offerings, the software licensing is free, and Logic Apps has half the cost compared to normal subscriptions. The rest of the services can be used at a normal rate.

Also, Windows 10 images can be used without any licensing charges, as well as Azure virtual desktops.

Huge cost savings can be achieved if this feature is activated in non-production environments. The only trade-off is that there is no financially backed SLA availability on these services guaranteed by Microsoft, and, of course, the need to have an active Visual Studio subscription.

Azure Visual Studio personal credits

If your organization uses Visual Studio subscriptions, each licensed user has the right to a personal account with Azure credits that are renewed monthly.

The amount of credits depends on which Visual Studio subscription is used:

- $150 on Visual Studio Enterprise
- $100 on MSDN
- $50 on Visual Studio Professional/Visual Studio Test Professional

The only thing the users need to do is activate their subscription for Visual Studio (`https://my.visualstudio.com`) and they will be able to begin using this special subscription with credits.

Having these subscriptions in place is no substitute for subscriptions used by projects, for example, but it can alleviate your costs slightly, as developers can test out services and learn in a personal, separate environment that is completely free.

Keep in mind, though, that with such limited credits, we advise you to shut everything down after use if you don't want your credits to be consumed in just a few days.

SQL Server Developer and SQL Server Express licenses

There are two special SQL Server licenses that are free and can be used for development environments:

- **SQL Server Developer** is a special SQL Server license offered by Microsoft that includes the same functionalities as SQL Server Enterprise but whose use is limited to development environments with no real data
- **SQL Server Express** is a special SQL Server license offered by Microsoft that provides limited functionality (limited database size, cores, and memory, for example)

The incredible thing about both licenses is that they *are completely free*. Our preference is for the Developer edition, as you can set up the same solutions, with no limitations, for Dev and Pre/Pro environments, with the only difference being the license type between environments.

You can imagine, if you check SQL Server Standard and Enterprise license pricing, how much you could save just by switching to these licenses in your projects that have IaaS SQL Server databases.

There is one important thing to be considered though, and that is that it is not currently possible to downgrade from Standard or Enterprise to these editions, so a complete reinstall or a migration is needed in order to transition to these licensing models. But it is possible to upgrade from these licenses to Standard or Enterprise.

In some cases, the effort of reinstalling or migrating may pay off, as the potential savings to be attained by doing so may be very big.

Summary

We did not expect it, but in the end, we finished up with yet another lengthy chapter. In this chapter, we have gone through the different types of databases and DBMSs available, as well as their strengths and weaknesses. We have also covered optimization initiatives and ideas that can be used for cost optimization in both IaaS and PaaS services. In addition to these points, we have also learned how to optimize database licensing and our options for Reserved Capacity.

We hope that our insights have been useful and shed some light on key topics to be considered for cost optimization in databases.

It is impossible to cover all the concepts and all the possibilities in the database domain, and it goes far beyond this book, but at least the basics are covered here as a starting point for FinOps practitioners' analysis.

To fully close the circle, in the next chapter, we are going to cover storage optimization, to review the concepts and initiatives we can work on to design more cost-optimized solutions in the cloud related to storage services.

9

Implementing Storage Optimization

Cloud Storage is one of the most powerful offerings in the cloud services portfolio. The possibilities are endless, as cloud providers offer almost unlimited storage with incredible performance on tap that we can leverage to run our workloads in the cloud. From disks to newer storage paradigms such as object storage, we have a broad offering of services to cater to all possible use cases.

However, this wide offering sometimes is confusing for cloud practitioners, as with so many options available, we may suffer from decision paralysis. From a cost perspective, this complexity affects cost drivers, making it really hard to understand how much we need and how much it is going to cost as well. This complexity, of course, also impacts the difficulty of applying cost optimization in a Cloud Storage domain, and this is what we are going to try and tackle.

In this last chapter dedicated to technical initiatives for cost optimization, we are going to explore different concepts related to different storage services in the cloud, fully closing the circle on what we will cover as part of the Optimize pillar.

We will begin with a short introduction of key concepts related to storage services, which will set the basis needed to fully understand how they work. After this short introduction, we will review these initiatives on the different types of storage available in the cloud (i.e., block, file, and object storage), as well as other key topics such as backup and log storage optimization.

In this chapter, we are going to cover the following main topics:

- Storage key concepts
- Block storage optimization
- File storage optimization
- Object storage optimization
- Other storage optimization initiatives

Storage key concepts

In this section, we are going to introduce some terms and architectural concepts that are essential to fully understand how storage costs work in the cloud.

Let's set the ball rolling by explaining the three main storage technologies that are used, as well as their advantages and the most suitable use cases for each one of them.

Types of storage in the cloud

There are mainly three types of Cloud Storage that you can choose from, depending on the use case:

- **Block storage**: This is analogous to **Storage Area Network (SAN)** or **Direct Attached Storage (DAS)** storage. Data is stored in the form of blocks and offers low latency and really good performance. It is used for virtual machines at a higher cost than other storage types.

- **File storage**: This offers the possibility to access storage through a shared filesystem or a file share with different clients. It works similarly to **Network-Attached Storage (NAS)** devices with common protocols such as **Server Message Block (SMB)** and **Network File System (NFS)**, used in Windows and Linux, respectively.

- **Object storage**: This is a modern take on storage, and its appearance has been fairly recent. Data is stored as objects, and this type of storage is really well suited for unstructured data. Basically, it offers a storage engine based on HTTP REST APIs for storage operations. It is really cost-effective, but not all applications support its use. It often offers advanced features such as object metadata and versioning, as well as object snapshots and data temperature tiers.

Let's summarize the key takeaways using a simple table:

	Block	**File**	**Object**
Use case	Databases or other HPC I/O use cases that need direct read/write access to storage with low latency Virtual machine disks Containers	Local network storage	Documents, backups, and IoT
Clients	One client	Multiple clients	Multiple clients

Examples	Azure managed disks AWS EBS disks GCP Persistent Disk	Azure storage account file storage Amazon EFS GCP Filestore	Azure Blob storage Amazon S3 GCP Cloud Storage
Strengths	High performance	Simple access management	Scalability and distributed access
Limitations	Scalability	Scalability	Not for frequently changing data
Performance	High	High	Low
Scalability	Low	High	High
Cost	High	High	Low

Table 9.1 – A summary of the types of storage in the cloud

In the upcoming sections of this chapter, we will develop this topic and cover the best use cases for each storage type, as well as different initiatives to build cost-optimized solutions around them.

We know that all this information can be a little overwhelming at first, especially for cloud newcomers. From our point of view, a good approach that can help you choose the most suitable storage paradigm for a specific use case is to make use of **decision trees**, which cloud providers make available to simplify and streamline cloud solution design relating to storage, and even other services such as load balancers. Understanding these decision trees can be a great starting point to understand the differences, advantages, and disadvantages of each type of storage in detail, as well as their unique features.

You can find decision tree examples to choose the best service to host your data at these links:

- Azure: https://learn.microsoft.com/en-us/azure/architecture/guide/technology-choices/data-store-decision-tree
- GCP: https://cloud.google.com/architecture/storage-advisor

As we are going to be covering a lot of storage services throughout the chapter, we also recommend that you check out the **available monitoring metrics** that each cloud provider offers for each storage service we cover. Understanding these metrics, and how they are related to storage costs and performance, will be the cornerstone to detecting non-optimized storage platforms. We recommend first reading the entire chapter and then researching this topic further after all the information covered here is fully understood.

These links can be a good starting point for this metrics research:

- Azure: `https://learn.microsoft.com/en-us/azure/azure-monitor/reference/supported-metrics/metrics-index`

- AWS: `https://docs.aws.amazon.com/AmazonCloudWatch/latest/monitoring/aws-services-cloudwatch-metrics.html`

- GCP: `https://cloud.google.com/monitoring/api/metrics_gcp`

With this in mind, let's examine an interesting concept that can be leveraged to optimize how we think about storage when designing cloud solutions.

Thick versus thin provisioning in disks

To begin our journey of storage cost optimization for virtual machine disks, it is essential to understand different approaches to provision disk capacity on machines.

These concepts come from traditional on-premises computing, where we had big virtualization environments that applied different approaches to present disks to virtual machines.

When we use **thick provisioning**, disk capacity in its entirety is pre-allocated to a machine. This means that the disk takes up all the capacity, regardless of whether it's used or not.

Conversely, **thin provisioning** allocates a part of the storage, only what is essential for the machine to function, and then dynamically adds more storage as it is needed.

The differences between these approaches are highlighted in the following figure:

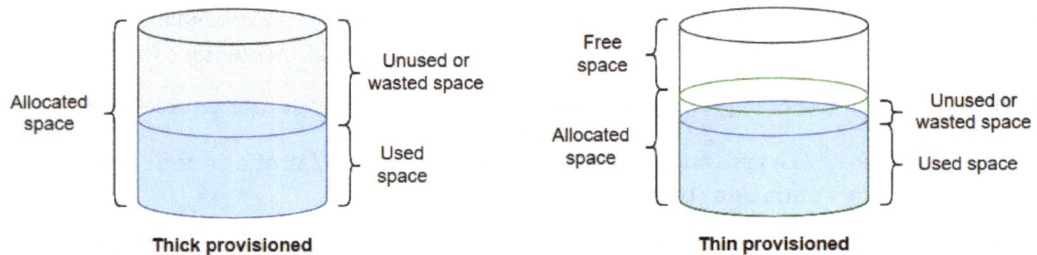

Figure 9.1 – Thick versus thin provisioning in disks

When working in the cloud, the virtual machine disks almost always use thick provisioning, as the entire capacity for the disk is taken up by a virtual machine regardless of how much of it is occupied.

It is important to note that thick and thin provisioning also apply to other disk parameters, such as disk bandwidth, which are often also thick-provisioned in the cloud. However, there are some disks that allow for bursting, essentially allowing for thin provisioning in some ways.

Regardless of the use of thick provisioning in most cloud IaaS disks, we think that the concept of thin provisioning is still valid and is a philosophy we can use when designing virtual machine solutions, which is *to just provision what you are going to use.*

It is always easy to grow from an adequately sized disk in any of the disk parameters, but it is always hard to downsize.

However, this philosophy does not only apply to disks, as it can also be applied to object and file storage as well. Make sure to just provision the space you need, and if the space gets filled up, make sure to free up some of it or move colder data to other storage services, instead of always thinking of expanding storage first, which is often the easy way out.

> **Important note – reducing disk size in cloud providers**
>
> Disk size reduction is currently not supported in either Azure managed disks, Amazon EBS, or GCP Compute Engine Persistent Disk.
>
> The only way to reduce disk size is by migrating data manually, using tools such as Robocopy, AzCopy (Azure), or other similar tools.
>
> This is another reason why it is essential to provision the disk size you need, no more and no less, when creating a virtual machine.

If disks are not properly adjusted to their correspondent use case, we are essentially paying for something that we are not using, which, in the cloud, means generating waste.

Let's move to another key concept that is essential for storage cost optimization, which is snapshots and how to effectively use them.

Disk snapshots

First, *a snapshot is not a substitute for a backup.* The idea of a snapshot is to capture a *photo* of a disk at a specific point in time. However, because it is easy to take a snapshot, a lot of technical teams use them as backup substitutes, which is not right in our view.

The main use of snapshots is to capture the previous status of a virtual machine before a change or planned intervention is performed on it. By having this snapshot at hand, we also have a rollback mechanism if things go wrong.

This use case means that snapshots should not be permanent but ephemeral, as eventually, snapshots should be deleted when the planned change is complete and validated, and we can ensure that they won't be needed anymore.

Snapshots are point-in-time copies of specific disks, while a backup protects a complete virtual machine, including all its disks.

However, with archive offerings of snapshots in AWS and GCP, the difference between snapshots and a backup is thinner than ever, as you can also store snapshots on a long-term basis at really good prices.

The following diagrams represent how total storage space varies from snapshots to a backup:

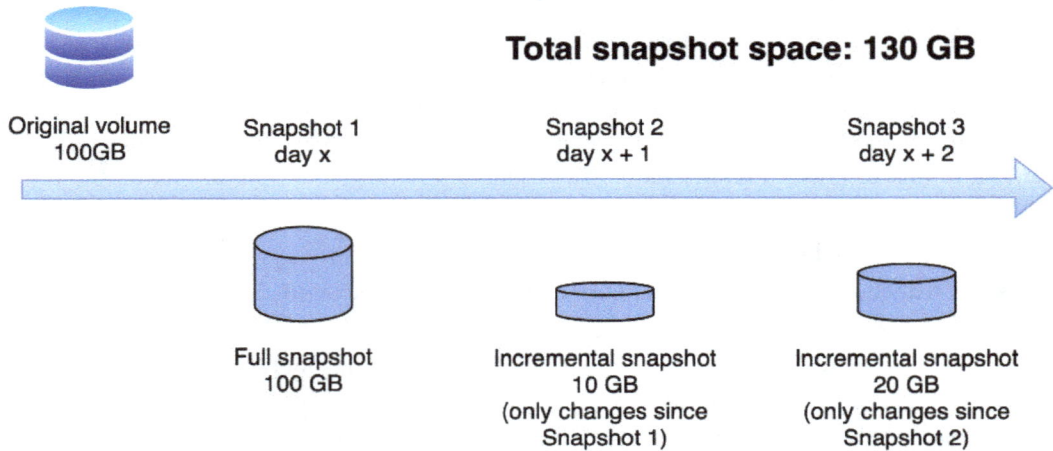

Total snapshot space: 130 GB

Original volume
100GB

Snapshot 1
day x

Snapshot 2
day x + 1

Snapshot 3
day x + 2

Full snapshot
100 GB

Incremental snapshot
10 GB
(only changes since
Snapshot 1)

Incremental snapshot
20 GB
(only changes since
Snapshot 2)

Figure 9.2 – Snapshot storage capacity usage

The first snapshot of a volume is always full and takes full capacity, while an incremental snapshot only takes up as much space as the changes that happened since the last snapshot taken.

Conversely, with backups, we always take a full backup of the virtual machine:

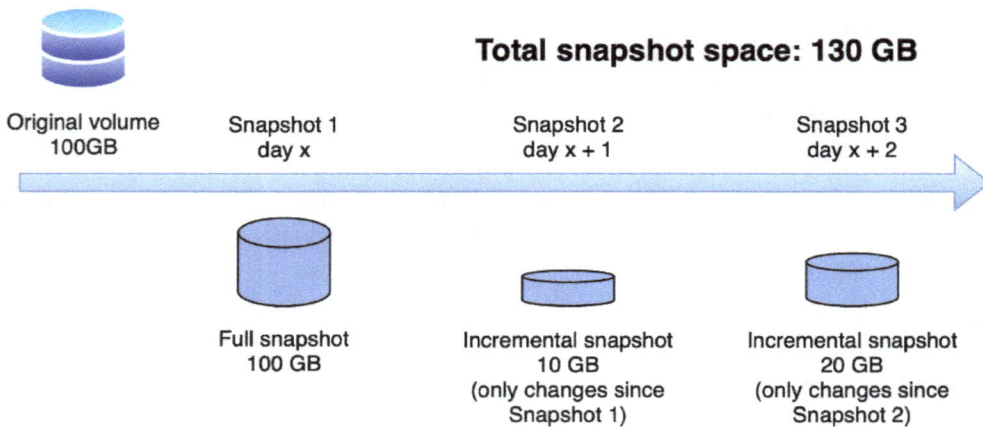

Total snapshot space: 130 GB

Original volume
100GB

Snapshot 1
day x

Snapshot 2
day x + 1

Snapshot 3
day x + 2

Full snapshot
100 GB

Incremental snapshot
10 GB
(only changes since
Snapshot 1)

Incremental snapshot
20 GB
(only changes since
Snapshot 2)

Figure 9.3 – Backup storage capacity usage

This table summarizes the essential differences between backups and snapshots:

	Snapshot	Backup
Capacity used	With incremental snapshots, the total used capacity can be greatly minimized	The sum of the total capacity multiplied by the number of days that data is kept as per retention policies
Level	Only disk	Virtual machine level
Use cases	Before changes in configuration or planned maintenance Development purposes	Production workloads with backup, retention, and business continuity policies Disaster recovery
Periodicity	On-demand Scheduled – this is not usual, but some cloud services allow you to schedule, such as the EBS Snapshot Scheduler	On-demand Scheduled using backup policies
Time it takes	Almost instant and can be done anytime	Takes some time and it is usually done out of hours
Scope of backup	A point-in-time restore of the disk	Complete virtual machine

Table 9.2 – Snapshots versus backups

Storage redundancy

Storage redundancy is another key benefit that the cloud offers. Storage redundancy means, in essence, that data is replicated/copied or backed up in different places. This feature is managed by cloud providers, and it is a great feature to ensure that, in the event of any contingency, there will be minimal or no data loss.

Building these replication mechanisms would be really costly without the cloud, as we would need to have different data centers to host our virtual machines, as well as strong network connections between them to guarantee consistent replication. In the cloud, it's just a box that we need to tick upon the creation of resources.

While this feature is great for high availability purposes, it can have a great impact on cost if it's not properly configured and adapted to each use case. Due to the impact of this setting on storage costs, we need to ensure it is necessary, and if that's the case, try to choose the setting that represents the sweet spot between costs and high availability.

Let's analyze how redundancy is implemented in each one of the cloud providers. Understanding this is essential to understanding optimization in these storage services.

Azure storage accounts

In Azure, when working with **storage accounts**, we are presented with different redundancy options. A storage account offers different data structures (i.e., blob, file, table, and queue) and redundancy applies to all of them.

The possible redundancy settings are the following, ordered from lowest to highest on cost:

- **Locally Redundant Storage (LRS):** This is the lowest cost redundancy option. If we use this option, data is stored as three different copies inside the same data center.

- **Zone-Redundant Storage (ZRS):** In this redundancy option, data is replicated asynchronously between three availability zones inside the same region. An Availability Zone has its own power, cooling, and networking to achieve segregation with other Availability Zones.

- **Geo-Redundant Storage (GRS):** For workloads that require high availability, we often require an additional copy of our data outside of the primary region. This is similar to LRS that is replicated to a LRS account in the secondary region.

- **Geo-Zone-Redundant Storage (GZRS):** In this last redundancy model, ZRS replication between availability zones is combined with an additional copy of our data that is replicated to another region.

The following diagram summarizes the different redundancy options:

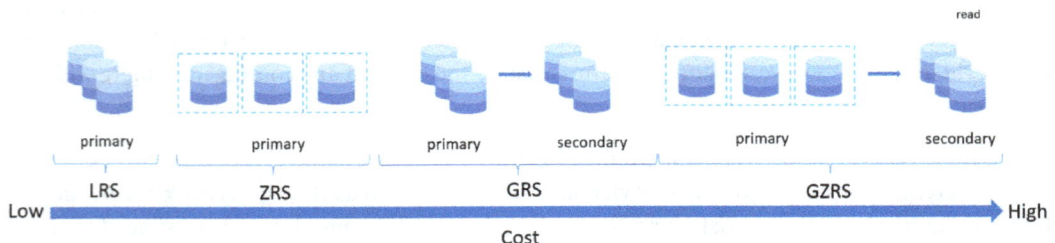

Figure 9.4 – Azure storage account redundancy options

Keep in mind that having geographical redundancy settings does not mean that data is instantly available in the secondary region. For data to be available, you need to perform a failover operation, which can take some time and does not fulfill high availability.

To overcome this limitation, there are two additional redundancy options on which geo-redundant copies are available at all times for read operations, which are as follows:

- **Read Access Geo-Redundant Storage (RA-GRS)**
- **Read Access Geo-Zone-Redundant Storage (RA-GZRS)**

Keep in mind that RA-GRS is the option with the highest price tag. Compared to non-read access counterparts, pricing is around **18% higher in Blob Storage per GB**, for example.

Azure managed disks

In Azure, when working with Azure managed disks, we have the following redundancy options:

- **Locally Redundant Storage (LRS)**: This is the lowest cost redundancy option. If we use this option, data is stored in three different copies inside the same data center. With this option, write latency is better than its ZRS counterpart.
- **Zone-Redundant Storage (ZRS)**: In this redundancy option, data is replicated asynchronously between three availability zones inside the same region. An availability zone has its own power, cooling, and networking to achieve segregation with other Availability Zones. This option has higher write latency due to async replication but is better for high availability purposes.

ZRS options are at around a **33% higher cost than LRS** if we compare their price per GB.

Azure managed disk snapshots

In Azure, snapshots of managed disks can be LRS or ZRS, the same as Azure managed disks.

An interesting point that is nice to know is that snapshot costs are the same in both LRS and ZRS, so our recommendation is to always store them with the ZRS setting if the option is available in your region.

Amazon S3

In AWS, when working with **Simple Storage Service** (S3) Storage, we create an S3 bucket inside a Region, and our data is automatically replicated within a region between a minimum of three availability zones. However, there are some exceptions, such as S3 One-Zone Infrequent Access and S3 on Outposts.

We also have here the possibility to set up replication, but in this case, we need to manually set it up in the destination Amazon S3 bucket, unlike Azure, where the process is handled by Microsoft.

We have different two options here, depending on whether the replication spans multiple Availability Zones or multiple regions:

- **Same-region Replication (SRR)**
- **Cross-region Replication (CRR)**

By combining Amazon S3 built-in replication with these features, we can achieve the same results as in Azure. It is even possible here to set up two-way replication, which can be a great solution for some workloads. Unlike Azure, we can also set up replication between production and UAT/pre-production environment data using this method (make sure that data is anonymized first, as you don't want to come into conflict with your compliance/GDPR team).

A great advantage of using this method of replication is that we can set up colder storage tiers in the destination bucket, allowing us to have a much cheaper copy of all the information that is stored in the primary bucket readily available, should we need it.

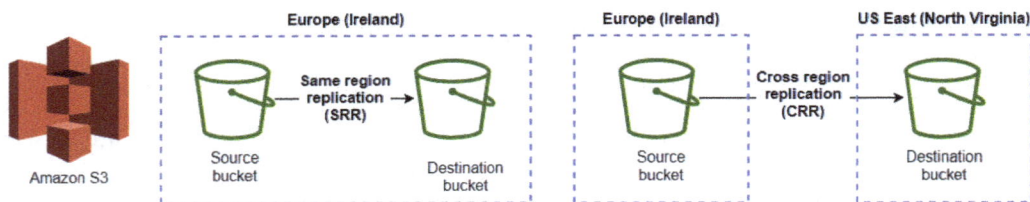

Figure 9.5 – AWS Replication (SRR-CRR)

Amazon EBS

By default, Amazon EBS volumes are created inside an Availability Zone. Within the Availability Zone, data is automatically replicated between different devices to prevent physical hardware failures.

For disaster recovery purposes, we have the option to use cross-region Amazon EBS snapshot copy replication, which we can combine with Amazon Data Lifecycle Manager for automated snapshots that are replicated to other regions. By having the snapshot duplicated in another region, we can spin up a virtual machine really fast in the event of any contingency.

In Amazon EBS, we have no further settings to set up any redundancy or replication, and therefore, it has no impact on costs.

GCP Cloud Storage

In GCP, we have different cloud locations or regions that consist of zones, while zones are an abstract representation of one or more data centers.

When a GCP Cloud Storage bucket is created, we are presented with different redundancy options:

- **Regional**: Here, our data is hosted in a specific region, such as São Paulo. Inside the region, our data is replicated between two different zones to ensure fault tolerance.

- **Dual-region**: Here, our data is hosted in a pair of regions. Data is replicated between regions.

- **Multi-region**: Here, our data is hosted in a pair of regions, but they are managed by Google Cloud Storage, which ensures that data remains within the region. Multi-region has lower costs than dual-region. This option is only available for Asia, US, and EU territories.

To replicate data between regions, there are mainly two settings:

- **Default replication**: Data is replicated within the hour but often takes just a few minutes.

- **Turbo replication**: Data is replicated within a target of 15 minutes but is only available on dual-region buckets. This replication is ideal for high availability scenarios, where the **Recovery Point Objective** (**RPO**) should be minimized. Keep in mind that turbo replication data transfer is billed at twice the price of default replication.

Here are some key cost considerations in the GCP Cloud Storage pricing model:

- Replication charges are only incurred for writes in buckets that span multiple regions.

- In dual-region and regional buckets, egress charges are not billed when reading data. Conversely, in multi-region buckets, egress charges of reads are billed.

- Make sure to host your GCP Compute Engine virtual machines in the same region as their associated storage, in dual-usage and regional buckets, to minimize data transfer charges.

GCP Compute Engine disks

When making use of Google Compute Engine disks, we have the following redundancy options:

- **Local SSD disks**: Ephemeral disks (they are persisted until the virtual machine is stopped or deleted) that are connected to our virtual machine. They offer no redundancy at a cheap price.

- **Persistent disks – zonal**: The default setting where storage is hosted within a zone inside a region.

- **Persistent disks – regional**: Data is replicated between two different zones in the same region. It is the type of disk that should be chosen for high-availability scenarios.

- **Hyperdisk**: A special type of disk that offers the fastest storage. In regard to redundancy, it offers the same as a persistent disk.

GCP snapshots

GCP Persistent Disk snapshots can also be hosted in one region (regional) or multiple regions (multi-regional) across a territory.

Having a multi-regional snapshot is a guarantee for high availability. Keep in mind that if you don't choose a specific region, the snapshot will be located in the closest region available for your disk.

Should you choose a specific default location for your snapshot, keep in mind that you will be charged for egress costs upon data transfer.

With redundancy options available across all three major public clouds already covered, let's jump in and see how block, file, and object storage work.

Block storage

Block storage is a technology used to store files and data in cloud and on-premises scenarios. Block storage divides data into blocks of equal sizes that are stored as separate pieces, each one with its corresponding identifier. The underlying storage then places these blocks where they are most efficient for fast retrieval.

For reference, these are the services in each cloud provider that we refer to when we speak of block storage:

- **Azure**: Azure managed disks
- **AWS**: Amazon EBS disks
- **Google**: GCP Compute Engine disks

In this section, we are also going to consider snapshots as block storage. Their use is closely linked to disks, as they "capture" the status of a disk at a specific point in time.

How does it work?

In this kind of storage, a file is stored across different blocks that are not sequential. Apart from the list of files, there is always a lookup table with the identifiers that correspond to each file. Unlike Azure, we can also set up replication between production and **User Acceptance Test** (**UAT**)/pre-production environment data using this method.

This lookup table is used in read and write operations to find the blocks where data is stored:

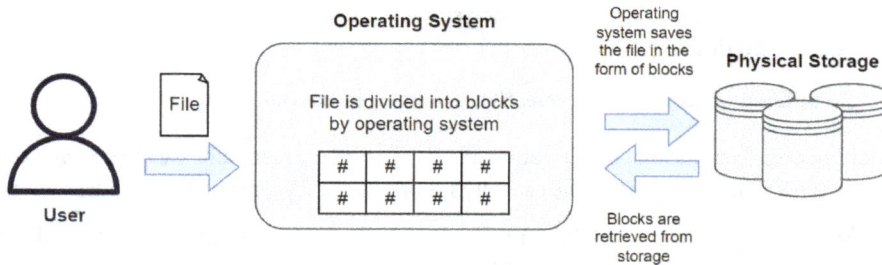

Figure 9.6 – Block storage

IOPS, throughput, and latency

What are IOPS, latency, and throughput? As we are going to discuss these terms in upcoming sections, let's give a brief explanation of what each term:

- **IOPS** is a unit of measurement that describes how many I/O operations per second a disk can handle as a maximum – for example, 10,000 IOPS is equal to 10,000 read/write operations per second as a maximum.

- **Throughput**, conversely, refers to how much bandwidth a disk has for these operations. It is measured in bytes per second.

- **Latency** is another key metric that describes how much time is needed for the storage to process a request or transaction.

The following diagram exemplifies how these parameters are related and what they represent:

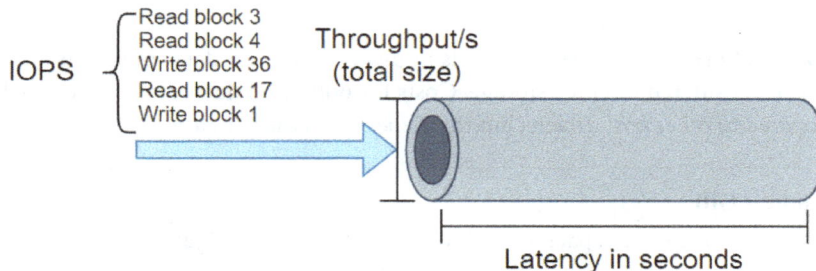

Figure 9.7 – IOPS, throughput, and latency

These concepts are key to understanding the needs of workloads in regard to storage performance.

Cost drivers

To begin our cost optimization journey with block storage, the first thing we need to do is to analyze what drives costs up and down.

Let's analyze the different factors, one by one, that can have an impact on storage costs:

- **Capacity**: This is a common cost driver on all storage types. Essentially, we are charged on a per **Gibibyte (GiB)** basis. We pay for the capacity we provision, regardless of how much of it is used.

- **Tier**: Most cloud providers use different disk types based on the kind of physical disk, mainly **Solid-State Drives (SSDs)** and non-SSDs, such as HDDs. They also offer a disk on which provisioned throughput and IOPS can be customized to each workload need, in high-performance scenarios. In summary, cost is proportional to performance.

- **Transactions**: In this regard, Azure is the exception, as transactions are billed as a separate cost driver. This is not the case in GCP and AWS, where the transaction price is included in the capacity pricing. This fact makes Azure managed disks' cost drivers harder to understand, with billing being higher as well, so review Microsoft pricing conditions and documentation carefully.

- **Provisioned IOPS**: Some disks allow you to set the IOPS for your disks, such as GCP Extreme or Amazon EBS io2. The problem with this setting is that it is difficult to estimate the needs of a specific workload. If needs are not clear but your workloads are highly demanding, our recommendation is to first ask the software vendor (if any) for IOPS requirements during the design phase. If this is not possible, make sure to monitor IOPS usage closely so that the provisioned IOPS setting can be adjusted in real time to meet the exact needs of your workload.

- **Provisioned throughput**: Following the same concepts as provisioned IOPS, this setting is used for high-performance or low-latency scenarios.

- **Redundancy**: As covered in the previous section of this chapter, redundancy in object storage applies to Azure managed disks and Azure snapshots.

- **Outbound data transfer**: If replication is in place, outbound data transfer is another cost driver to be considered in block storage. Costs for outbound data transfer are usually pretty low compared to other cost drivers, but this depends on the use case.

Important note – GiB

GiB is a unit used to measure capacity that is often confused with **Gigabyte (GB)**.

GB often applies to storage devices, such as pen drives or hard drives, while GiB is mostly used to measure capacity for operating systems such as Windows.

The equivalences are the following:

1 GiB = 2³⁰ bytes = 1,073,741,824 bytes ≈ 1.07 GB

1 GB = 10⁹ bytes = 1,000,000,000 bytes ≈ 0.93 GiB

As an overview, in the following section, we provide a cheat sheet in the form of a table of each service and their corresponding cost drivers, which will help us understand which parameters have an effect on cost when using each block storage offering in all three major public clouds.

Azure managed disk cost drivers

These are the different types of disk platforms available in Azure and their correspondent cost drivers:

Type of disk	Cost drivers
Ultra disks	<ul><li>Capacity</li><li>Provisioned IOPS</li><li>Provisioned throughput</li><li>Provisioned vCPU reservation charge</li><li>Shared disk setting</li></ul>
Premium SSD	<ul><li>Capacity</li><li>Redundancy</li><li>On-demand bursting setting and burst transaction fees</li><li>Outbound data transfer</li><li>Transactions</li><li>Shared disk setting</li></ul>
Premium SSDv2	<ul><li>Capacity</li><li>Redundancy</li><li>Provisioned IOPS</li><li>Provisioned throughput</li><li>Outbound data transfer</li><li>Shared disk setting</li></ul>
Standard SSD	<ul><li>Capacity</li><li>Redundancy</li><li>Outbound data transfer</li><li>Transactions</li><li>Shared disk setting</li></ul>

Standard HDD	• Capacity • Redundancy • Outbound data transfer • Transactions
Snapshots	• Capacity • Storage type • Redundancy

Table 9.3 – Azure managed disk cost drivers

Amazon EBS cost drivers

These are the different types of disk platforms available in AWS and their correspondent cost drivers:

Type of disk	Cost drivers
General-purpose **gp3**	• Capacity • Provisioned IOPS • Provisioned throughput • Outbound data transfer
General-purpose **gp2**	• Capacity • Outbound data transfer
Provisioned IOPS SSD **io2**	• Capacity • Provisioned IOPS • Provisioned throughput • Outbound data transfer
Provisioned IOPS SSD **io1**	• Capacity • Provisioned IOPS • Outbound data transfer

Throughput optimized HDD st1 Cold HDD sc1	• Capacity • Outbound data transfer
Snapshots	• Capacity • Tier

Table 9.4 – AWS EBS cost drivers

GCP Persistent Disk cost drivers

These are the different types of disk platforms available in GCP and their correspondent cost drivers:

Type of disk	Cost drivers
Standard SSD Balanced	• Capacity • Redundancy • Outbound data transfer
Extreme	• Capacity • Outbound data transfer • Provisioned IOPS
Hyperdisk	• Capacity • Outbound data transfer • Redundancy • Provisioned IOPS
Snapshots	• Capacity • Redundancy • Tier

Table 9.5 – GCP Persistent Disk cost drivers

File storage

File storage is another storage technology that is widely used in both on-premises and cloud scenarios.

It uses a structured hierarchy of files and folders that allows you to set up granular permissions, which can be given at the folder or file level in that structure, while allowing for permission inheritance. When a user requests a file, they provide the full path of the file that the user wants to retrieve or work on, as shown in the following diagram:

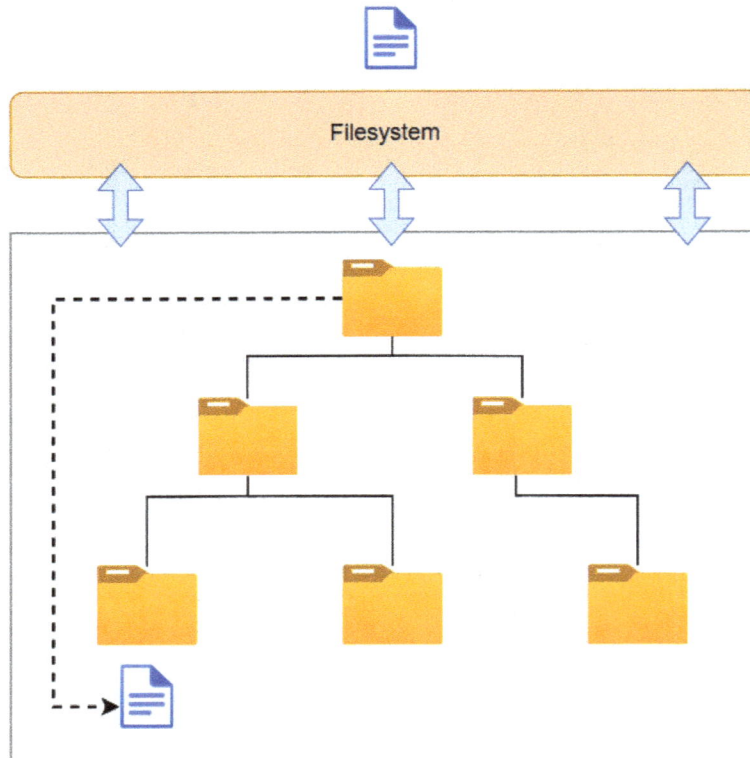

Figure 9.8 – File storage

This is the technology used in **NAS** devices on-premises, for example. This technology makes use of protocols such as SMB (Microsoft Windows) and NFS (usually used for Linux workloads) to access files over a network.

For reference, these are the services in each cloud provider that we refer to when we talk about object storage:

- **Azure:** Azure Files
- **AWS:** Amazon EFS
- **Google:** GCP Filestore

One thing that is worth mentioning is that AWS EFS is currently the only service that offers file storage on which you only pay for the actual capacity you are using, instead of provisioned capacity. In Azure and GCP counterparts, you specify the size of your storage shares, and you are charged for that, regardless of how much of it you are using.

How does it work?

In file storage, your disk needs to be mounted or accessed from a network location. When file storage is in the cloud, the network location can be an address on the public internet or a private address that we can access through our virtual network (for example, if we use Azure Files with a private endpoint while disallowing public access).

The files follow a strict hierarchical structure, on which we can set up permissions at every level to specify what users can do at every level of the hierarchy. After permissions are set up, we end up with a list of **Access Control Lists (ACLs)** on every file.

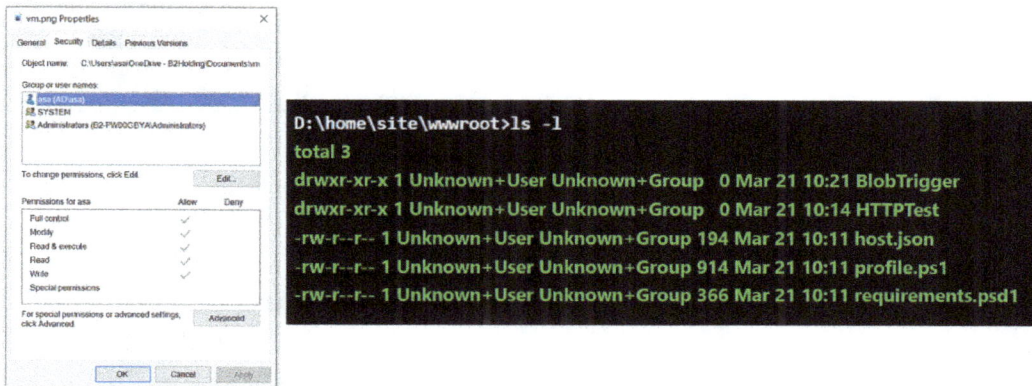

Figure 9.9 – Windows and Linux ACL examples

This filesystem allows you to set access rights at the user level, as well as set files to be read-only or password-protected in some cases.

Cost drivers

The cost drivers in file storage are a simpler subset of the ones in block storage, only with some additional considerations on the tier/data temperature cost driver.

In file storage in Azure and AWS, the tier concept changes to reflect data temperature instead of the physical disk type. Although the offering of data temperatures is not as broad as in object storage, there are essentially hot and cool options, which is an important feature for cost optimization. Conversely, in GCP, it is a much simpler approach, where there is no possibility to use different data temperature tiers, only different physical disk options such as SSD or HDD.

As an overview, in the following section, we provide a cheat sheet in the form of a table of each service and their corresponding cost drivers, which will help us understand which parameters have an effect on cost when using each file storage offering in all three major public clouds.

Azure Files cost drivers

Regarding file storage, these are the different tiers that Azure storage account file storage offers as well as their cost drivers:

Storage tier	Cost drivers
Premium*	<ul><li>Capacity</li><li>Redundancy</li><li>Snapshot size</li><li>Metadata at rest (ACLs and other properties)</li></ul>
Transaction Optimized	<ul><li>Capacity</li><li>Redundancy</li></ul>
Hot	<ul><li>Snapshot size</li><li>Number of transactions</li><li>Metadata at rest (ACLs and other properties)</li></ul>
Cool	<ul><li>Capacity</li><li>Redundancy</li><li>Snapshot size</li><li>Metadata at rest (ACLs and other properties)</li><li>Data retrieval**</li></ul>

Table 9.6 – Azure Managed Disk Cost Drivers

*Transactions are included in the Premium tier. For highly transactional
workloads, Premium can be cheaper than Hot/Transaction Optimized with much
better performance.*

**Data retrieval is a special fee that is charged whenever data is accessed in the
Cool Storage tier and is a fixed price per GiB.*

Amazon EFS cost drivers

Regarding file storage, these are the different tiers that Amazon EFS offers as well as their cost drivers:

Storage tier	Cost drivers
Standard	• Capacity
One Zone	• Outbound data transfer • Data and metadata read/write (if elastic throughput is activated)* • Provisioned throughput • Backup size
Standard Infrequent Access	• Capacity • Outbound data transfer
One Zone Infrequent Access	• Access requests (per GB transferred) • Data and metadata read/write (if elastic throughput is activated)* • Provisioned throughput • Backup size

Table 9.7 – AWS EFS cost drivers

*Elastic throughput is essentially a serverless option for AWS EFS, on which you
pay just what you use. In this mode, costs should be controlled more closely with
Amazon Cloudwatch, as EFS adapts to throughput spikes, and this may have a
reflection on costs.*

GCP Filestore cost drivers

Regarding file storage, these are the different tiers that GCP Filestore offers as well as their cost drivers:

Storage tier	Cost drivers
Basic HDD/Standard	• Capacity
Basic SSD/Premium	• Redundancy
High Scale SSD	• Outbound data transfer
Enterprise	• Backup size

Table 9.8 – GCP Filestore cost drivers

Object storage

In this storage technology, data is stored as objects. But what is a storage object? It is an abstraction of a data structure that contains information, which can be anything from a CSV or JSON file to an image or audio recording. These objects are stored in containers or buckets, depending on which cloud we work on. Apart from the information that is stored in the object itself, this storage technology allows for other features such as object metadata (which can be used for a lot of applications) or file versioning.

This storage technology is well suited for unstructured data, with its highlights being scalability, flexibility, and finally, cost.

For reference, these are the services in each cloud provider that we refer to when we talk about object storage:

- **Azure**: Azure Blob Storage
- **AWS**: Amazon S3
- **Google**: GCP Cloud Storage

How does it work?

When we want to fetch an object, we use a similar approach as in file storage, with the complete path (in this case, it is a URL address) to the file and the container where it is hosted, which allows us to uniquely identify it from others. This feature allows us to use a folder-like structure as well in object storage, which is similar to file storage.

We can see how object storage works in the following diagram:

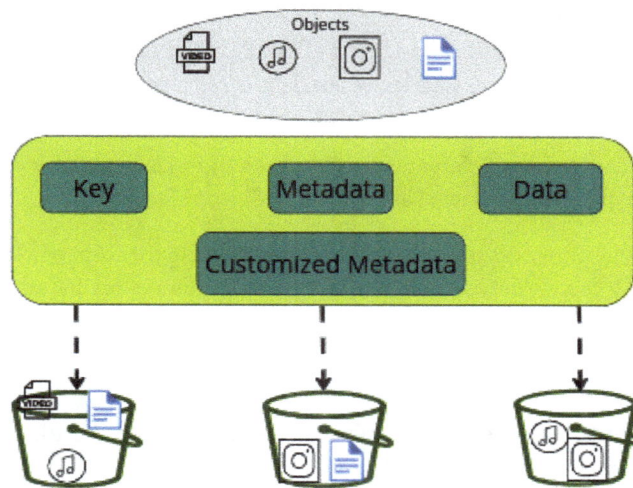

Figure 9.10 – Object storage

Another advantage of this storage technology is that all the storage operations, such as modifying a file or creating a new one, are performed with APIs, using HTTPS RESTful calls and not proprietary protocols, as was the case in file storage, or using the operating system as a proxy, as in block storage.

Due to object storage's nature of being API-based, we can access data from anywhere without any ties to virtual machines or operating systems. Cloud providers also offer **Software Development Kits (SDKs)** for the most popular programming languages to interact with their object storage offerings, so we can adapt our applications to this new storage paradigm.

Cost drivers

The cost drivers in object storage are a simpler subset of the ones in block storage, only with some additional considerations on the following cost drivers:

- **Capacity**: This is a common cost driver on all storage types. Essentially, we are charged on a per GiB basis, as with block and file storage. In this case, however, we only pay for the total space used by our objects, instead of the capacity that we provisioned in the block and file storage offerings.

- **Tier/data temperature**: We will delve deeper into this concept in the *Object storage optimization* section. Essentially, we are offered different data temperature tiers. On hotter tiers, the price per GB goes up, but the transaction price, such as reads and writes, goes down, while in cool tiers, it is the reverse, with a cheap price per GB and costly transactions. Archive tiers are for long-term storage, but there is latency when data is retrieved (ordered hourly), and reads/writes on data are really costly.

- **Requests**: Each data tier has different prices for reads and writes that vary a great deal between them. In archive tiers, data retrieval also has an additional fee per GB.

As an overview, in the following section, we provide a cheat sheet in the form of a table of each service and their corresponding cost drivers, which will help us understand what parameters have an effect on cost when using each object storage offering in all three major public clouds.

Azure Blob cost drivers

For object storage in Azure, these are the tiers offered for Azure Blob Storage and their correspondent cost drivers:

Storage tier	Cost drivers
Premium	• Capacity • Write/read/other operations • Redundancy
Hot	• Blob inventory (if creating rules) • Versioning (if this feature is activated) • Snapshots (if this feature is activated) • Index (GB/month) (exclusive to Hot storage)
Cool	• Capacity • Write/read/other operations • Data retrieval • Redundancy • Blob inventory (if creating rules) • Versioning (if this feature is activated) • Snapshots (if this feature is activated)
Cold (preview)	
Archive	

Table 9.9 – Azure Blob Storage cost drivers

Amazon S3 cost drivers

For object storage in AWS, these are the tiers offered for Amazon S3 storage and their correspondent cost drivers:

Storage tier	Cost drivers
S3 Standard	• Capacity
S3 Intelligent-Tiering	• Outbound data transfer • Redundancy • Request • Inventory (if creating rules) • Versioning (if a feature is activated)
	• Monitoring and automation (per 1,000 objects)
S3 Standard-IA	• Capacity
S3 One Zone-IA	• Outbound data transfer
S3 Glacier Instant Retrieval	• Redundancy
S3 Glacier Flexible Retrieval	• Request
S3 Glacier Deep Archive	• Data retrieval • Inventory (if creating rules) • Versioning (if a feature is activated)

Table 9.10 – Amazon S3 cost drivers

GCP Cloud Storage cost drivers

For object storage in GCP, these are the tiers offered for GCP Cloud Storage and their correspondent cost drivers:

Storage tier	Cost drivers
Standard	• Capacity • Redundancy • Write/read/other operations • Outbound data transfer • Insights inventory (if creating rules) • Versioning (if this feature is activated)
Nearline Coldline Archive	• Capacity • Redundancy • Write/read/other operations • Outbound data transfer • Data retrieval • Insights inventory (if creating rules) • Versioning (if this feature is activated)

Table 9.11 – GCP Persistent Disk cost drivers

Versioning, soft delete, and snapshots

Object storage offers another great feature, which is the possibility to keep multiple versions of the same files, as well as soft delete features, to avoid loss of data or unwanted deletion/modification of objects.

When **versioning** is activated, every time an object is changed, we can keep a number of versions that can be configured beforehand.

When **soft delete** is activated, deleted data is stored in the bucket or storage account for a period of time (retention).

By default, this feature is often not activated to avoid unwanted charges in storage. Let's provide a small example of how versioning works with the following diagram:

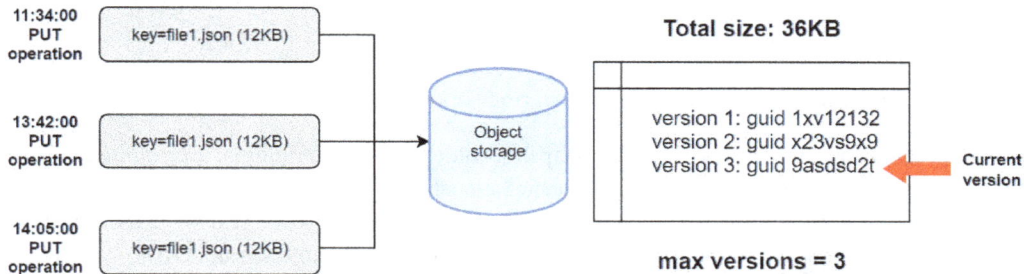

Figure 9.11 – An example of object storage versioning

Once there are multiple versions of an object in place, we can access any of them by using their corresponding unique identifiers, using standard methods such as REST APIs or an SDK.

From a cost perspective, every time we modify an object, we create a new version of it:

- If there are older versions already in place, the oldest one will be deleted and replaced by the next in line
- If there are no versions of an object yet, they will be created, and we will be charged for as many versions as we keep of every object

In addition to these features, in Azure, you currently create snapshots of an object stored in Blob Storage that is not stored as a version but as a separate object (i.e., a capture of that object at a point in time). These snapshots can be used in the same way as traditional snapshots, and we should follow the same life cycle rules that were discussed previously in this chapter.

Block storage optimization

In this section, we will cover different strategies and ideas on how to achieve cost optimization when using block storage services. Let's set the ball rolling by analyzing what we can do to optimize the size of our snapshots.

Snapshot optimization

With the differences between snapshots and backups already covered in the introductory sections of this chapter, let's move forward and analyze different strategies to optimize snapshot usage.

So, the **first point** to take into consideration is to *use snapshots when they are really needed.* By considering this and eliminating snapshot waste, we can generate some savings by deleting some old snapshots whose life cycle has ended or orphaned snapshots.

Old snapshots

There are different strategies we can apply to keep track and remove old snapshots:

- We could, in a **reactive** way, check for snapshots older than three months, for example (which is a reasonable time, taking into account that snapshots should be ephemeral). To do so, we can use different methods such as the CLI, SDKs, or cloud consoles, and we can even make this visible in dashboards so that we can track how we advance in this initiative.

- In a **proactive** way, we can set up alerts to be notified when snapshots are older than three months (or whatever retention we decide upon in our snapshot policy), so we can contact the owners of the snapshot and propose its removal. To do this, we can use Azure Policy or AWS Config, or even set up an automation in a serverless function to run with a schedule, detecting and deleting old snapshots.

The proactive approach is always the better choice, as we ensure that inefficiencies won't reappear in the future.

Orphaned snapshots

Another point that is important to cover is orphaned snapshots. Unfortunately, orphaned snapshots are more common than we'd like, as sometimes, when a virtual machine is deleted, its snapshots are not deleted along with it.

Snapshot storage optimization

As an alternative to a snapshot life cycle, we can also optimize snapshot storage by working on the main cost driver of snapshots in the cloud, which is the storage or capacity used.

The good thing about snapshot storage is that it only uses the used space on the disk and not its provisioned capacity, which reduces the total size of snapshots across our environments. Compared to other storage types such as disks, snapshot storage is cheaper.

In GCP and AWS, snapshots are incremental by default. This means that the first snapshot is always full, and subsequent snapshots only reflect the changes that happened from the first snapshot, further reducing the total size of snapshots. However, this is not the case in Azure.

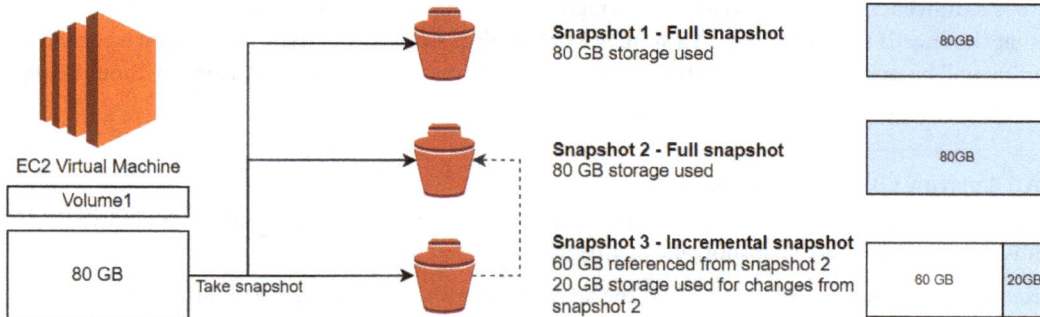

Figure 9.12 – A full and incremental snapshot example

> **Azure snapshots – incremental versus full snapshots**
>
> By default, subsequent snapshots on the same virtual machine are not incremental in Azure. However, it is possible to use incremental snapshots, but this comes with some implications. For example, incremental snapshots cannot be moved between subscriptions, and they cannot be stored in Premium SSD storage.
>
> Make sure that technical teams know of this feature, which will reduce the total size of subsequent snapshots a great deal.

We can also reduce the storage used by our snapshots in different ways:

- Separate OS and data disks, and just make snapshots on the data disk where stateful data is stored. Larger disks create larger snapshots, so make sure that you only take a snapshot of what is necessary.

- If snapshots are created on an OS disk, make sure that unneeded data is stored elsewhere – for example, non-critical logs or other files – and not on the OS disk itself, to reduce the size of the disk that is used as much as possible.

- Separate data across different disks. Multiple smaller disks can equal the performance of a big disk. Having separate disks will reduce the number of snapshots needed, as different data types may require snapshots less frequently.

Snapshot tiering

In AWS and GCP, you can have either standard snapshots or archive snapshots.

Archive snapshots are much cheaper and best suited for long-term retention or compliance requirements. With current pricing, the price per GB of archive snapshots in AWS is 75% cheaper, while in GCP, they are 50% cheaper compared to standard snapshot storage.

There is a drawback to archive snapshots though, and it's that we are billed per GB of data retrieved if a snapshot needs to be restored. In a similar way to object storage cold tiers, this feature is ideal for data that will be rarely accessed and that is kept for regulatory or compliance reasons, without being touched for a number of years.

An Amazon EBS archive snapshot note

In AWS, you can archive a standard snapshot. When an AWS incremental snapshot is archived, it is converted to a full snapshot automatically.

Keep in mind that snapshots that had the archived snapshot as a reference will increase their size to add the missing data that the archived snapshot had, which will increase the snapshot capacity total cost.

This is not the case in GCP, where you can have incremental archive snapshots. In GCP, snapshots can be created in the archive, so the process of archiving is not needed.

Using archive snapshots is a great way to close the gap between backups and snapshots. If a backup is not possible for some reason, we can use archive snapshots to have full copies of the disks we need in much cheaper storage, which is a really cost-effective way to preserve data on a long-term basis.

Ephemeral disks

Ephemeral disks are disks on which their content is not persisted when your virtual machine is shut down.

The main advantage of these disks is that their cost is included in the virtual machine price, which makes them much cheaper than persistent disks. Here are the names of these disks for each cloud provider:

- **Azure**: Ephemeral OS disks
- **AWS**: Amazon EC2 instance store
- **GCP**: Compute Engine local SSDs

These disks are ideal for some workloads, such as stateless solutions, on which it is not necessary to store any data in virtual machines for use cases such as batch processes or containerized solutions, where data is stored and persisted elsewhere and not inside the virtual machine. They also offer better latency than persistent disks, as they are directly attached to the virtual machine, which can be even more beneficial for these described use cases.

Every time a virtual machine is stopped, the disk is deleted, and therefore, its data is lost, including the operating system. However, when the virtual machine is rebooted, the data is persisted.

Consider using this feature for the aforementioned use cases; it is a great way to generate some savings when running stateless IaaS virtual machines. This feature can also be combined with the use of Spot Instances for even more cost optimization.

Disk rightsizing

As we did with virtual machines, we need to analyze whether our disks are correctly sized for their uses. For this rightsizing exercise, we need to consider two different elements:

- The first is based on the disk's **performance**. Is the disk performance tuned just right for the workload a virtual machine is running?

- The second is based on the disk's **capacity**. Are we making full use of the capacity that was provisioned for the disk? Should we offload some colder information to other storage types or other cheaper disks?

Let's discuss both of these elements in the following sections.

Is my disk correctly sized (performance)?

Azure, AWS, and GCP each offer different disk tiers based on the performance needs of our workloads.

On the one hand, we have disks that aim for high-performance workloads, where the lowest latency and highest performance are needed, and on the other hand, cheap HDD disks that can be used in scenarios where performance is not a key factor, such as development or sandbox environments.

For cost optimization purposes, the key to rightsizing is to always choose the right tool for the job. Our recommendation is to ensure that you fully use the provisioned throughput and IOPS that your disk offers. To do so, we can thoroughly key in disk metrics, such as the percentage of IOPS used or the percentage of throughput used, analyzing whether or not we are making reasonable use of the bandwidth of our disks.

If the results point to disk underuse, consider downgrading the disk to a cheaper one that will adjust better to your bandwidth needs.

> **Note – high-performance disks in sandbox and development environments**
>
> Our recommendation is to avoid using high-performance or Premium disks in development and sandbox environments.
>
> As these environments do not often require high-performance disks, having them in place should not be the norm. Setting up policies to ensure that these disks are not created is a great proactive way to prevent inefficiencies.
>
> Also, in such policies, we can add exceptions for services on which the use of high-performance disks is justified.

Is my disk correctly sized (capacity)?

Now that the performance part is covered, let's move on to capacity.

The first step is to determine whether the disk capacity is correctly sized for our workload (we take for granted the idea of thin provisioning, which has already been discussed).

We may have three different scenarios:

- **Scenario 1**: We have spare capacity
- **Scenario 2**: The disk is sized just right, with enough capacity to get by, and we have alerts in place to get notified when the disk is getting full
- **Scenario 3**: We don't have that much capacity left

In **scenario 1**, we don't have much room for improvement. As we already explained, cloud providers don't allow us to reduce disk size in any way, even if space is freed and returned from the OS layer to the virtual machine. Our only option here is to create a disk with a smaller volume size and copy data, which can be difficult and time-consuming in some cases. This is why it is so important to never overprovision disk capacity.

In **scenario 2**, things are just right, and our disk is working like a well-oiled machine. We have enough capacity for our current needs, and we have the means to identify, by using alerts, when more capacity is needed. If an alert fires, then we will be also in **Scenario 3**.

Lastly, in **scenario 3**, expanding the disk is not the only way forward. In fact, in our view, it should be the last resort. Before expanding the disk, ensure that it is not possible to move some data or free some disk space. As we will explore in the next section, a good alternative is to try to move infrequently accessed data to other storage types, such as file or object storage.

Offloading to file and object storage

As we already explained in the introduction of this chapter, block storage is the storage type that has the highest price tag.

Evidently, when the use case supports doing so, one of the biggest cost optimization initiatives we can implement is to replace block storage with file or object storage:

- If we consider a transition from **block to file storage**, it can be done in a somewhat transparent way, as the virtual machines will still be able to access data in file storage using standard protocols such as NFS or SMB. We just need to ensure that we do it for workloads that are not that dependent on disk performance, such as document stores.

- When considering moving from **block to object storage,** things are not that easy. The protocols used to access information are totally different (remember that object storage works with HTTP and RESTful APIs), so the only way to do it may be to refactor the applications and workloads to support object storage. In some cases, this can be totally impossible, especially for older applications.

Let's use a diagram as an example of how much impact this can have on our costs.

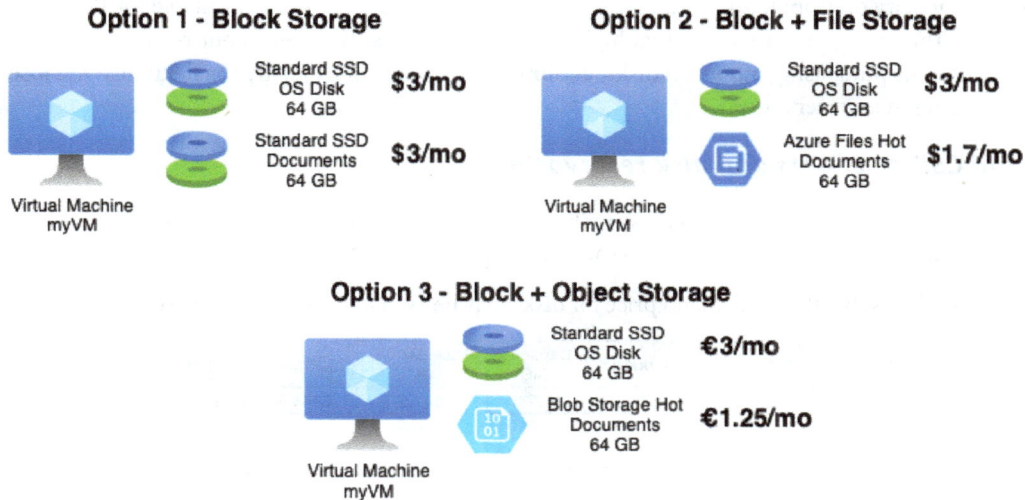

Figure 9.13 – An offload example in Azure Virtual Machines. The prices
were obtained from West Europe in June 2023 for LRS storage

The price differences are clear, and we have not even used the data temperature tiers that are offered in object and file storage, which would result in even more evident price differences.

Let's sum it up:

- Consider moving data to other storage types even if it requires some preparation or solution modernization, as it may be worth your while cost-wise. Using more modern storage services is never a bad idea as well.

- For data that is moved, try to separate the least accessed data from data that is accessed more frequently, and move cold data to colder storage tiers. This will result in amazing cost savings.

Reserved capacity

Currently, disk reservation is only available in Azure. In AWS and GCP, the possibility to reserve capacity in Amazon EBS and GCP Compute Engine Persistent Disks is not supported.

The possibilities of reserved capacity for managed disks is limited, as it is offered in disks instead of disk size. You cannot distribute a single reservation for 1 TB between disks on different virtual machines.

Reserved capacity only applies to disk SKUs and not to snapshots and is only available for Premium SSD disks. The savings are not really that high (around 5% in west Europe) so our recommendation is not to use this purchasing model on disks, as, for us, the savings don't justify the flexibility you lose by purchasing these reservations.

EA and CSP discounts and disk reservations

Also, in our experience, when a discount is applied to customers due to high volumes, it is only applied to on-demand prices and not to disk reservations.

The result of this is that the on-demand price for disks may be even lower than the price of reservations:

Service	Region	Redundancy	Capacity (GiB)	Price per GB OD	Price	Price RI 1 yr	Savings 1 yr
Azure Managed Disks P30	West Europe	LRS	1024	$0.15	$148.68	$141.25	5%

Table 9.12 – Azure Managed Disks reserved capacity example

Now that we understand the options that we have to use reserved capacity in Azure, we will close our section on block storage optimization. In the next section, we will explore different cost optimization ideas and initiatives that we can use on object storage services.

File storage optimization

In this section, we will cover different strategies and ideas on how to achieve cost optimization when using file storage services. To begin with, let's analyze how to rightsize our file shares not only in capacity but also in performance, in order to cut unneeded costs.

File storage rightsizing and data temperature

Following the same concepts as disk storage rightsizing, we need to make sure that our file storage services are correctly sized in regard to provisioned capacity.

The first step in this process is to analyze how much of our provisioned file storage we actually use. It is also important to have some visibility in how many transactions we are being billed for per bucket or storage account.

Having this information at hand allows us to project how much it would cost to have the same information in another storage class (for example, one that is better suited for highly transactional systems). It even helps us to decide to move the information to a smaller storage that is a better fit in terms of capacity for our needs. As we already covered in this chapter, applying the idea of thin provisioning is always the best approach when provisioning and deciding how big our storage services will be.

Apart from this point, we should also choose the better fit for the type of data we will be storing in file storage. If we don't do so, we can end up paying much more than we anticipated.

An Azure example – different tier costs per GB and transaction

This table shows how important it is to choose the right tier for our data to be stored in file storage:

Service	Region	Redundancy	Price per GB OD	Price per 10,000 write transactions	Price per 10,000 read transactions
Azure Files Premium File Share	West Europe	LRS	$0.192	N/A (included)	N/A (included)
Azure Files Transaction Optimized	West Europe	LRS	$0.060	$0.014	$0.002
Azure Files Hot	West Europe	LRS	$0.027	$0.066	$0.005
Azure Files Cold	West Europe	LRS	$0.015	$0.122	$0.012

Table 9.13 – Azure Files different tier pricing

Let's break down the preceding table:

- The per GB cost of a Premium file share is **12 times the cost** of per GB in Azure Files Cold
- The price per 10,000 write transactions in Azure Files Cold is **8.6 times the cost** of 10,000 transactions in Azure Files Transaction Optimized
- Due to Azure Files Premium having all transactions included, in some situations, it can be more costly to have highly transactional storage in Standard than in Premium

Regarding file storage and data temperature, this feature is not as powerful as in object storage due to a lack of granularity, as data temperature can only be set at the share level and not at the object level, and the temperature tiers offered are limited (the Archive tier is not offered in this case, for example), but even with these limitations, it can be used to distribute data, based on how it's going to be used for cost optimization.

With all these ideas covered, let's move on to reserved capacity for file storage options.

Reserved capacity

Currently, reservations for network storage are only available in Azure. In AWS and GCP, the possibility of reserving capacity in GCP Filestore and Amazon EFS is not supported.

Reservations are only available in units of 10 TiB and 100 TiB for 1 and 3 years, respectively. These are the tiers where reservations can be applied:

File share type	SMB	NFS
Standard file shares (general-purpose v2) LRS/ZRS	Supported	Not supported
Standard file shares (general-purpose v2) GRS/GZRS	Supported	Not supported
Premium file shares LRS/ZRS	Supported	Supported

Table 9.14 – Reserved capacity

Snapshots also count as eligible for these reservations, but not on Premium file shares.

Service	Region	Redundancy	Capacity (TiB)	Price per GB OD	Price	Price RI 1 yr	Price RI 3 yr	Savings 1 yr	Savings 3 yr
Azure Files Premium File Share	West Europe	LRS	10	$0.19	$1,966.08	$1,612.17	$1,297.62	18%	34%
Azure Files Transaction Optimized	West Europe	LRS	10	$0.06	$614.40	Not supported	Not supported	Not supported	Not supported
Azure Files Hot	West Europe	LRS	10	$0.03	$277.50	$227.59	$183.14	18%	34%
Azure Files Cold	West Europe	LRS	10	$0.02	$153.60	$125.92	$101.39	18%	34%

Table 9.15 – Reserved capacity

The possibilities in this regard are currently very limited, but it is important to highlight that we can achieve some cost savings if we use reserved capacity for file storage if we expect data to not change for years, as long as we can ensure that the reserved capacity purchased won't be wasted in any way.

Object storage optimization

In this section, we will cover different strategies and ideas on how to achieve cost optimization when using object storage services. In the case of object storage, we have multiple ways to optimize our costs, as it offers many features, but we must also be mindful of how these features can skyrocket our cloud bills as well. Let's begin by understanding how to make the most of one of its most important features, which is data temperature tiers.

Object storage tiering

As we already introduced for file storage, one of the key offerings of object storage is a set of storage classes, or data temperature tiers, that are offered to store our information.

In object storage, we have available *hot* storage classes on which we expect a lot of read/writes, as well as *cold* classes more suitable for storing files that are rarely accessed. The extreme of *cold* tiers are archiving offerings that we can use to store files for years at the lowest cost possible. These archiving tiers have the drawback that writes have a much higher cost, and there can be some latency (in terms of hours) from when we request a file to when we get it.

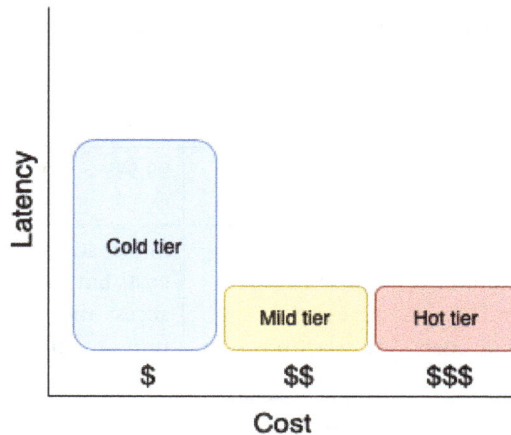

Figure 9.14 – Object storage cost versus latency

How do we achieve cost optimization in object storage? Essentially, it all comes down to exactly aligning the use case of our data to the storage class or tier where it is stored.

This can take a lot of work, as we need to separate data based on its temperature and how and when it is accessed. However, this work can prove to be fruitful, as cost savings that can be achieved after this process can be enormous.

As a small example, these are the pricing differences between tiers in Azure:

- **Per GB of storage**: Per GB pricing in the Archive tier is around **11 times** cheaper than the Hot tier

- **Write operations**: The write operations cost in the Archive tier is **222 times** the price of the Hot tier

- **Read operations**: The read operations cost in the Archive tier is around **1,277 times** the price of the Hot tier

As you can see, the differences can be significant, which reinforces our idea that if there is a mismatch between our temperature tier and the use case of the data, we are essentially creating waste, which will result in additional costs.

For reference, we will provide some tables to understand the different tiers offered in each cloud provider and the characteristics of each one.

An Azure example – Blob tiering

These are the different tiers offered in Azure Blob and their key features:

	Hot tier	Cool tier	Cold tier (preview)	Archive tier
Availability	99.9%	99%	99%	Offline
Availability (RA-GRS reads)	99.99%	99.99%	99.9%	Offline
Usage charges	Higher storage costs but lower access and transaction costs	Lower storage costs but higher access and transaction costs	Lower storage costs but higher access and transaction costs	Lowest storage costs but highest access and transaction costs
Minimum recommended data retention period	N/A	30 days	90 days	180 days
Latency (time to first byte)	Milliseconds			Hours
Supported redundancy configurations	All			LRS, GRS, and RA-GRS only
Life cycle transitions	Yes			
Retrieval charge	N/A	Per GB retrieved		

Table 9.16 – Azure Blob tiering

An AWS example – Amazon S3 tiering

These are the different tiers offered in Amazon S3 and their key features:

	S3 Standard	S3 Intelligent Tiering	S3 Standard-IA	S3 One Zone-IA	S3 Glacier Instant Retrieval	S3 Glacier Flexible Retrieval	S3 Glacier Deep Archive
Availability	99.99%	99.9%	99.9%	99.5%	99.9%	99.99%	99.99%
Usage charges	Higher storage costs but lower access and transaction costs	Higher storage costs in the Frequent tier and lower storage costs in the Infrequent tier, but lower access and transaction costs	Lower storage costs but higher access and transaction costs				
Minimum storage duration charge	N/A		30 days	30 days	90 days	90 days	180 days
Minimum capacity charge per object	N/A		128 KB			N/A	
Latency (time to first byte)	Milliseconds					Minutes or hours	Hours
Availability zones	≥ 3			1	≥ 3	≥ 3	≥ 3
Life cycle transitions	Yes						
Retrieval charge	N/A		Per GB retrieved				

Table 9.17 – Amazon S3 tiering

A GCP example – Cloud Storage tiering

These are the different tiers offered in GCP Cloud Storage and their key features:

	Standard	**Nearline**	**Coldline**	**Archive**
Availability	99.99%	99.9%	99.9%	99.9%
Availability (multi regions – dual regions)	99.99%	99.995%	99.95%	99.95%
Usage charges	Higher storage costs but lower access and transaction costs	Lower storage costs but higher access and transaction costs	Lower storage costs but higher access and transaction costs	Lowest storage costs but highest access and transaction costs
Minimum recommended data retention period	N/A	30 days	90 days	365 days
Latency (time to first byte)	Milliseconds			
Life cycle transitions	Yes			
Retrieval charge	N/A	Per GB retrieved		

Table 9.18 – GCP Cloud Storage tiering

Now that the concept of storage classes based on data temperature has been covered, let's move on to a great feature to automate object transfer between tiers, which is using life cycle policies.

Life cycle policies

Life cycle policies are a great feature that is offered in object storage services in the cloud. What they do essentially is to automate object transition between tiers, based on rules that can be configured beforehand.

By leveraging this feature, for example, we can set up so that files that have not been accessed for a year should be moved to a cold class.

These life cycle policies can not only be used to move objects but also to delete them if specific conditions are met, based on any of the information that we have about the object, such as last access, object size, object tags, and the folder or filename.

As an example, we could implement a life cycle policy that applies the following transitions:

- If an object is not accessed in the hot tier for X days, it will be moved to mild tier storage

- If an object is not accessed in the mild tier for Y days, it will be moved to cold tier storage

- If an object is not accesed in the cold tier for Z days, it will be permanently deleted

The policy can be summarized in the following figure:

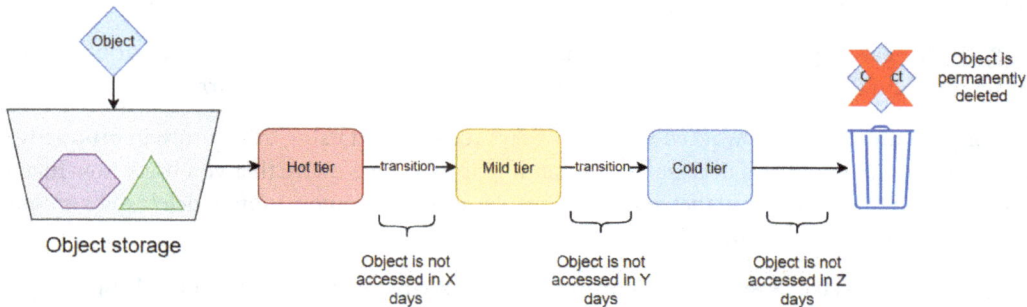

Figure 9.15 – An object storage life cycle policies example

Each cloud provider has defined a minimum period of time that a specific object needs to spend in each tier before a transition. For example, in GCP, if we put an object in Nearline storage, it needs to stay there for at least 30 days as a minimum. There are also retrieval fees in that kind of storage every time we want to access the objects stored there. You can check all these details in the tables that we created in the previous section for each cloud provider.

If we take into account these limits imposed by cloud providers, we can design any rules that we want for our objects' life cycles.

Our main recommendation is to make use of this key feature of object storage services, simplifying storage objects management and making use of colder tiers, which can be challenging for organizations that are not experienced in using the cloud. For bigger environments, it is also impossible to manage file transitioning manually, and it can be time-consuming for operation teams.

This is the perfect solution to have a standardized and automated solution that fosters cost optimization and proper data governance, based on rules that we can apply not only at the project level but also at the organizational level.

Limiting and tracking versioning, soft delete, and object snapshot usage

As we have already explained in the *Object storage* section, object versioning is an amazing feature that is part of every object storage service offered in the public cloud.

We just need to be mindful that with multiple versions of the same object, we are essentially paying for the storage space of each one of them in the correspondent storage class where the object is hosted. This can be difficult to track in cloud bills, which are not the easiest to read sometimes.

This can be the case especially if, due to automation, we generate multiple versions of our objects or even snapshots of them (a snapshot of object storage is offered only in Azure currently).

In the organizations we have worked with, we always recommend using this feature in productive environments and only if it is necessary. If versioning is used, make sure that you track how many versions of each file are kept and that you have a regular process or automation to delete old versions that can be disposed of after some time.

Versioning can have a big negative impact on our cloud costs and can be easily overlooked, especially if this specific feature is not known or fully understood by FinOps or cloud practitioners. Its implementation differs from one cloud to another, so make sure to review each cloud provider's documentation on this topic thoroughly to fully understand the implications, desired configurations, and best practices.

You can find all details regarding versioning in each cloud at the following links:

- Azure Blob versioning: `https://learn.microsoft.com/en-us/azure/storage/blobs/versioning-overview`
- Amazon S3 Versioning: `https://docs.aws.amazon.com/AmazonS3/latest/userguide/versioning-workflows.html`
- GCP Cloud Storage Object Versioning: `https://cloud.google.com/storage/docs/object-versioning`

Now that we understand the impact that versioning can have on our object storage costs, let's move on to the object storage inventory features.

Object storage inventory

There are multiple tools that are offered to obtain insights of storage use across our organizations. We can use these tools to have a centralized way to determine how many objects we have and how big they are, as well as how objects are distributed between data temperature tiers, for example.

These tools generate reports that can be analyzed by other services and shown in reports, using products such as Amazon Athena or GCP BigQuery.

To use these tools, we usually select a set of metadata that needs to be gathered from all the objects across our buckets or storage accounts – for example, the tier that an object is placed in, its size, and its tags.

We totally support the use of these tools to gain greater visibility on the use of object storage, but we also need to be mindful of the costs that these tools can bring to the table. As these tools essentially request data from our object storage services, we need to limit how much data is gathered to the bare minimum we need for analytics purposes, instead of including all the file metadata with non-relevant fields for analytics purposes.

Let's analyze an example in Google Cloud on how costs can be impacted by inventory reports.

GCP – Storage Insights inventory reports

Let's say we have a bucket with 10,000 million objects in the London (Europe-west2) region. The objects are evenly distributed into three prefixes – prod_01, prod_02 and prod_03.

We can create a rule that covers all three prefixes:

Region	Prefix	Objects	Price per million objects	Price per report (daily)	Price per month
europe-west2	*	10,000,000,000	0.0028	$28.00	$840.00

Table 9.19 – A GCP Storage Insights inventory report example

Alternatively, we can create a rule that only covers prod_1:

Region	Prefix	Objects	Price per million objects	Price per report (daily)	Price per month
europe-west2	prod_01	3,333,333,333	0.0028	$9.33	$280.00

Table 9.20 – A GCP Storage Insights inventory report example

As we can see, the cost difference between them is vast.

We can also further reduce the data generated by inventory reports by selecting the metadata fields to be included in the reports (here, there are three required fields, and the rest are optional).

Reserved capacity

As with other storage types offerings, reserved capacity for object storage is only offered in Azure.

It is currently offered for the Hot, Cool, and Archive tiers for one or three years, and the reservations can be purchased in packages of 100 TB or 1 PB per month.

As an example, here is how much you can save by using reserved capacity for 100 TB in LRS:

Service	Region	Redundancy	Capacity (TiB)	Price	Price RI 1 yr	Price RI 3 yr	% Savings 1 yr	% Savings 3 yr
Azure Blob Premium	West Europe	LRS	100	$19,968.00	Not supported	Not supported	Not supported	Not supported
Azure Blob Hot	West Europe	LRS	100	$1,986.56	$1,646.00	$1,325.00	17.14%	33.30%
Azure Blob Cool	West Europe	LRS	100	$1,024.00	$840.00	$676.00	17.97%	33.98%
Azure Blob Cold	West Europe	LRS	100	$460.80	Not supported	Not supported	Not supported	Not supported
Azure Blob Archive	West Europe	LRS	100	$184.32	$165.00	$152.00	10.48%	17.53%

Table 9.21 – An Azure Blob reserved capacity example

In our object snapshot usage opinion, reserved capacity is only significant for really big organizations where object storage is widely used and there is a stable storage baseline that is always used, and always for shared scopes to avoid reservation waste.

Other storage optimization initiatives

In this section, we will cover other initiatives that can be relevant to storage optimization but that don't fit neatly into the previous sections, such as backups and log storage optimization.

Log storage optimization

The idea of **observability** has made an impact in the IT world for a few years now.

Using observability platforms, we are able to collect performance telemetry such as metrics, as well as logs and complete traces of what happens on our systems. Once all the information is in place, we can then build dashboards and reports with all this information, providing great value for technical teams and stakeholders.

However, having these platforms is a double-edged sword, as this huge amount of data and information can cause storage costs to go through the roof if not controlled properly.

There are many flavors of observability platforms, including having IaaS solutions such as **ELK** (**Elastic Search + Logstash + Kibana**), but for the purposes of this book, we are going to focus on public cloud-native services that offer these functionalities.

The costs of these services depend, mainly, on the following:

- **Data ingestion**: The amount of data that is ingested into the solution
- **Data retention**: The amount of data that is retained
- **Data analysis**: The queries, analytics, and dashboards on top of the stored data

With these cost drivers in mind, performing cost optimization means trying to reduce as much as possible all the cost drivers. FinOps practitioners should ask the following questions:

- Are we using all the data that is ingested? Could we reduce some parts of it (e.g., limit logging from the virtual machines where logs are collected to just the logs that are essential for observability purposes)?

- If that is not the case, can we offload older data to other storage services such as object storage, where we can use colder tiers? Such data can be retained in a much more cost-effective way to be compliant with data retention policies.

- Do we need the data retention that is currently set up, or is it just a commodity to avoid setting up archival processes or data offloads to other storage services?

- Can we reduce in any way the analytical tools that we use on our data?

Optimizing the price of these services involves reducing their uses to what is exactly needed – nothing less and nothing more. This exercise is not easy at all and requires a great deal of effort.

Backup storage optimization

Using native for backup in the cloud is a great way to optimize the costs. Instead of purchasing costly licenses and hardware appliances (plus hosting and housing costs), you just need to create a few resources in the cloud so that your workloads are protected.

Conversely, as is the case with other services, costs can get out of hand easily if we don't fully understand backup cost drivers and work on them from a cost optimization perspective.

There is an additional problem with backup as well. Due to backup retention policies, we may need to retain backups for years, which can result in huge amounts of storage occupancy in our Cloud Storage services.

In this section, we will try to provide a few ideas to work on to try to optimize backup costs.

The recovery point objective, the recovery time objective, and adjusting backup policies

Let's now introduce two key concepts that are key to understanding how backup policies protect our workloads and how we can ensure business continuity:

- **Recovery Point Objective (RPO)**: This corresponds to the maximum period of time (i.e., the worst-case scenario) in which data can be lost without severely impacting business or operations

- **Recovery Time Objective (RTO)**: This corresponds to the period of time expected to recover from a contingency, disaster, or other event that can bring down operations

This diagram represents these parameters in a graphical form over time:

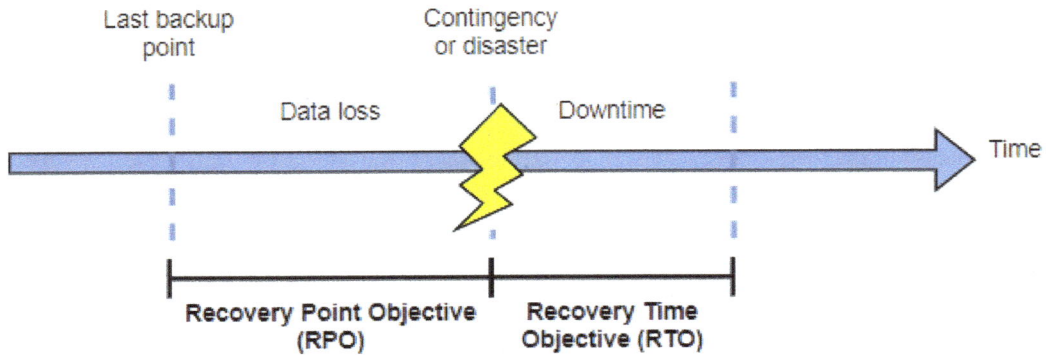

Figure 9.16 – RTO and RPO

Both parameters are agreed upon from a business perspective, and it is part of the IT department's duty to ensure that we can fulfill these needs. Apart from RPO and RTO, there is another key document that we need to review before going forward with our cost optimization exercise, which is a backup policy.

A **backup policy** describes different things:

- Which data needs to be retained?

- How long to retain the data?

- When do backups occur and how frequently?

- The process that is followed for data to be deleted

Once all these parameters are clear, we are ready to begin cost optimization for backups, which is essentially the following:

- Ensure that only resources indicated in an organization's backup policy are protected. We may eventually find development or sandbox virtual machines, or even proofs of concept, that have been included in backup policies by mistake. Deleting these can result in different degrees of cost savings, depending on the virtual machine and disk size.

- If our retention policies state that backups should be kept for X years, make sure that none of the backups that are hosted in backup storage are older than this period. Sometimes, organizations tend to avoid deletion of these, keeping them just in case they are needed in the future, which can generate massive amounts of storage.

- Ensure that backups are kept in the most cost-effective storage possible. Avoid disk storage for backups as much as possible, as the pricing of disks is way higher than object or file storage.

With this in mind, let's move on to the next section, which discusses selective disk backup.

Selective disk backup

This idea is simple in essence but sometimes difficult to implement. Basically, the idea is to include in a backup only the disks that need to be backed up.

For example, let's say that we have a virtual machine where all key data is stored on the data disk, and we have nothing at all in the operating system or other disks. Does it make sense to have a full daily backup of all those additional disks if data does not change and we don't need it at all? The answer is, probably not.

With this idea in mind, you can review the different backup policies and protected objects and try to exclude from daily backups specific objects where a backup is not needed.

This idea can also be used to exclude backup objects that have already been backed up using other backup software or products, thus avoiding storing backup duplicates across different backup products.

> **Azure Backup – a selective disk backup example**
>
> In Azure, we can implement this idea by using its selective disk backup feature.
>
> This allows us to protect and restore, if needed, a subset of the disks of a virtual machine for cost optimization purposes. To do so, you can include or exclude disks from a backup by using **Logical Unit Numbers** (**LUNs**) assigned to the disks.
>
> Also, there is a great Excel tool available from Microsoft to calculate backup costs; make sure to check it out: `https://view.officeapps.live.com/op/view.aspx?src=https%3A%2F%2Fdownload.microsoft.com%2Fdownload%2F0%2Fb%2F7%2F0b7c4140-24b4-4eff-9b2b-64ecee97d667%2FAzureBackupDetailedEstimatesV10.xlsx&wdOrigin=BROWSELINK`.

Backup redundancy

As we have described throughout the chapter, using redundancy options for storage has an impact on costs, and this is also the case with backup storage.

This applies to both backup in IaaS and PaaS resources.

Our recommendation on this topic is to only use redundancy options when it is required. For example, in Azure PaaS resources such as Azure SQL, we often use ZRS for backup storage, which is a trade-off between having some redundancy options and costs. You ensure that in the event of a contingency in a data center, there are additional copies spread over other availability zones, so your backup will be available anyway. Redundancy options that span regions have a higher price tag and only make sense in highly critical multi-regional solutions.

As an additional measure, be mindful of these redundancy settings when creating resources. In Azure, when you create an Azure SQL database from the portal, the default redundancy setting is GRS, which is the priciest of them all.

It is part of a FinOps mindset to be aware of all the cost drivers of different resources and ensure that you only use what's needed every step of the way.

Reserved capacity

As with other storage services in Azure, there is also the option to use reserved capacity for Azure Backup.

You can buy reserved capacity in units of 100 TB or 1 PB a month for 1-year and 3-year periods, respectively, and it is only available on the Standard tier, not Archive.

Service	Region	Redundancy	Capacity (TIB)	Price	Price RI 1 yr	Price RI 3 yr	% Savings 1 yr	% Savings 3 yr
Azure Backup 100 TB Standard	West Europe	LRS	100	$2,293.76	$2,018.50	$1,835.00	12.00%	20.00%
Azure Backup 100 TB Archive	West Europe	LRS	100	$245.76	Not supported	Not supported	Not supported	Not supported

Table 9.22 – Azure Backup reserved capacity

Our recommendation is the same as with other reserved capacity offerings in Azure; only use it when high volumes are assured and you can guarantee no reservation will be wasted.

Summary

With this chapter, we conclude the most technical part of this book and all the topics included in our Optimize pillar.

We have reviewed all the different kinds of storage that can be used in the cloud and how they work, as it is often difficult to fully understand it from cloud provider documentation in our experience. Apart from the introductory section in this chapter, which was full of storage concepts and options, we delved into how to apply different optimization initiatives and ideas for cost optimization related to Cloud Storage, highlighting the key topics to be aware of and showing examples at each step of the way.

We hope that, although technical and sometimes deep and lengthy, these chapters have proven to be useful for your future FinOps endeavors.

Having closed this big part of the book, we will cover the Operate pillar in the next chapters, where we will focus on proposing new KPIs to work on related to FinOps, as well as proposing how to set up new organizational processes to help in our journey to iterative cost optimization.

Keep in mind that, in *Part 5* of the book, specifically in *Chapter 13*, we will be applying all these ideas and concepts into actual examples, where we will implement the ideas covered in this part of the book.

Part 4: Operate – How to Set Up a Governance Model around Cloud Costs

This part describes in depth the **Operate** pillar, which is based on setting up a governance model, creating new KPIs and processes around FinOps practices, and improving the visibility of FinOps initiatives' statuses and obtained savings, to highlight the value of FinOps teams to stakeholders.

This part has the following chapters:

- *Chapter 10, Designing and Implementing FinOps KPIs*
- *Chapter 11, Defining New FinOps Roles and Processes*

10

Designing and Implementing FinOps KPIs

The need to keep track of our business objectives is not exclusive to FinOps practices. All companies need a way to quantify or measure their results and achievements in line with their previously established targets. One of the best ways to do this is to use **Key Performance Indicators (KPIs)**, which help us obtain key insights for our business.

In this chapter, we will introduce what a KPI is, and how KPIs can be used to fuel and give more visibility to our FinOps practices. KPIs are the key to discerning whether we are on the right track with our FinOps practices, providing us with quantifiable means to measure progress and success, which is essential to FinOps sponsors, management, and key stakeholders to keep the practice alive (as well as its budget).

In addition to this, we will describe a methodology to design new KPIs from ground zero that will help you create your own indicators that will add value to multiple areas of your FinOps practice. To top it off, we will also cover some detailed examples of KPIs in each category there is that can be used to boost and raise visibility of FinOps work and progress.

KPIs also represent all the information needed for our reports and dashboards, as we already described in *Chapter 5*, of this book.

In this chapter, we will cover the following topics:

- What is a KPI?
- FinOps and KPIs

What is a KPI?

There is a lot of literature related to KPIs. From finance to HR to marketing and sales and, of course IT, they are widely used in almost all industries.

A KPI is a quantifiable measure that allows us to evaluate performance over time, providing us a means to gauge progress and generate additional insights for our business. In the case of FinOps, it helps evaluate a lot of key aspects of the practice, such as its maturity, the degree of implementation of a specific initiative, and purely financial measures such as the savings generated compared to last year as a result of FinOps.

Before continuing to go deeper into the use of KPIs, we want to also introduce the differences between a KPI and another measure that is used to create KPIs, a **metric**.

A metric is a specific measure that is recorded to track the status of a specific business activity in order to evaluate the success or failure of the performance of that activity. In the FinOps world, an example of a metric is the *CPU consumption of a specific virtual machine over time during business hours*. Let's imagine we collect the following information on our set of virtual machines:

Name	CPU % (avg)
vm1	50
vm2	70
vm3	10
vm4	35

Table 10.1 — Rightsizing KPI example 1

On the other hand, a KPI specifically measures how close we are to a business goal. We would not be able to evaluate how successful a metric is for the business if we don't establish clear goals and targets.

Continuing the example, we could create a KPI by setting up a specific business goal: *All virtual machines should have average CPU consumption of 50% percent or more during business hours to be considered properly rightsized.* Once we have this goal, we can go back to our virtual machines' CPU consumption metrics to evaluate each virtual machine and determine whether they are compliant with our policies.

Let's come back to our imaginary data and add an additional column with the aforementioned information:

Name	CPU % (avg)	Rightsized?
vm1	50	YES
vm2	70	YES
vm3	10	NO
vm4	35	NO

Table 10.2 — Rightsizing KPI example 2

With this information, we can compose a KPI that measures the *percentage of rightsized virtual machines, which in this case would be 50%.* This KPI will give interesting insights about how a specific FinOps initiative, in this case the rightsizing of virtual machines, is evolving over time, providing us with a way to objectively measure the success and impact of this specific initiative across the organization.

We can consider all KPIs as metrics, but not the other way around. Metrics are the foundation that supports KPI creation.

A metric by itself provides no value without further interpretation. We could record all the data in the world across our cloud environments, but without interpretation, it would be completely useless. Part of the job of FinOps teams is to identify which metrics we should focus on to get the insights related to cost optimization that we want.

As an example, you can check the available metrics per cloud resource at the following links:

- **Azure Monitor**: `https://learn.microsoft.com/en-us/azure/azure-monitor/reference/supported-metrics/metrics-index`

- **AWS CloudWatch**: `https://docs.aws.amazon.com/AmazonCloudWatch/latest/monitoring/finding_metrics_with_cloudwatch.html`

- **GCP Cloud Monitoring**: `https://cloud.google.com/monitoring/api/metrics_gcp`

With all this metrics information, we can define new KPIs as long as we are able to define which metrics are key for given purposes.

As a general overview, coming back to KPIs versus metrics, we can highlight the main differences between them by consulting the following table:

KPI	Metric
All KPIs are metrics	Not all metrics are KPIs
High level and strategic	Low level and operational
They give key business insights, telling you about current performance with respect to a specific business goal	They don't give any business insights on their own without interpretation
Calculated using a combination of relevant metrics	Quantitative or qualitative measurements
Example: Sales revenue growth	Example: Sales revenue per month

Table 10.3 — KPI versus metrics

We can also represent their relationship in a graphical view:

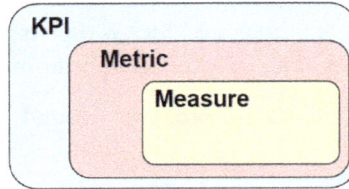

Figure 10.1 — KPI components

Lastly, we should consider, both for metrics and KPIs, the time frame in which they are measured or their refresh rate. In other business areas, such as stock investment trading, real-time data is essential and can give the edge over the competition, while in FinOps we can work with non-real-time information.

With this brief introduction to what a KPI is complete, let's describe step by step how we can define our own and put these concepts into practice.

KPI creation process

The KPI design phase should follow a specific process consisting of multiple steps to ensure the consistency and quality of our KPIs, which is what we will be covering in this section. We must remember that KPIs represent key business information that is invaluable for the organization. Due to this fact, a KPI should be totally impartial and unbiased, which is why its definition and creation should follow a specific structure. In addition to this point, we must keep in mind that KPIs are *living things* and subject to change many times over their life cycle due to business objective changes, the way they are measured, or just purely because of changes in an organization's goals. Because of this dynamic nature, it is always preferable to be formal in the definition and documentation of KPIs.

A KPI can be implemented by working through these different phases:

1. Definition
2. Implementation
3. Analysis
4. Actions

Definition

As we already explained, a KPI should be totally impartial and should not present any ambiguity about its purpose and the goal it aims to achieve. Reaching this non-ambiguity requires continuous refinement, which means an iterative process until the desired objectives for the KPI are met.

The metrics on which a KPI depends must have a clear measurement method and should be agreed with the different teams that will benefit from it. For example, if we measure a *sale*, or an *active client*, these terms should be clearly defined and specified in advance.

Each KPI should have an **owner**, who is the point of contact whenever anyone has any query or question regarding the given KPI. The owner is also responsible for ensuring that this KPI provides the right information. The owner does not have to be the person implementing it, but they are accountable for it and take responsibility for this specific KPI, to avoid accountability being diluted across multiple teams.

In the definition phase, we need to document and explore *which calculations and metrics will support this KPI*, as well as other key aspects such as the data refresh rate, data sources, KPI value range, and the format of the indicator, among others.

For each KPI, it is also a good practice to assign each one with a **unique identifier**, which will allow us to identify the domains and categories that the KPIs belong to. For example, we could define the KPI that we used in the previous example (virtual machine rightsizing) as *FINOPS_INFRA_001*.

Once the definition of the KPI is clear, we must assign the indicator a **priority** level, based on its importance or the value it represents for the business.

Last but not least, once the previous steps have been completed, we need to also decide the **audience** for this KPI, which means who should be granted access to this KPI information. Keep in mind that some KPIs represent highly sensitive data that should not be widely available, and this is why its audience should be clearly defined.

Implementation

Once the base components of the KPI are defined, we need to proceed to the implementation phase, in which we gather the information needed for this KPI. To do this, we must *extract* the information from the data sources. After this, the information will be *processed, transformed, and validated* so it can be presented in the defined prerequisite format.

The implementation phase is usually carried out by technical teams that set up a process (preferably automatic, but unfortunately, this is not the case much of the time) to gather this information and process it. Once the data is extracted and processed, it will be shown in a report or dashboard.

As a part of the implementation phase, we can involve **key users** to ensure that the process and information the KPI provides properly follow the principles of its definition toward the business goal that we want to achieve.

Finally, we need to also implement the **required permissions** so the expected audience will be able to access these dashboards and reports whenever necessary.

Analysis

Once we have all the information supporting our KPI and the calculation is ready as well, there comes another key phase of the process, which is the analysis and interpretation of the data we have obtained.

In this phase, we need to double-check whether the *behavior* of our KPI is as expected, ensuring that its values are within the thresholds and data range, matching the format that was defined.

During this phase, we also need to *interpret* the information to check that it serves its purpose of representing the performance status toward reaching a specific goal. We can also analyze the tendencies and patterns hidden behind our data.

To do so, we can use multiple analytical methods to review the data that we have gathered, including the following:

- **Descriptive**: Aims to review and summarize the patterns that we see in our data
- **Prescriptive**: Aims to provide recommendations and key insights to improve performance
- **Diagnostic**: Examines the data to understand the root cause for specific events or issues
- **Predictive**: Intended to predict future trends or tendencies beforehand

Once our dataset interpretation is ready, we are ready to move to the next phase.

Actions

Once our data has been properly and deeply analyzed and validated, we can consider the KPI ready for consumption by our target audience. At this point, what is usually the hardest phase begins, which is to *take actions* to improve the performance that this KPI shows over time.

We can identify two different types of action:

- **Corrective**: Corrective actions seek to return to its track unwanted patterns or specific issues we observe in the KPI data, that we can tackle on a short-term basis. These tasks can be carried out by smaller or specific teams and are easier to track.
- **Strategic**: These actions are more focused on the long term and derived from observed issues that require a great deal of time and work to be corrected, a process that usually involves a lot of people. These strategic tasks may, for example, end up in the form of a specific high-impact project or company restructuring across the organization.

We can summarize the process in the following figure:

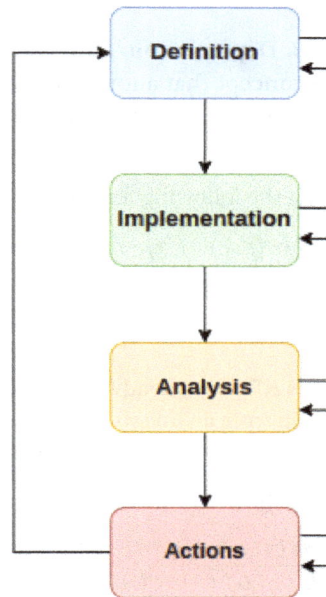

Figure 10.2 — KPI life cycle

All these phases help us shape our KPI, ensuring that its purpose is met and maximizing this value. We recommend revisiting frequently the purpose and definition of each KPI as, over time, they can be lost in translation. If a KPI is not important enough or if it loses its value, it should be retired or replaced. Having too many KPIs in scope can overwhelm and overload, losing its true purpose of bringing value. Now that we understand the phases that should be followed for a KPI definition, let's analyze the different types of KPIs that we can identify.

Types of KPIs

There are different categories of KPIs, based on the value they bring and the way they are calculated. In this section, we aim to define each category and provide some examples of each one.

Strategic KPIs

Strategic KPIs are more high-level and usually the most important ones for organizations. These KPIs usually target a C-level or management audience and are defined in close collaboration with them. They represent the performance and health status of organizations and usually use a bigger time scope, spanning months, quarters, or semesters. Strategic KPIs are often more about monitoring progress or trends toward reaching long-term business objectives.

Some examples of strategic indicators widely used in most organizations across the world to evaluate financial performance are the following:

- **Earnings Before Interest, Taxes, Depreciation, and Amortization (EBITDA):** This KPI represents a complex financial concept that allows us to evaluate the profitability of our organization. EBITDA is represented by a monetary figure that can also be shown as a percentage if it's divided over the total revenue (EBITDA margin).

  We can calculate its value by using the following formula:

$$EBITDA = Net\ Income + Taxes + Interest\ Expense + Depreciation\ \&\ Amortization$$

Figure 10.3 — EBITDA formula

- **Return of Investment (ROI):** This KPI was already covered during our introductory chapter, and it represents the profit we obtain from an investment. ROI is represented using a percentage that shows the profit per EUR, $, or whatever currency we used for the investment.

  We can calculate its value by using the following formula:

$$ROI = \frac{Net\ investment\ gain}{Cost\ of\ investment} \times 100$$

Figure 10.4 — ROI formula

- **Revenue:** Revenue represents the income brought to a company by its business activities, before deducting other expenses such as taxes and operational costs. This KPI is represented as a monetary figure.

  It is calculated using the following formula:

$$Revenue = number\ of\ customers \times price\ of\ services$$

Figure 10.5 — Revenue formula

- **Cost of Goods Sold (COGS):** This KPI represents the costs that are derived from producing a product or service, such as materials used or personnel cost, among others. This KPI is represented as a monetary figure.

  We can calculate its value by using the following formula:

$$COGS = Beginning\ Inventory\ value + Value\ of\ purchases\ during\ the\ period - Ending\ Inventory\ value$$

Figure 10.6 — COGS formula

- **Gross profit margin**: To put it simply, this KPI represents the profit a company makes after deducting the cost of doing business. It is often shown as a percentage of net sales, and it can be calculated using the following formula:

$$Gross\ Profit\ Margin = \frac{Net\ Sales - COGS}{Net\ Sales} \times 100$$

Figure 10.7 — Gross profit margin formula

All these examples represent financial performance. If we analyze them closely, each one serves a specific purpose that is different from the rest, as they focus on specific components that are part of the overall financial performance of an organization.

Let's now jump to our next category: operational KPIs.

Operational KPIs

Operational KPIs are tools to evaluate the business's day-to-day operations and are usually more focused on the short term. These KPIs represent performance in specific operational areas. Instead of having bigger time frames, these KPIs are more granular, focusing on time windows of days, hours, or minutes.

Having this information to hand about specific areas eases the process of identifying and remediating possible issues or challenges that may surface during day-to-day operations or business processes.

Some examples of operational KPIs focused on different business areas are the following:

- **Average ticket resolution time**: For companies that rely on ticketing systems to provide support to their clients, this KPI represents how much time it takes for a support agent to solve a ticket, from its creation date to its resolution. This KPI allows us to evaluate some key aspects of our operations such as agents' performance, the efficiency of our business processes, and customer satisfaction, among others. This KPI is expressed in minutes, hours, or days and only requires that the creation dates and resolution dates of tickets are recorded somewhere.

We can use the following formula to calculate this KPI:

$$Avg\ ticket\ resolution\ time = \frac{\Sigma\ resolution\ time\ per\ ticket}{number\ of\ tickets}$$

Figure 10.8 — Average ticket resolution time formula

- **Server downtime**: This KPI represents, in the IT infrastructure domain, the total time that a server is inactive or unavailable due to incidents, issues, and maintenance processes. It is represented in hours or minutes, and is used as a measure of the health of our infrastructure. This parameter is usually part of a **Service-Level Agreement (SLA)** that service providers agree with clients when providing infrastructure services.

- **Time-to-productivity**: This is another example of a widely used KPI. This KPI measures how much time it takes new hires to adapt and become fully productive in their new roles. It is a key HR KPI that is essential for hiring processes as it shows how well business onboarding processes and training processes work in organizations. It can be calculated as an average, using the data that is available with the following formula:

$$Avg \, time\text{-}to\text{-}productivity = \frac{\Sigma \, time\text{-}to\text{-}productivity \, per \, employee}{number \, of \, employees}$$

Figure 10.9 — Average time-to-productivity formula

- **Cost per lead**: For marketing companies, it measures how much is spent to acquire new leads; that is, potential new clients that could be converted into paying customers. Spending more should mean more leads and clients, and this KPI aims to identify these patterns in sales and business generation. This KPI is expressed as a monetary figure.

It can be calculated using the following formula:

$$CPL = \frac{marketing \, spend}{number \, of \, new \, leads}$$

Figure 10.10 — Cost per lead formula

As you can see from the examples, all these KPIs highlight specific information that is aimed at a specific business process. They still provide a lot of value but on a much lower, more operational level.

KPIs by functional area

Apart from the type of KPI, we can also group KPIs per **functional area**. These KPIs focus on areas such as finance, human resources, or marketing. Using this approach to grouping KPIs simplifies the interpretation for the targeted audience, making data from the same domain easier to read.

Coming back to the examples of strategic and operative KPIs, they could be categorized into the following functional areas:

- **Customer experience**: *Ticket resolution time*
- **IT Infrastructure**: *Server downtime*
- **Human resources**: *Time-to-productivity*
- **Marketing**: *Cost per lead*
- **Financial**: *Cost of goods sold, ROI, and gross profit margin*

We can also identify **cross-functional KPIs** that touch multiple areas and are usually the ones that have a bigger impact on organizations. An example of a cross-functional KPI is the strategic KPI *EBITDA*.

Leading/lagging

Apart from the KPIs that we have already presented, we can analyze our KPIs from a time-bound perspective.

We can consider **lagging KPIs** the ones that represent what has happened in the past; facts that cannot be changed.

But having this information at hand may change the future, as this information allows us to react and correct any problematic patterns we have detected.

As an example of this, let's imagine a possible scenario: we ran a marathon two years ago, and arrived in the second last position. The year after that, we also participated in the same race and came in last position. Having this information, the most probable outcome, should we decide to participate this year, is that we will arrive in one of the last positions of the race. For this analysis, we only employed lagging KPIs.

But let's again imagine and say that we decide to act on this information, and for this year's marathon we train properly, rest more than eight hours a day, and plan a diet to maximize our results. Putting this plan into motion does not automatically mean that we will get a better position in the race, but the chances are higher.

Let's say we begin measuring additional metrics, such as body weight, muscle, and fat percentages and ingested kilocalories per day, to track our progress toward this objective to be a better and healthier runner. By using leading KPIs, we eliminate the potential of viewer misinterpretation on possible future outcomes by providing quantifiable and measurable information.

What we've just discussed is a perfect example of a **leading KPI**. Leading KPIs seek to anticipate where we are going by adding additional information that can be used to detect patterns and tendencies behind our data.

As some examples of leading KPIs that represent the health of a population, we could identify the following KPIs, among others:

- *Average body mass index per person* (should be between healthy levels between underweight and overweigth)
- *Average hospital stays* (should be minimized)
- *Average fat ingestion per week in grams* (should be around 30% or 35% of total daily calories)

After having covered leading and lagging KPIs, we now move on to a methodology that we can use to align our goals with the company strategy.

Objectives and key results

Apart from covering KPIs, we also wanted to take this opportunity to describe a methodology that can be helpful for FinOps practices, which is **Objectives and Key Results (OKR)**. This methodology can be used to set goals in a collaborative way that encourages engagement and alignment between teams.

We consider that the concept of OKR can be easily applied alongside FinOps practices, as well as in other related projects such as journeys to the Cloud or migration projects.

The idea behind OKR methodology is for different teams to set out high-level objectives and refine them over time. This can foster collaboration and cross-team alignment to push organizations to the next level.

To introduce this methodology, we must first define the concept of OKR. An OKR consists of two components:

- **Objective**: A description of what we want to achieve. Objectives should be short and ideally engaging and motivating.

- **Key results**: A set of metrics or KPIs that can help us measure the progress toward the objective.

The idea of this methodology is to define *two to five key results per Objective*. Having more than three is not recommended to support engagement and make them easier to remember. Another point to keep in mind about the number of objectives is that it should never dictate or influence the objectives themselves. Having a lot of objectives may lead to losing focus and team burnout, while making prioritization a challenge. It is also perfectly fine to have one or two objectives as well, as long as they dictate really clear goals for teams or individuals to reach.

We can illustrate this idea in a graphical way via the following figure:

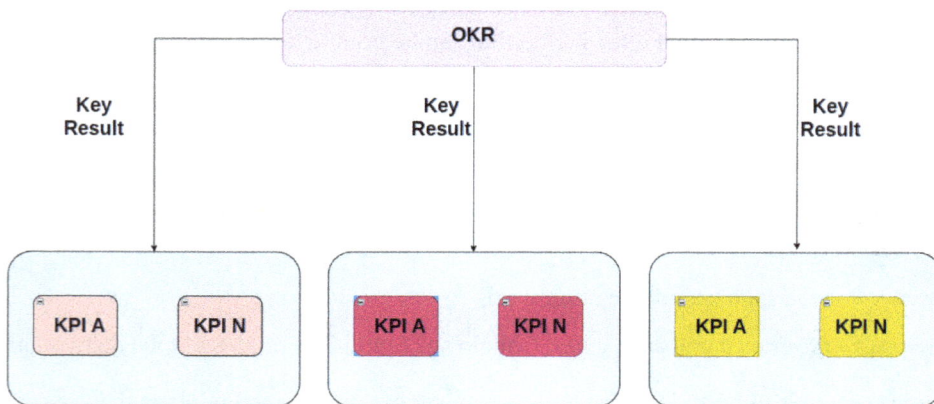

Figure 10.11 — OKR methodology

As we can see, each key result can use KPIs or metrics to allow us to measure our key result in reaching a goal.

As an example, let's say we define the following objective:

- **O**: Increase cost optimization in our IaaS Azure environments

For us, the high-level objective is clear. But we need to define what we consider as a *cost optimization increase*. We need to define key results for this Objective to do that. We could set out these goals:

- **KR1**: Implement *80%* of Reserved capacity in IaaS Production environments
- **KR2**: Implement *100%* of Hybrid benefit use in IaaS Production environments
- **KR3**: Reach *€50,000* yearly savings from cost optimization initiatives

The figures in the preceding example could be just the first iteration, and could be refined in the future by periodically revisiting and adjusting our Key results. Further in the future, even revisiting the Objective itself could lead to different interpretations or drive the key results to different focus areas, should we need it.

OKRs are often categorized into the following types:

- **Committed**: OKRs that are set to get a commitment from different teams to reach a goal. Committed goals often imply a stretch from the team trying to reach them, and they should be challenging but realistic. These OKRs can only be reached by achieving 100%. Commitments can be adapted and changed over time through a change process that should be decided beforehand. Priorities change in every organization, so having such change processes will help us align our OKRs with organizational objectives and keep them achievable.

- **Aspirational**: These KPIs are set to serve as an inspiration to reach a goal that is difficult but aspirational, setting the bar for success way higher. These OKRs can be more long term and are intended to motivate and generate engagement.

- **Learning**: In these OKRs, the goal is focused on learning instead of acting. For example, learning more about a specific business process, or increasing the information that we have on a specific topic. As an example, a thorough research in the topic of Unit economics could give a lot of interesting new insights about cloud financial behavior if we invest in it.

OKR methodology goal definition is often used alongside another goal-setting concept, which is **SMART** objectives:

- Specific: Target a specific goal that leaves no room for ambiguity.
- Measurable: We need to be able to measure and track our progress in objective terms.
- Attainable: Goals should be feasible. If they are not, teams will lose motivation fast and we won't get any closer to our goals.

- **Relevant:** We need goals that make an impact in our organization.

- **Time-bound:** We should define our objectives to be limited in time, or the teams engaged in them will lose interest and motivation. As an example, most companies define OKRs on a yearly basis.

Having these rules can help us set goals that will be more interesting and valuable for both the organization and the teams involved.

As you can see, the OKR methodology can be a great ally for FinOps practices. During the long road of FinOps adoption, we will need to keep different teams engaged and motivated, while keeping clear track of our progress toward specific goals. *OKRs can act in this case as a way to iterate over goals that we can set in advance, gluing teams together while increasing the visibility of the practice to the organization.* It will also help to put to practice KPIs and metrics already available from cloud providers (or newly defined ones) into attainable goals that will be more tangible for organizations.

With this basic introduction to OKRs in mind, let's reflect on the key differences between OKRs and KPIs, as summarized in the following simple table:

	OKRs	**KPIs**
Purpose	Motivational and more directional to reach a specific ambition	Increase performance or learn new insights
Focus	Based on organizational goals and aspirations	Based on historic results and organizational goals and aspirations
Timeline	Revised on a fixed period (e.g., quarterly or yearly) or fixed short term	Long-term measurements
Approach	Aggressive and bold	Measures operations and should prompt specific actions if the results are not on track
Scope	Broad	Usually narrow and specific

Table 10.4 — OKRs versus KPIs comparison

Taking into account these differences, our recommendation is to use a mix of OKRs and KPIs to both track and drive FinOps practices. By combining both approaches, we can engage the teams while measuring everything we need to the last detail, which will result in better results and FinOps visibility improvement overall.

With all these ideas in mind, let's jump to the next section, where we will cover how we can apply all the ideas we've covered in the FinOps domain.

Using KPIs for FinOps practices

We have already prepared the ground by introducing what a KPI is and presented, as an example, a FinOps KPI dedicated to rightsizing initiatives. We have also defined a methodology to design and implement our own KPIs alongside some more examples of KPIs in each category.

It is time to put all this knowledge to use in the FinOps domain. In this section, we will go through the whole process of defining a FinOps KPI, following the phases that were described in the previous sections. With these examples, we seek to achieve the most difficult part of the learning process, which is to put the ideas and concepts we have covered on paper into practice to make it truly useful for our readers.

To close this section, we will also provide other KPI examples in each of the categories to serve as a starting point to build up your own KPIs in the future.

Without further ado, let's jump into it by setting up the context for our KPI example.

Example of a FinOps KPI in Azure – region placement

In this example, we are going to define a KPI from start to finish, following the process that we previously defined. Let's begin with some short context.

Our FinOps team has detected a worrying increase in indirect costs, specifically data transfer costs between regions. This does not add up for the cloud operation teams, as our services should all be hosted in the same region, being a smaller company with not so many cloud resources.

To act on this issue that we detected and reduce these costs, we decided to set out a new cloud policy that states that all our cloud resources should be allocated in one specific region, let's say West Europe (Azure). Having this policy in place will ensure that we don't incur additional data transfer costs and will simplify our cloud operations overall. We will cover how we can enforce these policies in the next chapter, so we will leave that specific part aside for a moment.

After we have got an agreement on this policy with the cloud operations team and other involved parties, we need to try to find a KPI to measure our progress toward reaching this goal and achieving full compliance with our newly defined policy. Let's begin with the KPI definition.

Definition

We should start the definition phase by establishing a clear **owner** for this KPI. In this case, we can consider that this metric falls under the **FinOps lead** role responsibilities for simplicity.

This KPI will be part of our **operational KPIs**, as it is a KPI that our technical teams will use to oversee the compliance of this new policy we have defined.

The FinOps lead will be accountable for this KPI, even though they will be relying on FinOps technical teams to gather the required data and information to implement this KPI. He will also act as a driver to push this initiative across business units as much as possible.

This KPI will be tracked **monthly**, as we have decided it does not require a more granular time frame. Tracking this KPI monthly will also allow us to analyze its evolution over time.

Regarding the **format**, we can define this KPI as a percentage that can be calculated using the following formula:

$$\frac{Number\ of\ resources\ west\ Europe - Number\ of\ resources\ outside\ of\ west\ Europe}{total\ number\ of\ resources}$$

Figure 10.12 — Example of KPI formula

The first point to consider is how to get the data that we need. To do that, we need to identify the data sources and define the process required to extract this information at the necessary level of granularity. In some cases, we will need to initiate a discovery process to learn and analyze the possible ways that we could use to get this information. In this case, we could use either of the following data sources:

- **Azure Cost Management Invoices**: These have all the charges for all the services we use in the given region where these services are hosted. We could process this data to count the number of resources that are/are not in the desired region.
- **Azure Resource Graph**: We can get the information easily by doing a simple query in Azure Resource Graph that counts the resources in each region.

We will deep dive into how we can do this in the implementation phase.

In this case, we have decided to make the information of this KPI public for all the stakeholders and interested parties to see, as it does not include any sensitive information. This choice will simplify the implementation as well.

Last but not least, we should assign this KPI a unique identifier that follows our KPI naming convention, which is as follows: *{KPI Area}-{KPI Internal Category}-{Numeral}*.

To be able to assign a name to this KPI, we need to define its category and priority as well:

- The *priority* will be medium in this case. It is an important KPI for FinOps practices, but it does not have a high level of impact on our organization.
- We can categorize this KPI as an *infrastructure* KPI.

In addition to this, we can use a numeral based on how many KPIs are in this category; in this case, just this one. Taking this information into account, we could assign this KPI the following identifier: **FINOPS_INFRA_001**.

Implementation

Once our definition is ready, we need to begin the implementation of this KPI. The complexity of this phase can vary drastically between one KPI and another.

This implementation is usually carried out by a technical team with expertise in the data sources and/or resources involved in this KPI.

Let's come back to the options we identified and ponder how much effort each of them will require:

- Cost Management Invoices
- Azure Resource Graph

If we opt for **Azure Cost Management Invoices** processing, we'd need to do the following:

- Get the information from the cloud bills in Azure Cost Management. In this case, the best option would be to use Azure Cost Management Cost Exports to export this information to a Storage Account.
- Once this information is stored in an Azure Blob container, we need to process and filter the data to get the specific information required.
- After the information is there, we can publish it via Power BI or any other BI platform or report that can connect to our storage platform.

On the other hand, if we opt for **Azure Resource Graph** we would have a much simpler solution, with the following requirements:

- Define an Azure Resource Graph query to get the necessary information from our cloud resources. The query could be something like this:

```
resources
| summarize resourceNumber=count() by location
| project compliant
=iff(location=='westeurope','Yes','No'), resourceNumber
| summarize sum(resourceNumber) by compliant
```

This information can be gathered programmatically and put in any storage or database that can be connected to our BI platforms. Because we are also using Azure Resource Graph, we could consider a **Log Analytics Workbook** as well, which would be much easier to develop and for our KPI information to be presented. The downside of this Log Analytics Workbook is that we won't be able to keep any historic information on this KPI, as it only shows the current value of the KPI.

Keeping in mind both options, we opt for Azure Resource Graph as it seems to be simpler of the two to implement, choosing to save the KPI data in an Azure Storage Account with the timestamp of the data extraction that will be run monthly. Once the information is there, we can connect our BI platform to this storage platform and put this KPI in a newly created FinOps report with which we will track this KPI's progress.

The technical team implementing this KPI should also ensure, by using **platform and process monitoring**, that there are no failures in these data extraction processes.

The last step in this part of the process is to validate that the audience of this KPI has access to the reports and dashboards where it is presented.

Once all of this is done, we can move on to the analysis phase.

Analysis

Once the data is in place and we are to present it via dashboards or reports, we need to enter the analysis phase. In this phase, we need to analyze the data and the trends of the information presented by this specific KPI, to see if it makes sense and to ensure there are no errors.

The owner, in this case our FinOps lead, is the person ultimately responsible for this whole process and is the one validating that the information of this KPI is correct.

This particular KPI is as straightforward as it gets, but with other KPIs this phase is often essential to get additional insights and conclusions from the information that the KPI highlights, following the same idea of a learning OKR, of which the important aspect is learning new things about our businesses and processes.

For this specific example, our analysis is going to be simple and descriptive, and we can use this phase to analyze the region placement of our cloud resources and how many of them are not located in the cloud region that they should be.

Actions

Depending on the outcomes of the analysis phase, we may need to plan for **corrective actions** to act on what we have learned from this KPI information.

In this specific example, we could identify resources that do not follow our policy in one or more business units, due to miscommunication or any other reason. Regardless of the reason, we will need to collaborate with and support them to correct the situation.

In this specific example, moving resources from one region to another can be challenging, as not all resources support this. This will require proper analysis and some effort from technical teams, which should be prioritized according to this KPI's criticality. It may imply some application downtime and careful planning to minimize the impact on customers and/or users.

After the actions have been implemented, we need to collect fresh KPI information and demonstrate the impact of our corrective actions using this newly defined KPI.

All these phases can be revisited iteratively to improve our KPI definition and implementation and get more optimal and better information from our cloud resources and processes, again applying the ideas of **crawl, walk, and run**.

With that, we close this small example of how a KPI can be defined and implemented. To close this chapter, we will provide additional examples of each KPI category based on our experience and an analysis of the pain points, which we hope will serve you as a great starting point for your future business practices.

More FinOps examples

In this section, we will provide KPI examples in all the different categories and types of KPIs that we have covered throughout the chapter, from strategic KPIs, to operational, functional, and leading/lagging KPIs, that can be used to reflect the health of FinOps practices in many areas.

In addition to these FinOps-specific KPIs, FinOps practices can also help define other important KPIs related to cloud architecture and operations that will also support our cost optimization journey.

Keep in mind that you can also use the concept of **unit economics** to propose and define new KPIs to discover additional insights that can be hidden inside our cloud billing data.

Remember that, as a general requirement to be able to progress in KPI definition and implementation, it is essential to have defined a proper **naming convention and tagging strategy**. Without those in place, the tools you can use to show this key information will be much more limited, as we will lose key granularity levels such as business unit, region, and other parameters.

Without further ado, let's begin with strategic KPIs.

Strategic KPIs

From a FinOps perspective, strategic KPIs should show on a high level the progress of FinOps practices, measuring the savings that we generate and how optimal our cloud resources are throughout the business units of our organization.

Regarding non-FinOps KPIs, we are going to also name a few that can help increase the resource visibility of other key aspects such as modernization and cloud migration.

Some examples of strategic KPIs are as follows:

- **Cloud adoption**: When we are going through a journey to the cloud, one of the most common points we need to track is our progress in shifting our workloads to the cloud. Following this idea, we could define a KPI that represents the extent of cloud adoption with the following formula:

$$\frac{Total\ cloud\ resources}{Total\ cloud\ resources + Total\ on\ prem\ resources}$$

Figure 10.13 — Cloud adoption

This KPI format takes the form of a percentage. We can count as a resource any unit that is measurable, such as virtual machines, containers, and even processors and CPUs. The important aspect of this KPI is to have a measurable indicator to track cloud adoption. Logically, this KPI won't be useful at all for cloud-native organizations.

- **Modernization**: To get most out of the cloud, as we have explained throughout this book, fewer and fewer managed services should be used, which means prioritizing PaaS, serverless, and cloud-native services over IaaS. What we truly want to express with this idea is the cloud maturity or the level of modernization when working in the cloud. To convert this abstract idea into an actual measurable indicator, we could use the following formulas:

$$IaaS\ usage = \frac{Total\ IaaS\ resources}{Total\ cloud\ resources} \qquad PaaS\ usage = \frac{Total\ PaaS\ resources}{Total\ cloud\ resources}$$

$$SaaS\ usage = \frac{Total\ SaaS\ resources}{Total\ cloud\ resources}$$

Figure 10.14 — IaaS, PaaS, and SaaS usage

In the preceding formula, **Total cloud resources** represents the sum of IaaS, PaaS, and SaaS resources. We could also consider dividing PaaS into its multiple flavors: BaaS, FaaS, CaaS, and any other way you like. Once we have this information, we will be able to create a pie chart that aggregates this information to summarize the modernization KPI:

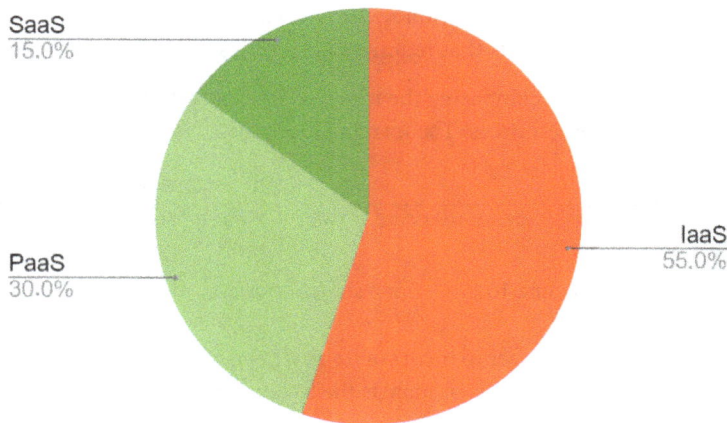

Figure 10.15 — Modernization KPI

We know that this KPI doesn't cover all aspects of modernization, but it can be used as a starting point to discover better ways of representing this concept.

- **Use of Infrastructure-as-Code**: Continuing with the modernization idea, we may want to know the degree of application of the IaC methodology, especially in big companies with multiple and diverse business units and departments. Automation means reducing administrative overhead and optimizing the time taken for our IT operations, so it is definitely an important indicator.

A possible way to implement this KPI is to flag our cloud resources with a specific tag , such as `source: Terraform`, to indicate resources that have been deployed from Terraform, Bicep, CloudFormation, or any other IaC language. We could also flag resources deployed manually from the CLI or Cloud Console with `source: manual`. Having this essential tag will allow us to calculate the following formula:

$$IaC\,adoption = \frac{Resources\ deployed\ with\ IaC}{Resources\ deployed\ from\ IaC + Resources\ created\ manually}$$

Figure 10.16 — IaC adoption

- **Open source adoption**: There are organizations where open source software is the norm by policy, to try to reduce license costs as much as possible. Open source is also very common in cloud-native, containerized, and Linux environments, with software such as OpenShift and Kubernetes being used widely in a lot of organizations. For organizations with such a policy, it is useful to track how many resources there are that use open source software compared to licensed software. This KPI can serve two purposes: on one hand, it will track the progress toward implementing this company policy, and also helps track the cost optimization that results from eliminating products that require paid licenses over time.

We can calculate this KPI by comparing the number of resources that make use of proprietary licenses over open source resources. The formula looks like this:

$$Open\text{-}source\ adoption = \frac{Open\text{-}source\ resources}{Open\text{-}source\ resources + Proprietary\ licensed\ products}$$

Figure 10.17 — Open source adoption

In addition to these examples, remember to incorporate as strategic KPIs the financial KPIs that were covered in *Chapter 5* into your FinOps KPIs, such as the following:

- **Total costs, savings, and cloud waste**
- **Year-over-year, month-over-month, month-to-date, year-to-date**
- **Potential savings per initiative**
- **Unit economics: cloud costs per customer**

With these ideas and examples in mind, let's move on to operational KPIs, which measure much more specific and day-to-day metrics.

Operational KPIs

Operational KPIs in the FinOps domain can provide more granular information that different teams can use in their day-to-day operations to monitor non-optimal configurations across our cloud resources, as well as enabling the tracking of different initiatives carried out as part of FinOps practices.

Apart from KPIs that purely belong to FinOps practices, we will also propose other key KPIs that should be considered, as they will provide additional information and insights on specific technical parameters and configurations.

Some examples of strategic KPIs are the following:

- **Virtual machines per family, vCPUs, operating systems**: For IaaS-centered environments, there are some parameters that are essential to oversee throughout our business units and environments:

  - **Instance family**: Awareness of the distribution of virtual machines across the different families that are offered (memory-optimized, compute-optimized, and other similar families) is essential to analyze the usage of our virtual machines across environments. As described in *Chapter 6*, we should have a governance plan including **virtual machine family standardization**, which should include the recommended families per environment depending on the use case. Having this governance plan will enable both cost optimization and appropriate sizing of virtual machines. This KPI can help to provide the requisite information to oversee the distribution we have and identify any room for improvement if present.

- **VCPU**: Following the same idea as in the previous point, knowing how many virtual machines we have with 2, 4, or 8 vCPUs will give us information on the types of workloads we run on our virtual machines, which can be also used as an indicator in analyzing **virtual machine rightsizing**.

- **Operating system**: In the same manner, the operating system used on our virtual machines, or any other compute resource that has an OS runtime, can be invaluable for **licensing optimization** exercises, while adding a high-level view that is also essential from an architectural point of view.

- **Use of a BYOL licensing model**: As we also explained during *Chapter 7,* the **Bring-Your-Own-License (BYOL)** model is key to attaining cost optimization in the cloud. With this KPI, we can count the number of resources that are making use of this key feature when available, which will help us track this license optimization initiative if proposed.

- **Usage of resources (CPU, RAM, network, etc.)**: Analyzing how much of the provisioned compute we are actually using across our cloud resources is another key essential piece of information we need to bring to light in any **resource rightsizing** exercise. Having this information together in one place will allow us to detect the virtual machines with the highest cost and the lowest resource usage, which are the ones where we should focus the rightsizing exercise first to get the highest savings possible. This information can also be useful to analyze whether any of the virtual machines has been assigned a virtual machine size that is not adequate, resulting in almost constant full CPU, memory, or network usage, which can hinder the performance of running applications and workloads.

- **Uptime of virtual machines**: During *Chapter 6,* we reviewed how the **shutdown of virtual machines during off hours** can have a huge impact on overall costs. Using this KPI, we can track the total hours that every virtual machine is powered on, which will allow us to determine the extent to which teams shut down their virtual machines when not in use. From our past experience, this information is not available from cloud providers by default, but it is easily obtainable by doing a few calculations using the information we get from monitoring tools such as **Azure Monitor** or **AWS CloudWatch**.

- **Use of autoscaling in resources that support it**: Autoscaling initiatives are also key to creating elastic, cost-optimized environments that adapt to demand over time and only use the resources that are needed.

- **Reserved Instances**: Purchasing Reserved Instances or Saving Plans can yield greater savings if done right, as we covered in *Chapter 6*. It allows us to get a great discount on some services in exchange for a long-term commitment, but can also be a double-edged sword and generate waste rapidly if not properly managed. To fully benefit from this purchasing model there are some key KPIs that we need to incorporate into our operational dashboards. Let's take their definitions from *Chapter 6*:

  - **Reserved Instance utilization**: This KPI reflects, as a percentage, out of all the hours that the reservation has been paid for, how many hours a virtual machine benefited from it. This KPI should be 100% at all times.

  - **Reservation coverage**: This KPI reflects the percentage of Reserved Instances use out of all the hours of virtual machines used throughout your environments. This value strongly depends on your reservation strategy (what environments to reserve, on which conditions, and to what extent), and it is really useful to track how Reservations are used across an organization.

  Both of these KPIs should have as high values as possible (especially utilization, which should be 100% to avoid cloud waste in the form of paying for RIs that are not essentially used). Both KPIs can be tracked directly from cloud providers and are essential for Reserved Instances governance.

- **Tagging and convention usage**: It should be already clear how important it is for organizations to have a proper naming convention and tagging strategies in place. Considering this point, it makes sense to also define a KPI to track each resource's adherence to both tagging and naming convention strategies. Once we have this information available, we should use this KPI to try to correct non-compliant resources and achieve full alignment with our defined strategies.

With these examples, we complete this section on operational KPIs. As we recommend throughout the book, please also review the different operational KPIs that are offered by cloud providers as part of built-in dashboards or reports such as **AWS CUDOS**, which will help to complete your understanding of these examples. Let's jump to our next section, which focuses mainly on functional KPIs.

KPIs by functional area

This concept can be used in the FinOps domain to create different subcategories aligned with FinOps functions (we will cover this concept in the next chapter) to properly divide and distribute our KPIs in a logical way. This will help to better organize the information, creating an optimal experience for the audience that consumes our dashboards and reports.

Some examples of possible areas we could use to organize our KPIs, alongside some KPI examples that can be included in each category, are as follows:

- **Financials**:

  - Invoiced spend

  - Amortized spend

- Budget versus actual spend
- Savings attained
- Cloud waste
- Potential savings per initiative

- **Cloud governance**:

  - % of tagging policy adherence
  - % of naming convention adherence
  - IaC adoption
  - Open source adoption
  - Modernization

- **Cloud Operations**:

  - Number of orphaned resources
  - Number of resources per region
  - Number of virtual machines per family
  - Number of virtual machines per OS
  - Number of virtual machines per CPU
  - % of BYOL in cloud resources

- **Reserved Instances**:

  - Reserved instance utilization
  - Reserved instance coverage

With these categorization ideas in mind, let's move on to some examples of leading and lagging KPIs.

Leading/lagging

As an example of leading and lagging KPIs, we can extend the cloud adoption KPI that we defined at the beginning of this section.

This KPI is represented with the following formula:

$$\frac{Total\ cloud\ resources}{Total\ cloud\ resources + Total\ on\ prem\ resources}$$

Figure 10.18 — Cloud usage formula

This KPI only represents *how many of our resources we have migrated to the cloud*, but that is not the full definition of what cloud adoption is. Cloud adoption should also include other aspects such as cloud awareness and training to fully reflect how cloud-ready a company is.

This KPI is also a lagging KPI, as it presents data that represents what has already happened. Following the idea of leading KPIs, we could add some additional KPIs to help us improve overall cloud adoption, as follows:

- **% of IT budget dedicated to the cloud**
- **% of applications hosted in the cloud** (this will add another point of view not purely focused on resources)
- **% of new projects hosted in the cloud**
- **Number of cloud certifications obtained across the different technical teams**
- **Cloud training investment**

All these new inputs of information will add new dimensions and insights about cloud adoption. All of these inputs and the KPI itself will help us detect patterns and analyze the impact of the actions we take related to cloud adoption, which can help provide a more complete view to measure how this key KPI will behave in the future.

With these examples, we hope to have laid the basis for your future work in this area. In our experience, each FinOps journey is different, and it is up to you to select the initiatives that are most appropriate for your organization. Based on the initiatives selected, you also then define the KPIs to be used to track their progress.

Summary

In this chapter, we have covered what a KPI is and how it can help us convert perceptions into measurable indicators that we can use to track initiatives not limited to cost optimization. From high-level KPIs to low-level operational KPIs, we have learned how we can get key insights into many critical pain points for organizations that will help us on our journey to cost optimization by defining a step-by-step process to properly design and implement long-lasting KPIs that will make an impact.

Alongside KPIs, we have also covered the OKR methodology, which can help us set proper goals tied to our KPIs to foster team collaboration and raise motivation.

To close the chapter, we have tried to provide as many examples as possible to give our readers a practical approach that can actually be used in real life.

In the next chapter, we will venture into FinOps roles and processes, which is an essential aspect to grasp to prevent FinOps practices becoming a one-time exercise. Throughout the next chapter, we will also explain how to define a FinOps operating model from ground zero that will take the practice to new levels of quality and visibility.

11

Defining New FinOps Roles and Processes

Throughout the previous chapters, we have provided a lot of content to help you achieve success in FinOps domain across its different pillars: **inform**, **optimize**, and **operate**.

But there is one key question that remains unanswered, and that is: *how do I implement all of this in my organization?*

We have seen so many cases of companies that have tried to get into FinOps practices by performing an assessment or implementing a proof of concept in a specific business unit, getting different degrees of success. After the assessment is completed, due to lack of momentum, resources, or sponsorship, FinOps practices often become a one-time exercise instead of an iterative process. This situation usually leads to organizations falling into the same cost-optimization bad practices over and over again, repeating mistakes that were made in the past.

There is only one answer to this big challenge, and that is to overcome it by implementing proper governance by defining and implementing a proper operating model for FinOps practices, which will ensure that best practices and initiatives permeate into the day-to-day workings of organizations for long-lasting cost optimization and success. This is what we are going to cover in this chapter.

In this chapter, we will cover the following topics:

- Target operating model and FinOps
- FinOps operational model

Target operating model and FinOps

A **Target Operating Model (TOM)** is a strategic framework that seeks to define what a company is, essentially describing the vision of an organization and how it achieves its business goals.

Even though there are different approaches and definitions, a TOM usually includes the following core components:

- **Processes**
- **Technology**
- **Organization structure**
- **People and skills**
- **Governance and decision-making**
- **Culture and leadership**
- **Customer experience**

FinOps practices have an impact on a lot of areas that are defined in the TOM, such as processes, technologies, people, and skills. To properly implement FinOps in an organization, we have to change how an organization behaves in some areas, mainly the parts of the TOM that are related to the cloud and technology. As you can already imagine, this is no small task, and entails many challenges along the way.

As an example, we will use **Deloitte Luxembourg operating model**, which is often presented as an example of what a TOM looks like. It is represented in the following diagram:

Figure 11.1 — Target operating model example

In the preceding figure, we see the operating model represented in the form of a pyramid, in which the lower layers are the means to implement and achieve the higher layers. We can identify the following layers:

- **Vision and strategy**: What does the company do? It should include the products and services offered by the organization and the channels that are used, as well as other key strategic aspects of how the company operates and does business.

- **Target operating model**: The principles on which the company is based, as well as how it operates to implement the vision and strategy.

- **Organization design**: Organizational structures, as well as roles and responsibilities. Other key aspects such as KPIs are also included here:

  - **Technology**: Technologies and platforms supporting the products and services that the company offers

  - **Process**: Processes that support the business objectives

  - **People**: The level of capabilities and skills of the people required to reach the given objectives

- **Governance and reporting**: Reporting requirements and other governance measures used to run the organization in an efficient and effective manner.

All these components essentially transform the abstract concept of what an organization is into how it is implemented. For the purpose of this book, we are not going any deeper into how TOMs are defined and implemented. But we can take the same concept here and ask ourselves some key questions regarding FinOps to try to also transform the abstract idea of FinOps into reality.

In the introductory chapter, we explained **why** FinOps is important and **what** it is, even though on this second question we only scratched the surface. In this chapter, we aim to give a more complete answer to this question and also work on another fundamental question, **how?**

Before considering the way to do all of this, we should think about how a concept as abstract as FinOps is going to be implemented in reality. Different organizations have different needs and challenges, and therefore there is not one single way to implement FinOps in an organization.

Apart from thinking about how we are going to do it, we should also think about what the points are that we aim to work on with FinOps practices and how it's going to impact our organization.

All these questions can help us build a smaller-scale operating model purely focused on FinOps that describes what the practice is for us and how it works. We will cover this in the following section.

FinOps operational model

FinOps success is tied to the questions raised in the preceding section. We need to keep in mind that our stakeholders and sponsors will need those questions answered, and many more, before they agree on any serious commitment (in the form of budget, resources, and support) to the practice.

By thoroughly thinking about and answering these questions, we make sure to properly design and define the foundations of our practice, and essentially create a **FinOps operating model** that dictates how FinOps delivers value. This is the work that we are going to be covering throughout this chapter. Therefore, our FinOps operating model should at least include the following:

- **Organization model**
- **Rollout and execution plan**
- **Functions, capabilities, and processes**
- **Roles and responsibilities**
- **Governance**

Let's begin dissecting all of these components one step at a time. To start, in the next section, we will describe the different models of FinOps implementation that can be used.

Organizational model

There are many flavors of FinOps and many approaches we can take to implement the practices across an organization.

In this section, we aim to cover different organizational models that can work from our experience. We have primarily identified the following:

- **Centralized FinOps**
- **FinOps as a shared service**
- **Decentralized FinOps**
- Other models:
  - **FinOps assessments**
  - **FinOps as a proof of concept**

Keep in mind that there are other FinOps models that can work in specific situations, as every organization has their own specific needs.

Let's analyze how these different models are implemented and their key takeaways.

Centralized FinOps

In a centralized FinOps model, FinOps is a central function responsible for the governance, control, and implementation of all FinOps initiatives and capabilities across an organization.

This model is common in organizations that already have a **Cloud Center of Excellence (CCoE)** or are willing to build one. CCoE functions are mainly the following:

- Centralized cloud governance and enterprise architecture functions
- Creation of cloud guidelines, best practices, and governance policies
- Assistance and oversight of group projects, ensuring that best practices and standards are followed
- Furthering of cloud adoption and modernization across the organization

Organizations can benefit a great deal from having this central CCoE function to ensure standardization, governance, and adherence to company policies. Taking into account the role that a CCoE plays in a company, in our opinion, it would make total sense for FinOps practitioners to be part of this central team, but solely focused on cloud cost optimization.

It is also possible for organizations that lack sufficient resources in their business units, or smaller companies in general, to adopt this model, as it requires fewer resources than having specific FinOps resources in each business unit. We can summarize this model with the following diagram:

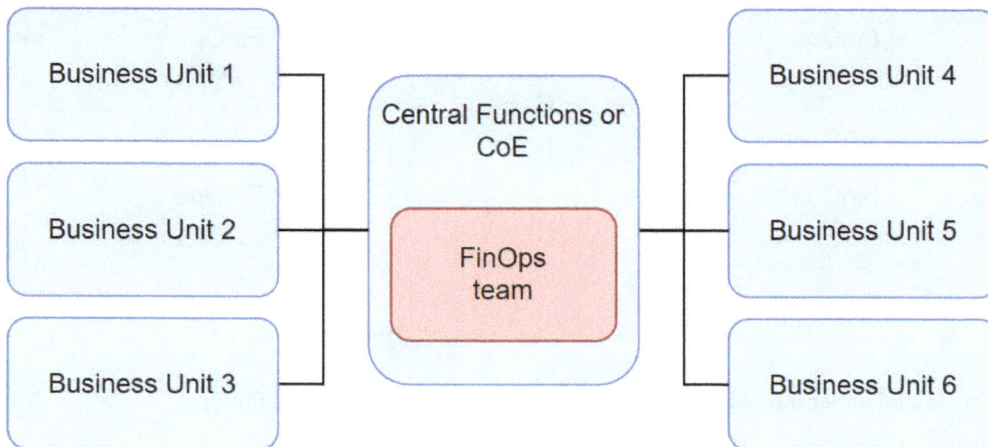

Figure 11.2 — Centralized FinOps

Now the centralized FinOps model is understood, let's jump to the next one, which is FinOps as a shared service.

FinOps as a shared service

In this model, FinOps is implemented in each business unit with local resources. The capabilities are distributed in the following manner:

- **Local FinOps teams** are in charge of FinOps practices in their business unit. This means that they implement FinOps initiatives and handle cost optimization from a local perspective.

- The **Central FinOps function** handles the centralization and standardization of FinOps practices. It ensures governance, control, the setting up of central cost optimization policies, and asset generation, including documentation or training, for business units to use. This is the part of FinOps that is considered a shared service across business units.

In this model, the central function usually also reports on FinOps adoption throughout the organization to management and stakeholders and is also in charge of pushing different business units for cost optimization. We can summarize this model with the following diagram:

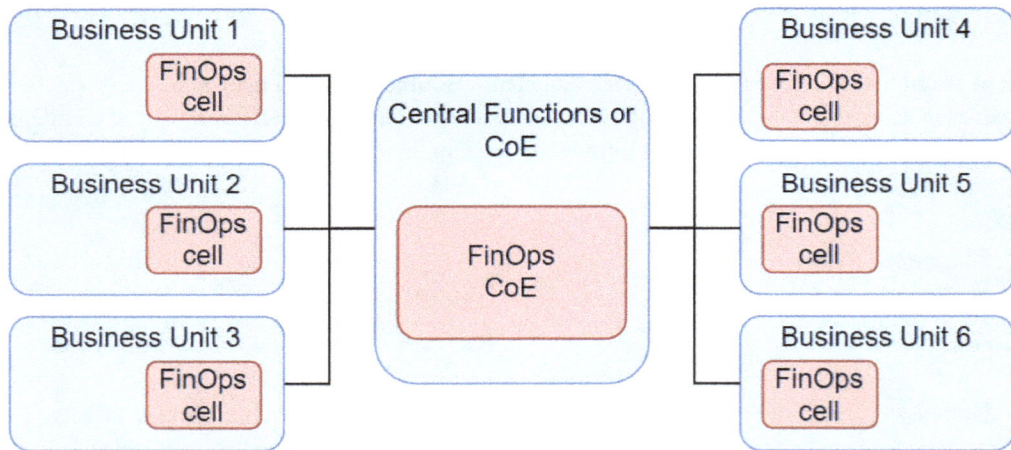

Figure 11.3 — FinOps as a shared service

With this model under our belts, let's examine the next one: decentralized FinOps.

Decentralized FinOps

Lastly, in a decentralized model, FinOps is part of each business unit without a central function to harmonize the practice. The capabilities are distributed across different operational teams.

In this model, standardization, governance, and tracking of overall FinOps adoption is definitely a challenge. On the other hand, this organizational model adapts better to each business unit's needs and it's easier to implement in organizations whose business units act more independently. We can summarize this model with the following diagram:

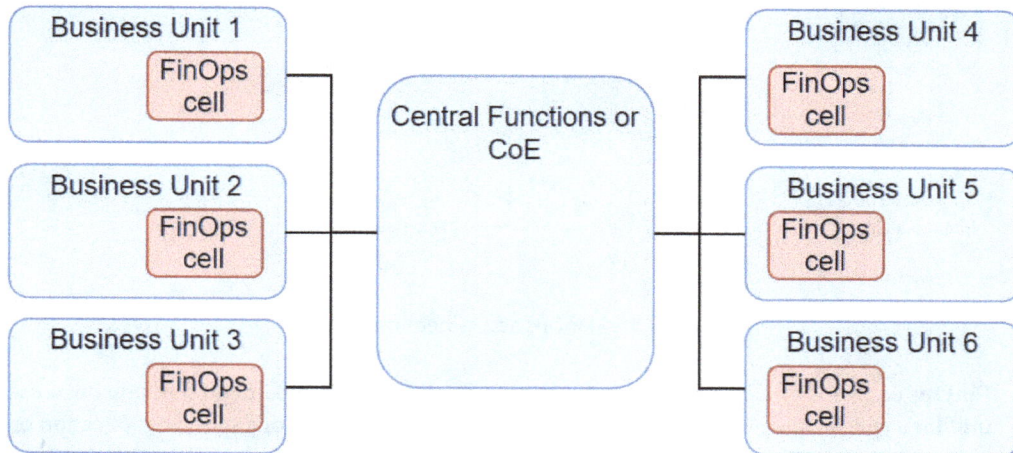

Figure 11.4 — Decentralized FinOps

In addition to these models, there are other models that can also work in more specific situations. This is what we will cover in the next section.

Other models

Apart from the standard organization models, we have other flavors of FinOps that can work in specific situations. These models are ideal for organizations that are not ready to fully commit to FinOps practices due to a lack of resources, sponsorship, or commitment from management.

In these situations, the best way to fuel FinOps practices and get funding is to demonstrate what FinOps brings to the table and how the organization will benefit from its application.

We have the following models for these specific situations:

- **FinOps assessment model**: In this model, a group of FinOps experts perform an assessment on the current status of an organization. After the assessment is done, the results are presented to management for consideration. In this organizational model, it is essential in our experience for FinOps practitioners to identify and propose applicable FinOps initiatives to optimize and improve cost control, and the corresponding estimated potential savings. This model is often offered by consultancy companies almost for free to sell FinOps to their different clients and generate buzz around the practice. This model is represented in the following diagram:

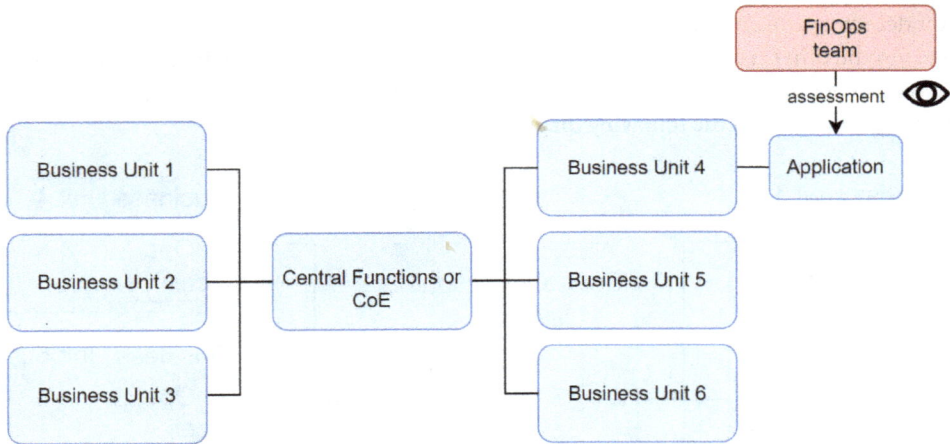

Figure 11.5 — FinOps assessment model

- **FinOps as Proof of Concept (PoC)**: In this model, FinOps is implemented in one business unit for a specific purpose, such as to optimize and reduce the costs of a specific application or system. The idea behind this model is to test-drive the practice before its full implementation in a company. After the PoC is done, results are presented to management. Please keep in mind that, if after an assessment, the initiatives proposed are not implemented and followed through, the PoC won't contribute in any way to FinOps maturity and cost optimization, and the resources that have been dedicated to it will mostly be lost. In addition to this, a single assessment will only be valid on a short-term basis, as it captures a specific moment in time and therefore the observations that are part of the assessment will no longer be valid after some time has passed. We can represent this model using the following diagram:

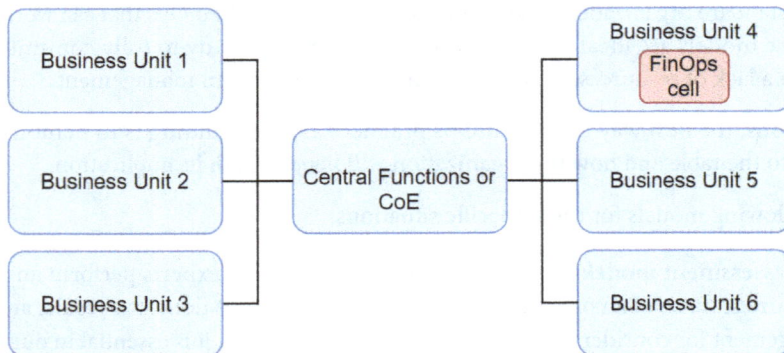

Figure 11.6 — FinOps as PoC

In both of these models, the ultimate objective is usually to bring down the barriers preventing the full implementation of the practice, or to get stakeholders' sponsorship for an implementation project, which can be challenging in some companies without some tangible proof or clear value proposal with actual results.

All these models that we have covered can be summed up in the following table:

Organizational model	Advantages	Disadvantages
Centralized	• Fewer resources needed with a centralized function • Easier standardization of policies and initiatives	• Lack of local FinOps resources increases governance complexity, as it depends on local and central relationships
Shared service	• Easier standardization of policies and initiatives • Easier governance and communication by having local and central FinOps resources	• More resources needed
Decentralized	• Fewer resources needed • Could lead to faster results, as having no central functions may be more straightforward	• Difficult FinOps governance, cost optimization policy, and initiative standardization
FinOps as PoC	• Easier to implement to demonstrate FinOps possibilities • Reduced costs and fewer resources required	• Limited scope can lead to less impact • The future of FinOps can depend on the results of the PoC, which can be dangerous if the business unit has no room for optimization
FinOps assessment	• Easier to implement to demonstrate FinOps possibilities • Reduced cost and resources required	• Limited scope can lead to less impact • Assessment with no implementation could lead to lack of follow-up

Table 11.1 — FinOps organizational models

Taking all these models into consideration, let's now describe how to define a rollout plan for the practice to take off.

Rollout and execution plan

Now that the different organization models of FinOps have been examined, we need a plan to roll out the practice.

There are multiple phases of this rollout that are usually categorized into three big areas:

- **Plan and design**: In this phase, we think about the operating model and tailor the practice to our organization. We need to properly design a rollout plan for FinOps to be implemented, which we refer to as the *FinOps implementation plan* and treat like any other project, with its own milestones and different phases. Having this rollout plan defined as a formal and official project will increase its visibility. In this phase, we also need to think about how we can measure FinOps maturity across different business units. This is usually done via the definition and design of key KPIs and deliverables. Apart from these topics, we also need to consider the functions that are going to be part of the practice, as well as the processes and governance policies that we need to define to ensure its implementation.

- **Mobilization**: In this phase, we begin to implement the practice by mobilizing the different teams through training and proper communication. FinOps teams are introduced and begin work alongside financial and technical teams to build the practice. FinOps initiatives tracking processes, which will be covered in depth later in this chapter, should be already in place, even in a basic form, and showing results by this phase. During this phase, FinOps teams are incorporated into operational teams by adopting ITSM tools such as Jira, for example.

- **Scale and evolve**: Once the practice is rolling and the dust from its initial implementation has settled, we need to ensure that the concept of **Crawl, Walk, and Run** is followed. We need to continuously improve and iterate by using the different tracking methods that we should already have set up, reflecting on everything we have done up to now, and trying to improve our approach to cloud financial optimization. As the practice evolves and matures, we are able to plan for more ambitious initiatives that require greater resources and budget, not needing to limit FinOps work to short-term initiatives of cost optimization.

Covering all these phases and creating a detailed plan is essential for long-lasting impact. Make sure to work properly on each of them to tailor the rollout plan to your organization's specific needs.

Let's now jump to the next stage of defining our FinOps operating model, which is to properly design FinOps functions.

Functions, capabilities, and processes

During the first chapter of this book, we explained why FinOps is important and the pillars of FinOps practices we should work on.

After going through all three pillars' initiatives and details, we should also have a clearer picture of all the initiatives that comprise a successful FinOps practice.

To complete our FinOps operating model, we now need to properly establish and define the responsibilities of our FinOps teams, as well as other parties that may be involved in FinOps work.

For this process, there are three different components we need to define:

- **Functions**: Functions are the building blocks of our operating model. They describe, at a high level, the different organizational entities that are the actors of our FinOps practices.

- **Capabilities**: Capabilities are more like use cases, and they define operating processes that are comprised of people, tools, technologies, information, and other resources to provide a service. We can consider that a function encompasses one or more capabilities.

- **Processes**: Processes are a set of FinOps activities that enable a specific purpose in the FinOps domain. Processes define the steps required to achieve a specific goal.

Let's provide an example of each one:

	Function	**Capability**	**Process**
Name	FinOps central billing team	Showback management	Monthly showback report meeting
Description	In charge of showback and chargeback, as well as invoice processing to ensure there are no errors in any cloud bill across all the different business units.	Perform the required analysis of cloud costs for cost allocation to different business units by processing billing data and distributing the costs in a granular way.	Perform a monthly meeting to present the results of each month's showback exercise, where the different business units are presented a report. In this meeting, the different business units can ask questions related to billing to better understand what they are charged for and why.

Table 11.2 — Function, capability, and process example

Remember that **FinOps will define its own set of functions, roles, and processes, but it is also going to impact existing functions, roles, and processes in the organization**.

For example, let's say we define a new FinOps process to check if new resources being created are following our cost optimization initiatives. We propose that, after anything is deployed in our cloud environments, our central FinOps team needs to check whether or not it follows the organization's cost optimization policies and approve or deny them accordingly. By setting up this new process, we will impact existing capabilities (*cloud resources deployment*, for example) and functions (*cloud operations*).

This is why, in order to properly define a FinOps operational model, we need to also analyze and learn about the current *target operating model* to develop an understanding of the relationships between these components and the potential impacts.

With all this information, you should already grasp the magnitude of what we are proposing. Working on all these components and properly defining, designing, and documenting them can take months or years.

As always, a good approach to tackle all this work is to begin applying **Crawl, Walk, and Run** here as well. We can begin the journey by describing some functions at a really high level and iterate over our work to provide more detail on which capabilities and processes are part of each FinOps function, clearly separating these between *FinOps-owned* and *FinOps-impacted*.

We can't do this work for you, but we can provide some examples that have worked for us in the past to serve as a starting point for your journey onward. This is what we are going to do in the following sections.

Function examples

The following are some examples of FinOps functions that we could define:

- **FinOps lead**: This function should be the main driver for FinOps practices. It should bring teams together through collaboration and align FinOps practices with organizational needs and goals. This function is usually also accountable for FinOps governance, roles, and processes defined within the FinOps methodologies in the organization.

- **Cloud central billing**: This is the function that we used in the preceding example. This function's main responsibilities could be the following:

  - Handle the processing and reviewing of invoices

  - Define and implement chargeback and showback processes to calculate the corresponding costs per business unit

  - Create cost evolution dashboards and reports and collaborate with financial teams to design and implement FinOps KPIs

  - Cloud contracts and agreement management

- **Cloud optimization**: This function is more technical, and could take care of the following tasks:

  - Ensure that cost optimization initiatives and concepts are implemented throughout all the different business units. Through training, documentation, and the creation of assets such as reports, dashboards, and specific tools, this function should standardize and support the different business units in their journey toward cost optimization.

  - This function can also act as a watcher that assesses the different environments and detects non-optimized workloads.

  - Prioritization of cost optimization initiatives based on potential savings or any other criteria.

  - Design, implementation, and operation (if needed) of FinOps assets such as dashboards, reports, automation, or specific FinOps-developed applications.

  - Reserved Instances and Capacity central management.

  - Cost optimization assessments of specific business units on request.

- **Cloud forecasting**: This function is fully focused on estimating and forecasting and could take care of the following:

  - Cost estimation of cloud solutions and projects before their implementation. This information can be used to enrich the project proposal before the project begins.

  - Calculation of the potential savings of different FinOps initiatives.

  - Supporting the cloud optimization function in performing cost optimization assessments.

  - High-level cloud cost forecasting and budget management.

With these examples as a starting point, let's now jump to some capabilities that could be part of these functions.

Capabilities examples

Some examples of FinOps capabilities can be defined as follows:

- **Central Reserved Instances and Capacity management**: For organizations where Reserved Capacity and Instances are managed and purchased centrally, this function could take care of determining when to purchase new Reserved Instances or Saving Plans, and properly adjust and exchange existing ones if needed, based on existing organizational policies.

- **Central license management**: For organizations where licenses are managed centrally, this function could take care of determining the licensing needs, purchasing the licenses that are needed for all the business units, and handling the required administrative overhead related to this, such as renegotiation, renewals, and so on. Having centralized license management often results in better discounts due to volumes being higher. Having this function will also help with licensing cost optimization initiatives such as instituting the *bring-your-own-license* model.

- **Tagging strategy management**: This function is in charge of the tagging strategy, ensuring that is properly defined and implemented across the organization and notifying the corresponding teams about any resources or workloads that are not compliant. This capability should also take care of updates and changes in the strategy over time.

- **Naming convention strategy management**: This function is in charge of naming convention strategies, ensuring that they are properly defined and implemented across the organization, and notifying the corresponding teams about any resources or workloads that are not compliant. This capability should also take care of updates and changes in the strategy over time.

With these capabilities in mind, let's review some examples of FinOps processes that can help to ensure FinOps practices are followed up.

Processes examples

The following are examples of FinOps processes that we could set up to ensure that the capabilities are carried out:

- **Cloud cost evolution monthly review**: A monthly meeting in which a cost evolution report is shared, explaining the financial highlights (savings, cost increases or decreases, and so on) that are observed in the report. In this meeting, each team involved could explain the main milestones and relevant changes in their cloud workloads that explain shifts in cloud costs.

- **Cost optimization quarterly assessment**: The central FinOps team could carry out a quarterly assessment of the status of each business unit, analyzing possible cost optimization initiatives and determining the potential savings that could be attained by implementing these initiatives.

- **FinOps tracking weekly or bi-weekly meetings**: Operative meetings in which the cloud operations and FinOps teams can review the progress of the different ongoing FinOps initiatives. FinOps teams can also propose, as part of this meeting, new initiatives and offer cost optimization insights to support other teams.

- **FinOps workshops**: FinOps training sessions explaining specific FinOps initiatives, given by FinOps teams to other technical teams including the cloud operations and infrastructure teams.

Let's now move on to how we can put these functions, capabilities, and processes into practice by setting up clear responsibilities to existing or newly defined organizational roles.

Roles and responsibilities

Once our FinOps functions and capabilities are defined, we need to close the circle by appointing people to newly created **FinOps roles,** or add some new **responsibilities** to existing roles in the organization. These FinOps roles will be the vehicle to implement the functions and capabilities and will also be in charge of the processes that are defined within.

To assign responsibilities to our FinOps capabilities, even though there are other valid ways, we almost always use a **RACI matrix** due to its simplicity. A RACI matrix is a method to define and document the responsibilities attached to a role, and it comes in the form of a chart that represents which of four aspects a given role falls into in terms of their relationship to capability. These four aspects are **Responsible, Accountable, Consulted, and Informed (RACI)**:

- **Responsible (R)**: Designates a person or group that is assigned directly to a specific task. They are the one that does the work to make it happen. There can be multiple parties responsible for a task and every task should have at least one responsible party.

- **Accountable (A)**: Designates a person or group that delegates and reviews the work involved in a specific task. The role of an accountable person is to make sure that the task meets the expectations and goals defined for it. There is only one accountable person per task.

- **Consulted (C)**: Designates a person or group that is consulted to provide input, feedback, and opinions on a specific task. This party usually represents someone impacted by the task in one way or another. Consulted parties are sometimes considered additional approvers as well.

- **Informed (I)**: Designates a person or group that should be briefed on the status or progress of a specific task. An informed party does not decide in any way, but they need to know the task status because they are impacted by it in one way or the other. This role is usually taken on by management or leadership roles.

There are other alternatives to a RACI matrix, but this method is so well known across different teams and disciplines that it is generally the easiest to use in any company.

As for the roles that are going to be part of this RACI matrix, we should add *FinOps-owned* roles and *FinOps-impacted* roles here as well, just as we did with the functions for our operating model. For example, in our RACI matrix we could add *FinOps lead (FinOps-owned)* and other roles such as *cloud operations/financial teams/project teams (FinOps-impacted)*.

Keep in mind that, in organizations where FinOps resources are really limited, **one person or group can bear multiple roles**. This is normal and expected, and does not excuse the fact that all roles should have a proper definition, be tied to one or more capabilities, and have a list of responsibilities attached to them.

As an example, let's go back to our imaginary capability, *Showback Management*, and try to represent the different responsible parties using a RACI matrix:

Task	Central FinOps CoE technical team	Central FinOps lead	Central financial teams	Local cloud operations teams
Invoice processing for showback analysis	R	A	C	I
Results validation and report creation	R	A	I	I
Meeting set up	C	A, R	I	I

Table 11.3 — RACI matrix example

The following is a breakdown of some of the key takeaways from the preceding table to illustrate key decisions that we have made:

- The central financial teams are *consulted* on our invoice processing task to ensure that financial best practices are followed and all required information is present.

- Invoice processing is a Central FinOps CoE technical team *responsibility*. *Accountability* for this task is held by the central FinOps lead, who owns this task.

- The local cloud operations team is sent the report before the meeting when it's ready. Due to this fact, they are an *informed* party for this task.

- Setting up the meeting with the local cloud operations team is a task for which the Central FinOps lead is *accountable and responsible*, as they both set up the meeting and are the party ultimately responsible for this process as well. The Central FinOps CoE technical team is *consulted* to ensure availability on their side for the meeting.

This is how this process should look. We should now be able to do the following:

- Set up the list of tasks required from each capability

- Determine the different parties involved in FinOps work, including new FinOps roles that may need to be defined

- Assign and jointly agree onwhich party will be responsible of each of the tasks we identified on each capability, as well as how these tasks will be carried out

Now that we know how to properly set clear responsibilities and assign them to each mapped role, we can cover other aspects of governance that we can use to ensure the long-lasting impact of our FinOps work.

Governance

After everything we've covered in the previous sections, we should have a strong foundation on which to build our FinOps practices. After all this work, we need to ensure, by setting up additional controls, that all the initiatives that we have worked on in the *inform*, *operate*, and *optimize* pillars will continue.

We need our initiatives to be part of the organizational policies that are enforced and controlled in order to avoid falling into the same cost optimization pitfalls of the past.

To ensure proper governance, we should work on the following topics:

- **Tagging strategy enforcement**
- **Naming convention enforcement**
- **FinOps initiatives enforcement**
- **FinOps initiatives tracking and maturity**
- **Overall cloud strategy alignment**

All of these topics have been already mentioned throughout the book in different sections and chapters, but we want to reinforce the message here to improve our overall control and visibility of all these points.

Tagging strategy enforcement

The tagging strategy applied in your organization has a profound impact on FinOps practices, as we have learned from this book.

It is a key initiative that is used as the basis of key business processes related to cost allocation, such as showback and chargeback. Our invoices, cost reports, and dashboards are most probably using it as well.

Due to this fact, we need to enforce its use across all the cloud resources throughout our organization. We cannot let resources have incorrect or missing tags.

To this effect, we can use multiple tools offered by cloud providers to enforce specific conditions on our resources, including the following:

- **Azure Policy**
- **AWS Tag Policies and AWS Service Control Policies**
- **GCP Organization Policies**

We can combine these tools with **tagging inheritance** to avoid auditing all resources and just focus our policies on the tags that are entered by users or operational teams.

We recommend implementing these policies on the highest level possible on the hierarchy tree of our accounts, subscriptions, or projects, so they apply to as many resources as possible. As an additional recommendation, we also recommend these policies to be deployed using **Infrastructure as Code (IaC)** through **Terraform, AWS CloudFormation, Azure Bicep**, or **Azure Resource Manager**, a practice we often refer to as **policies-as-code**, to ensure consistency and automation in this process. After the policies are applied, we must ensure that all resources are compliant and act to follow up on non-compliant resources, which should be remediated as soon as possible.

As a side recommendation, if IaC is used to deploy resources, we could even *add additional checks as part of our CI/CD processes to validate whether the tags entered in the code of the resources to be deployed are correct and aligned with our tagging strategy.*

With this idea in mind, let's also tackle another challenge, which is how to enforce our naming convention strategy.

Naming convention enforcement

Naming convention is, in a similar way to tagging, another of the cornerstones of FinOps work and is essential for many things.

It is often used when tagging strategies are not yet in place, as a temporary method to more easily identify key information on our cloud resources. Naming conventions can be used in reports, dashboards, and other cloud assets that filter cloud resources. Having an incorrect naming convention may filter some resources out from these assets, leaving some resources invisible on them, which is not ideal.

We can also enforce its use by using the same cloud services and ideas that we presented in the previous section:

- **Policies for naming convention enforcement**
- **Naming convention checks as part of IaC CI/CD**

Naming conventions are way more immediate as they dictate how we see our resources even when listing them. It is essential not just for cost optimization, but for cloud governance in general.

Tagging and naming convention enforcement should be a package that goes together, and should be part of any FinOps implementation in any organization.

FinOps initiatives enforcement and tracking

When a FinOps initiative that is part of the optimize pillar is implemented in an organization, it usually yields some results in the form of savings. However, we need to consider the implementation as only a part of the story. From that moment on, this initiative will need to live on as an organizational policy that must be followed when new resources are created.

Let's say we propose a new initiative to eliminate high-performance disks in development and sandbox environments. We go through its implementation and get some savings along the way by changing some high-performance disks we found for cheaper ones that are better suited for use in non-production environments. After this implementation process, we need to ensure that no high-performance disks will be created in the future in sandbox or development environments.

The easiest way to prevent this from happening, as we have already covered in the tagging and naming convention sections previously, is to enforce specific policies targeting key conditions that our resources must meet in the form of policies that will deny the creation of resources not compliant with these conditions and raise non-compliance flags on existing resources.

All the cloud providers offer the possibility to create and set up policies to deny or audit the creation of resources that are not compliant with a set of rules. We can set up specific rules for our FinOps initiatives, preferably using **policies-as-code** as we already described, to track compliance with our FinOps initiatives across all of our resources.

In addition to policies, which we will use to enforce these organizational policies for cost optimization, we will also need to have **dashboards or reports to follow up and track progress in every FinOps initiative** via the different operational KPIs that we should already have defined and calculated.

As an example, we could combine the information from our KPIs and create a simple weekly report such as this one:

KPI id	KPI common name	Scope	% of compliance last week	% of compliance this week	Weight
FINOPS_INFRA_VM_1	% of virtual machines rightsized	ALL	10%	15% (+)	0.33
FINOPS_INFRA_VM_2	% of non-Premium Managed Disks	DEV/SBX	80%	70% (-)	0.33
FINOPS_INFRA_VM_3	% Hybrid benefit for SQL Server virtual machines	PRO	50%	50% (=)	0.33
Total			46.6%	45% (-)	

Table 11.4 — FinOps tracking report example

The following are some key takeaways from the table:

- We define the initiatives of FinOps as well as the scope they should apply to. This allows us to be more cautious and test-drive initiatives in environments that are not that critical.

- We measure the percentage of application or compliance of each initiative by dividing the following

    - **compliant resources/total number of resources**

- We need to **incorporate historic data** in some way to track the advance in each initiative. In this example, we only show the progress from last week, so we can compare. Ideally, we should store the evolution of these KPIs over time somewhere to analyze trends and be able to do proper tracking. Note that we **highlight each initiative's progress**, showing the ones that have improved, the ones with no advance at all, and the ones where our compliance has reduced from last week.

- This report can be used to measure our initiative's progress in each business unit and globally, creating the most granular view possible.

- Once our KPIs are calculated, we assign a **weight** to them. In this case, we have assigned the same weight to each initiative, but **we could assign different weights based on how big the potential savings are in each one**. Having a weight allows us to create a new KPI that represents all of our initiatives in one, measurable, objective KPI. We can use this idea to score the FinOps progress of each business unit by calculating their percentages separately.

With these ideas, we should be able to properly track and publish the adoption of FinOps initiatives across our organization in a more specific and technical way as we define and implement them.

From our experience, these dashboards and reports can really benefit from incorporating **potential savings** information into them. If we can calculate and add this information alongside each initiative in an automated manner, we make prioritizing initiatives a much easier exercise, and give ourselves more tools to push for specific initiatives that have bigger potential savings attached to them.

We must remember that, in addition to this more technical view, we should also include the **unit economics** concept into these FinOps tracking dashboards and reports, in order to add additional informational inputs that track specific key indicators such as **cost per vCPU, cost per GB, and IaaS and PaaS unit costs**. These additional inputs will provide a more abstract yet specific point of view that will be invaluable to our reports and dashboards audience.

With these different inputs and ideas, we should be able to create our own **FinOps operational and technical dashboards and reports** that should be shared across more technical teams and followed up on a weekly basis to assess the impact and progress of FinOps practices.

Summary

In this chapter, we tried to provide a complete governance methodology around FinOps, fully focused on trying to integrate cost optimization practices into the DNA of organizations, with long-lasting measures that will make an impact not only in the short term, but also in the long term.

We defined what a target operating model is and how it can help us understand where FinOps fits in organizations. We also saw how to align these ideas to create our own FinOps operating model, which will include all the components that are needed for the practice to work like well-oiled machinery.

We described in depth various FinOps organizational models as well as how to define a proper rollout plan for the practice. In addition to this, we went through the process of designing and implementing new functions, capabilities, processes, and roles to support the practice, as well as understanding the impact that cloud financial management can have on existing functions, capabilities, and processes. Furthermore, we added other governance initiatives that we can put in motion to enforce best practices defined within the practice.

With this chapter fully dedicated to governance, this book is finally coming to an end. In the next chapter, which will be purely hands-on and more on the technical side, we will put into practice the different concepts, best practices, and ideas that were presented throughout *Chapters 6, 7, 8,* and *9*. By using case studies based on our previous experiences, we will highlight and describe how to apply different cost optimization initiatives. During this process, we will also analyze the resulting impact on costs, as well as the savings that could be attained by applying each initiative.

Part 5:
Hands-On Cost Optimization with Real-Life Use Cases and More

Throughout this part, we will apply all that we've learned in previous chapters on real architecture designs and case studies for all the different management levels for cloud services offered in Azure, AWS, and GCP.

We will cover **Infrastructure as a Service (IaaS)**, **Platform as a Service (PaaS)**, serverless and containerized workloads, as well as **Software as a Service (SaaS)** solutions and cost optimization examples, to make use of and apply all the concepts introduced in previous chapters as part of FinOps practices.

In addition to this, we will wrap up the book by summarizing what we've learned so far, as well as analyzing key synergies with new emergent trends such as artificial intelligence, machine learning, and sustainability.

This part has the following chapters:

- *Chapter 12, Case Studies for Mastering Cost Optimization*
- *Chapter 13, Wrapping Up and Looking Ahead*

Case Studies for Cost Optimization

This chapter will be a practical implementation of some of the initiatives that we covered in Optimize foundation chapters, which are Chapters 6, 7, 8, and 9. This chapter requires you to have gone through these chapters to fully understand the concepts that we are going to use here as the basis of our cost optimization initiatives.

The idea is to present two fictional case studies or solutions, and then illustrate the train of thought and the proposal process of cost optimization initiatives that FinOps practitioners should follow when analyzing workloads to be optimized.

We have selected two cases that are very much like real-world scenarios, to cover IaaS and PaaS workloads as well as storage and database services.

Without further ado, let's get into it. In this chapter, we will cover the following topics:

- IaaS case study – multi-tiered application migrated to the cloud
- PaaS case study – storage, serverless, and database optimization

IaaS case study – multi-tiered application migrated to the cloud

In this case study we are going to analyze in detail a standard web application IaaS architecture, with a multi-tiered solution:

- A *web layer* dedicated to the frontend of the application, where the users connect to.
- An *app layer* dedicated to the backend of the application, where the business logic of the application runs.

- A *database layer* where the application stores all its data in a clustered database setup. The database is hosted in a two-node cluster.

- A *jumpbox* that is needed for the project development team to work on the servers.

For this application, we will need three different environments:

- *Development*: Used to test new application features and to work on new releases and bug fixes

- *Preproduction*: A copy of production that is used to perform contingency tests and key user testing before new releases

- *Production*: The environment used for real users connecting to the application

We assume that this solution has been implemented in an *on-premises setup*, and we are requested to migrate it to the cloud.

Let's add as an additional premise that *we are not able to modernize the application* (we cannot use PaaS resources) in any way, as the software vendor won't support any PaaS modernization for this solution.

This application is intended for *50 users*, and it's a business-critical application. This fact means that we need to have multiple servers in each layer that can guarantee the service will keep running in the event of a failure or an issue in any of the servers, which means providing the solution with **high availability** or making it **fault tolerant**.

Solution description

In our on-premises setup, we have the following servers in use:

Server	Layer	Cores	RAM	OS Disk Size	Data Disk Size	Operating System
Weblayer0	Web layer	2	8	128 GB	128 GB	Windows Server 2022
Weblayer1	Web layer	2	8	128 GB	128 GB	Windows Server 2022
Applayer0	App layer	4	16	128 GB	128 GB	Windows Server 2022
Applayer1	App layer	4	16	128 GB	128 GB	Windows Server 2022
Dblayer0	Database layer	4	64	128 GB	2 TB	Windows Server 2022 + SQL Server Enterprise 2022
Dblayer1	Database layer	4	64	128 GB	2 TB	Windows Server 2022 + SQL Server Enterprise 2022
Jumpbox0	Jumpboxes	2	8	128 GB	64 GB	Windows 10

Table 12.1 – Case Study 1 – on-premises servers

Let's begin by presenting a diagram depicting the complete solution:

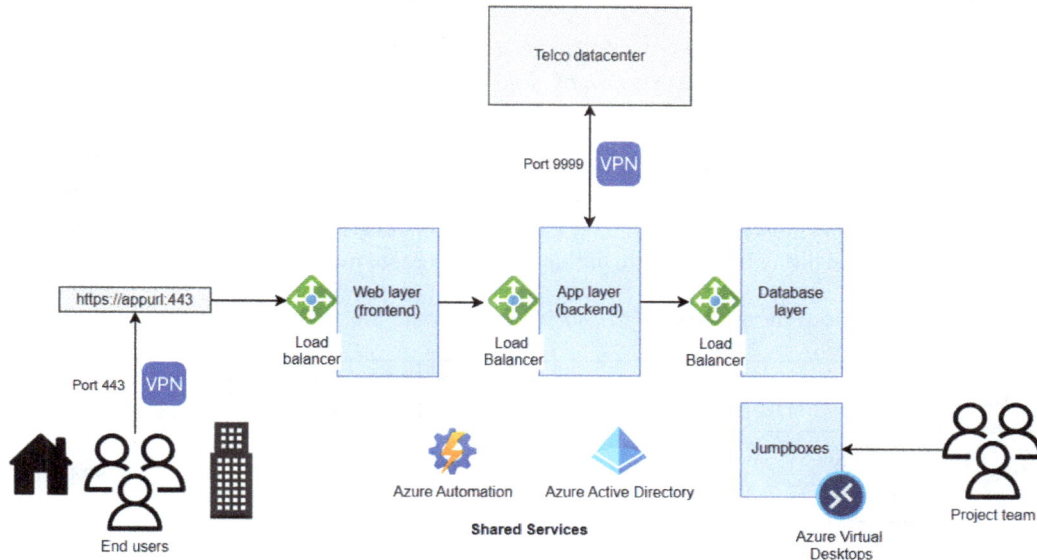

Figure 12.1 – IaaS Solution diagram

Let's do a quick one-to-one matching with Azure Virtual Machines counterparts. Here are some considerations:

- We have used Azure Pricing calculator (`https://azure.microsoft.com/es-es/pricing/calculator/`) using retail prices in July 2023 for the West Europe region. All cloud providers offer their own calculator so you can calculate the TCO of your solutions beforehand.

- Always choose the latest generation (as of now, it is version 5) VMs. They have the best price and performance as they use the latest generation processors. A good starting point for almost all workloads without specific needs is the general-purpose family. If we begin with this family and then review how our application is behaving, we can iteratively optimize by selecting the right size and checking how well it fits the solution, from both performance and cost perspectives.

- We have selected *Standard SSD Managed Disks* (mid-priced) since we think it will be sufficient for this solution. We have selected the **s** option of the VMs, which allow us to use Premium Disks if we need them in the future (this does not add any additional costs).

- RAM is often oversized in on-premises servers, especially in scenarios with SQL Server, where the database takes all the available RAM. We have selected a VM from the memory-optimized family (this is recommended for relational databases, among other things), E4s_v5, as the best match and starting point with 4 vCores and 32 GB of RAM.

- As we are going to set up a SQL Server cluster, we have selected the *SQL Server Enterprise license* as the starting point, as it was the license used on-premises as well.

- We assume that the users that are going to access our virtualization environment to connect the jumpbox have M365 Enterprise licenses, so the Windows 10 client license and AVD are free of charge in this case. For now, we assume we will be only using one jumpbox.

- Load balancers and VPN costs are going to be omitted from this exercise for simplicity because their associated costs are low and difficult to estimate in advance (e.g., how much traffic are the load balancers going to process?). We have also left out egress costs and data transfer costs because they are difficult to estimate in this imaginary case study.

With these considerations in mind, our VM lists in Azure would look like this:

Resources	Number of VMs	vCPUS	RAM	VM Size	Disk Size (OS Disk)	Disk Size (Data Disk)	Operating System	Price $/mo
Web layer	2	2	8	D2s_v5	E10 (128 GB)	E10 (128 GB)	Windows Server 2022	$340.62
App layer	2	4	16	D4s_v5	E10 (128 GB)	E10 (128 GB)	Windows Server 2022	$642.84
Database layer	2	4	32	E4s_v5	E10 (128 GB)	E40 (2,048 GB)	Windows Server 2022 + SQL Server Enterprise	$3,228.88
Jumpbox	2	2	8	D2s_v5	E10 (128 GB)	E6 (64 GB)	Windows 10	$98.35
						Total		$4,310.69

Table 12.2 – Case Study 1: Azure VMs selected

Of course, this is the price of our would-be production environment. If we assume the price will be the same for preproduction and development, we are at roughly **$13,000 per month**.

To calculate the savings that we obtain every step of our optimization process, we will use the following formula:

$$\% \, savings = \left(1 - \frac{final \, cost \, after \, optimization}{initial \, cost}\right) \times 100$$

Figure 12.2 – Savings formula

Using this as a starting point, we are going to try and optimize the solution by applying different initiatives that were described in Chapters 6, 7, 8, and 9 of this book.

Initiatives covered

We are going to begin with initiatives that are related to solution architecture and, when this part is clear and complete, we will proceed to other initiatives that let us generate some savings without changing the solution design, such as Reserved Capacity or Hybrid Benefit.

Let's begin by discussing what we can consider regarding cluster architecture.

Database cluster architecture

We could dedicate a lot of pages of analysis to this topic, and even consider moving to other DBMSs, but for the sake of simplicity we are going to implement one key initiative for database clusters, which is to use **Shared Disk cluster architecture**.

Our application services 50 users, and we need to ensure that if one server fails, it will be able to recover. With this number of users, we can be pretty confident that we don't need to distribute requests between servers to alleviate the load, so we could settle on an **active-passive** cluster if needed. This means that only one node will act as the primary at all times.

We can also check the requirements of SQL Server clusters regarding licensing:

- *Option 1 – Shared Nothing:* If we want to use active-active, we are limited to SQL Server Enterprise. SQL Server Standard only supports using SQL Server Basic Availability Groups that, among other things, have no read access on the secondary replica. This fact, and the limitation to one database, forces us to use SQL Server Enterprise.

- *Option 2 – Shared Disk:* If we choose to go with active-passive, we can use a **Windows Failover Cluster (WCF)** with a Shared Disk. By choosing this cluster, we can avoid the big price tag of SQL Server Enterprise, and we will also save by having only one big disk for our cluster.

Taking into account this fact and, looking at the size of the database data disk (2 TB), we consider it a good idea to propose a Shared Disk architecture for this cluster. If we apply this idea, instead of having one 2 TB disk per node and replicating data between nodes (active-active), we will have just one 2 TB disk and at any moment in time, only one node will be active, thus saving the cost of one entire 2 TB disk and the data transfer charges between nodes.

The result of this initiative can be seen in the following table:

Option	Disk Size (OS Disk LRS)	Disk Size (Data Disk LRS)	Operating System	Price $/mo
Option 1: Shared Nothing	2x E10 (128 GB)	2x E40 (2,048 GB)	Windows Server + SQL Server Enterprise	$3,228.88
Option 2: Shared Disk	2x E10 (128 GB)	**1x E40 (2,048 GB)**	**Windows server + SQL Server Standard**	$1,469.28
			Savings obtained	**$1,759.6**

Table 12.3 – Shared Disk and Shared Nothing comparison

By changing the cluster to this Shared Disks architecture, we have been able to reduce the number of large disks of the solution from 2 to 1, as well as downgrade to SQL Server Standard licensing. In doing so, we have been able to reduce the total cost of the database layer by *54.5%*.

For sure, we are losing some features by selecting SQL Server Standard licenses, but as per our assessment, the features offered are enough for this application, and we can still ensure high availability while having a cost-optimized solution.

This is how TCO looks after the changes to the environment:

Resources	Number of VMs	Cores	RAM	VM Size	Disk Size (OS Disk)	Disk Size (Data Disk)	Operating System	Price $/mo
Web layer	2	2	8	D2s_v5	E10 (128 GB)	E10 (128 GB)	Windows Server 2022	$340.62
App layer	2	4	16	D4s_v5	E10 (128 GB)	E10 (128 GB)	Windows Server 2022	$642.84
Database layer	2	4	32	E4s_v5	**E10 (128 GB)**	**E40 (2,048 GB)**	**Windows Server 2022 + SQL Server Enterprise**	**$1,469.28**
Jumpbox	2	2	8	D2s_v5	E10 (128 GB)	E6 (64 GB)	Windows 10	$98.35
							Total	**$2,551.09**

Table 12.4 – VM costs after Shared Disk clustering

Let's do a quick recap of the original and current total costs:

Environment	Original Cost after VM Sizing	Cost with Shared Disk DB
Development	$4,310.69	**$2,551.09**
Preproduction	$4,310.69	**$2,551.09**
Production	$4,310.69	**$2,551.09**
Total Cost	$12,932.07	$7,653.27
Total savings compared to original costs		**40.82%**

Table 12.5– Total costs of VMs after Shared Disk clustering

As we can see from the table, by applying this cluster architecture change, we have been able to reduce the total costs dramatically. With this big change in TCO, let's move on to rightsizing.

Rightsizing

Rightsizing VMs is an iterative exercise that should be performed continuously in iterations. It basically consists of analyzing whether our VMs are properly sized for the use we are making of them. As resource use may vary over time due to changes in user demand, we should have dashboards or automated scripts in place to help us provide more visibility of VM resource usage for this purpose.

To evaluate the resources that our workloads consume, we need to mainly look at memory, CPU, disk IOPS, and disk throughput metrics, and with this information assess whether our VM choice is right, or if it is too much or insufficient for our workloads.

As we cannot imagine in this case study how our imaginary application is going to behave in advance, for now, we are going to focus on something that we know we can do beforehand, and that is to properly size the preproduction and development environments.

For the preproduction and development environments, there are two choices:

- Leave it as it is *and use Power Scheduling to reduce its cost during off-hours,* or even shut down VMs when not in use. We do this to avoid being billed for this we are trying to use it as little as possible.

- *Reduce the size of the VMs by* at least one tier to reduce the cost while keeping the same VMs/ family for solutions similar across environments.

For this case study, we are going to leave preproduction as it is and apply Power Scheduling, while reducing development environment VM sizes.

Apart from rightsizing, we can consider limiting high availability to production and preproduction, making some tweaks in our development environment, with the following reasoning:

- In *preproduction*, we are going to have the same number of VMs per layer as in production. Contingency tests and failover tests are often done in preproduction so it makes sense to leave as it is, so it is essentially a copy of our production environment.

- For the *development* environment, high availability is not needed as it is only used for technical teams and developers to test before release into production. This fact effectively means that we can remove one VM per layer and leave only one, which results in even more cost savings.

This leaves us with the following changes in *development*:

- One server per layer.
- App layer VMs are downsized to D2s v5.
- Database layer VMs are downsized to E2s v5, which still has 16 GB of RAM but only 2 vCores. Being memory-optimized, it should perform well for development use.
- We leave the disks as they are. If we changed to non-SSD (standard HDD), the performance of the application, especially on the database, could be impacted.
- We leave the jumpbox as it is.

This is how the pricing picture in development will look after the preceding steps:

Resources	Number of VMs	cores	RAM	Virtual Machine Size	Disk Size (OS Disk)	Disk Size (Data Disk)	Operating System	Price $/mo
Web layer	1	2	8	D2s_v5	E10 (128 GB)	E10 (128 GB)	Windows Server 2022	$170.31
App layer	1	2	8	D2s_v5	E10 (128 GB)	E10 (128 GB)	Windows Server 2022	$170.31
Database layer	1	2	16	E2s_v5	E10 (128 GB)	E40 (2,048 GB)	Windows Server 2022 + SQL Server Standard	$633.32
Jumpbox	1	2	8	D2s_v5	E10 (128 GB)	E6 (64 GB)	Windows 10	$98.35
						Total		**$1,072.29**

Table 12.6 – VM costs after rightsizing

As in the previous initiatives, let's do a quick recap on how costs are evolving from the initial scenario:

Environment	Original Cost after VM Sizing	Cost with Shared Disk Database	Cost with Rightsizing
Development	$4,310.69	$2,551.09	**$1,072.29**
Preproduction	$4,310.69	$2,551.09	$2,551.09
Production	$4,310.69	$2,551.09	$2,551.09
Total	$12,932.07	$7,653.27	$6,174.47
Total savings compared to original costs		**40.82%**	**52.25%**

Table 12.7 – Total costs of VMs after rightsizing

By applying these changes and properly sizing development, we have reduced considerably (*19%*) the total cost of our solution, all environments included. Let's proceed to apply Power Scheduling.

Power Scheduling for non-productive environments

The idea of **Power Scheduling** revolves around the shutdown or scaling down of unused VMs or services during off hours. In production, this is usually not feasible as the service must always be available, while there are often background jobs running during off hours.

But we can apply this idea to the *preproduction and development* environments. Let's bring back the table that was shown in *Chapter 6*:

Schedule	Total Hours of Use	Savings
24 x 7	730	N/A
8 x 5	180	75%
12 x 5	270	63%
12 x 5 + 48(weekends)	318	56%

Table 12.8 – Savings depending on the rightsizing schedule

We are going to be cautious and not too aggressive about this, so let's say we agree with the project team and apply a *12x5 schedule on our preproduction and development VMs*, which will allow us to generate *63% savings* on compute costs. Please remember that storage costs (Managed Disks) are not affected by this change.

Let's reflect and review the total costs before and after all the initiatives that we proposed:

Environment	Original Cost after VM Sizing	Cost with Shared Disk Database	Cost with Rightsizing	Cost with Power Scheduling
Development	$4,310.69	$2,551.09	$1,072.29	$829.27
Preproduction	$4,310.69	$2,551.09	$2,551.09	$1,288.35
Production	$4,310.69	$2,551.09	$2,551.09	$2,523.03
Total	$12,932.07	$7,653.27	$6,174.47	$4,668.71
Total savings compared to original costs		40.82%	52.25%	64.12%

Table 12.9 – Total costs of VMs after Power Scheduling

By shutting down VMs during off-hours in development and preproduction, we were able to achieve more than *24% savings on the overall TCO of the solution in all environments.*

Now, another key question: assuming that the schedules have been agreed to with technical teams, how do we apply this using automation?

- For the VMs in the database, web, and app layer, we can use an **Azure Automation Runbook** that is able to shut down VMs on schedule. We can assign a specific tag to the VMs that we want to shut down using the runbook and it will run on schedule.

- For Azure Virtual Desktop jumpbox, there is a great feature called **Start VM on Connect** that can be used to leave the machines powered off when not in use. By using this, savings on jumpbox costs can be even higher than the costs we have calculated. Additionally, we can use the same runbook to power them off on schedule.

Reserved Capacity for production

Once rightsizing and Power Scheduling are in place, we are in a position to consider purchasing **Reserved Capacity** for production VMs.

Before even considering purchasing Reserved Instances, we need to ensure that the life cycle of the application will be aligned with the duration of the Reserved Instances (e.g., if the application is decommissioned after 1 year and we reserve for a 3-year period, how do we use this Reservation so it is not wasted?).

Here are some considerations:

- We have chosen a *1-year period Reserved Instance* in this case. As these Reservations are just scoped for this solution, we can purchase the exact amount of Reserved Instances needed to cover production. We have chosen 1 year to avoid long-term commitments.

- The *Jumpbox VM isn't covered by the RIs*, as we will be using AVD's Start VM on Connect feature, as well as Power Scheduling using the same Azure Automation Runbooks used for preproduction and development.

Let's analyze the potential cost savings that we'd obtain if we purchase these Reserved Instances:

Resources	Number of VMs	Cores	RAM	Virtual Machine Size	Disk Size (OS Disk)	Disk Size (Data Disk)	Operating System	Price $/mo PAYG	Price $/mo RI 1 yr
Web layer	2	2	8	D2s_v5	E10 (128 GB)	E10 (128 GB)	Windows Server 2022	$340.62	**$271.72**
App layer	2	4	16	D4s_v5	E10 (128 GB)	E10 (128 GB)	Windows Server 2022	$642.84	**$505.20**
Database layer	2	4	32	E4s_v5	E10 (128 GB)	E40 (2,048 GB)	Windows Server 2022 + SQL Server Enterprise	$1,469.28	**$1,440.88**
Jumpbox	2	2	8	D2s_v5	E10 (128 GB)	E6 (64 GB)	Windows 10	$70,29	$70.29
						Total		$2,523.03	$2,288.09

Table 12.10 – Total costs of VMs after Reserved Capacity

If we compare the result with the previous cost of production, we have saved around *10%* just by purchasing these Reserved Instances. In the TCO for all environments, we have saved around *5%* as well.

After all the optimization we went through, this is the final picture of each environment cost:

Environment	Original Cost after VM Sizing	Cost with Shared Disk Database	Cost with Rightsizing	Cost with Power Scheduling	Cost + PS + Reserved Instances
Development	$4,310.69	$2,551.09	$1,072.29	$829.27	$829.27
Preproduction	$4,310.69	$2,551.09	$2,551.09	$1,288.35	$1,288.35
Production	$4,310.69	$2,551.09	$2,551.09	$2,551.09	$2,288.09
Total	$12,932.07	$7,653.27	$6,174.47	$4,668.71	$4,405.71
Total savings compared to original		40.82%	52.25%	63.90%	65.93%

Table 12.11 – Total costs of VMs after Reserved Capacity

As we can see, applying Reserved Capacity in a limited scope yielded some additional savings. Yet we still have a high licensing cost in all three environments from the use of SQL Server. We will try to reduce these costs as much as possible in the next section by using the Hybrid Benefit feature.

Hybrid Benefit for Windows Server and SQL Server

Let's analyze the licensing costs for these VMs in all environments. As explained before, when we pay licenses using a PAYG model directly in the cloud, we are incurring surcharges that we wouldn't incur when purchasing the same licenses from Microsoft. To avoid these surcharges, we can leverage the **Hybrid Benefit licensing** model to use BYOL.

Keep in mind that cost optimization using *Hybrid Benefit is maximized when computers are not shut down*. The reason behind this is that, when we use PAYG and we shut down virtual machines during off-hours, we are no longer charged for either compute or licenses until the machine is started again. On the other hand, when we have a license purchased directly to Microsoft and we use Hybrid Benefit, we pay for the license on monthly, yearly, or three-year periods.

For this purpose, our proposal is going to be *to only apply Hybrid Benefit on production VMs that are not shut down* (this leaves the jumpbox out as well).

To cover production, take into account this table, which was presented in *Chapter 7*:

EA license (Software Assurance required)	Azure	Equivalence	Other benefits
Windows Server Standard Windows Server Datacenter	Azure VM	2 processor EA license = 2 Azure instances up to 8 vCPU or 1 Azure instance up to 16 vCPU	Windows Server Standard: 180 days coexistence on-prem-azure
			Windows Server Datacenter: dual usage on-prem-azure
SQL Server Standard SQL Server Enterprise	SQL Server on Azure VM (minimum eligible: 4vCPU)	1 EA core license STD = 1 vCPU (STD) 1 EA core license ENT = 1 vCPU (ENT) 4 EA STD licenses = 1 vCPU ENT	180 days coexistence on-prem-azure
SQL Server Standard SQL Server Enterprise	Azure SQL Single Database / Elastic pool (only in vCPU-Provisioned compute) Azure SQL Managed Instance	1 EA core STD license = 1 vCore GP 1 EA core STD license = 1 vCore HYP 4 EA core STD license = 1 vCore BC 1 EA core ENT license = 4 vCores GP 1 EA core ENT licenses = 4 vCore HYP 1 EA core ENT license = 1 vCore BC	180 days coexistence on-prem-azure

Table 12.12 – Hybrid Benefit on-premises and Azure equivalences

In production, we have the following VMs:

- 2x E4s v5 – 4 vCores
- 2x D2s v5 – 2 vCores
- 2x D4s v5 – 4 vCores

And we have these considerations:

- If we purchase on-premises Windows Standard Licenses we can use Azure VMs with Windows Server Datacenter
- We can exchange one Windows Server Standard license for two Azure VMs using **Azure Hybrid Benefit (AHB)** under 8 vCores

This means that we could cover all of them with the following licenses:

- 3x Windows Server Standard 2 processor licenses
- 8x SQL Server Standard per core license

Let's get the prices for these licenses, taken from a CSP contract pricing that we will use as reference. The exact pricing will depend on the type of agreement that each organization has with Microsoft:

- **SQL Server Standard 2-core 1-year annual license**: $1,825, with a price per month of $152
- **Windows Server Standard 2 core 1-year annual license**: $61, with a price per month of around $5

If we got 3-year licenses the price will be slightly lower, but again, we would be committing to much more money.

Let's compare the license costs of the Hybrid Benefit and PAYG models:

Environment	Cost + PS + Reserved Instances	Licensing Costs PAYG	Licensing Costs Hybrid Benefit
Development	$829.27	$207.36	$207.36
Preproduction	$1,288.35	$489.24	$489.24
Production	$1,782.89	$1,280.44	**$623**
Total	$3,900.51	$1,977.04	$1,319.6

Table 12.13 – Total license costs after Hybrid Benefit

Now that we have the license prices, let's update our table:

Environment	Original Cost after VM Sizing	Cost with Shared Disk Database	Cost with Rightsizing	Cost with Power Scheduling	Cost + Reserved Instances	Cost + Hybrid Benefit
Development	$4,310.69	$2,551.09	$1,072.29	$829.27	$829.27	$829.27
Preproduction	$4,310.69	$2,551.09	$2,551.09	$1,288.35	$1,288.35	$1,288.35
Production	$4,310.69	$2,551.09	$2,551.09	$2,676.63	$2,288.09	$1,630.65
Total	$12,932.07	$7,653.27	$6,174.47	$4,794.25	$4,405.71	$3,748.27
Total savings compared to the original		40.82%	52.25%	62.93%	65.93%	71.02%

Table 12.14 – Total VM costs after Hybrid Benefit

As we can see, by applying Hybrid Benefit only in production, we have been able to reduce the total license costs of the solution in all environments by *33%*, even though we have only applied it in production.

Of course, on the other hand, we now need to manage licenses and make sure that they are renewed from year to year and stay current.

With this last initiative, let's move to the conclusions of this exercise.

Summary of initiatives and final results

In this case study, we have applied the most common initiatives on IaaS workloads.

By doing so, we have been able to reduce the initial costs by *71%*, which would be really good news for our financial teams.

Keep in mind that the order in which we have applied the initiatives is key here as some initiatives depend on others. Here are some closing thoughts and considerations that are specific to this case study:

- Shared Disk setups for clusters are a good idea if the solution requirements are aligned and it can generate greater savings, especially if data disk are high.

- We should only consider using Reserved Instances and Hybrid Benefit with VMs that have been already rightsized.

- Hybrid Benefit should only be considered for VMs that are not shut down. If we use PAYG and shut down VMs, license costs will go down and Hybrid Benefit will not improve costs anymore.

- We should only consider Reserved Instances with VMs that we are not able to shut down because shutting them down is easier to implement and can lead to more savings.

- In this example, we have only applied Reserved Instances in production. We should consider its use in preproduction when VMs are not shut down. If VMs are shut down, then Reserved Instances should cover only the ones that we are unable to shut down.

- Regarding reserved Capacity, we always recommend starting with 1-year commitment periods. As Reserved Instances need to be aligned with each workload life cycle, in some cases committing to three years can be too risky. If your VMs are not yet rightsized or fully optimized, you can choose to purchase Saving Plans instead to have more flexibility.

- Don't purchase Windows Datacenter licenses for VMs hosted in the cloud. Even with the dual usage benefit, the high price tag is not worth it for cloud workloads as they are intended to be used for virtualization hosts in on-premises scenarios.

- Think twice before blindly keeping the same licenses that have always been used for SQL clusters. Carefully review the solution's requirements and consider downgrading the licenses if possible.

With all these valuable considerations, we will now proceed with the second case study that we are going to go through in this chapter.

PaaS case study – storage, serverless, and database optimization

In this section, we are going to use a standard **Extract, Transform, Load (ETL)** architecture for our case study, which is the gold standard of data and analytics architectures for data extraction, processing, and visualization.

Our solution consists of different tiers that are used to extract and process data:

- An *ETL layer* dedicated to extracting and transforming the data that is fed to our solution from different data sources
- A *data warehouse (DW) layer* where the application stores all its data in a database or storage services
- A *visualization layer* on which we use data stored in the data warehouse layer to be the basis of dashboards and reports on top of our data

For this application, we imagine we would need three different environments:

- *Development*: Used to test new features in our data processing process. We can also use this layer to build new dashboards or to apply changes to solve issues reported by end users.
- *Preproduction*: This environment will be as close as possible to production, and it is often used for User Acceptance Testing by a subset of key users.
- *Production:* The environment used by real users to connect to the reports and dashboards offered to users in the visualization layer.

This solution implements the complete process that is executed *once a day* (data is not present in real time).

The process is divided into smaller subprocesses that are executed from our ETL VMs. *The total number of subprocesses is 10 and has an average execution time of 60 minutes.*

The processes are distributed across *two VMs in our ETL layer*, and after the processing, *data is loaded into a database that is hosted on one server*. The processes to ingest and transform data before its visualization will be executed in batch mode, that is, run on schedule and without any interaction from the user side.

We will not be covering the visualization layer in this example because we have enough ground to cover the rest of the layers and this is usually the easiest part, with a lot of market solutions available.

Solution description

We assume for our case study that this architecture is working from a data center hosted on-premises, and we are asked to migrate it to the AWS cloud.

Let's begin by presenting a diagram depicting the complete solution:

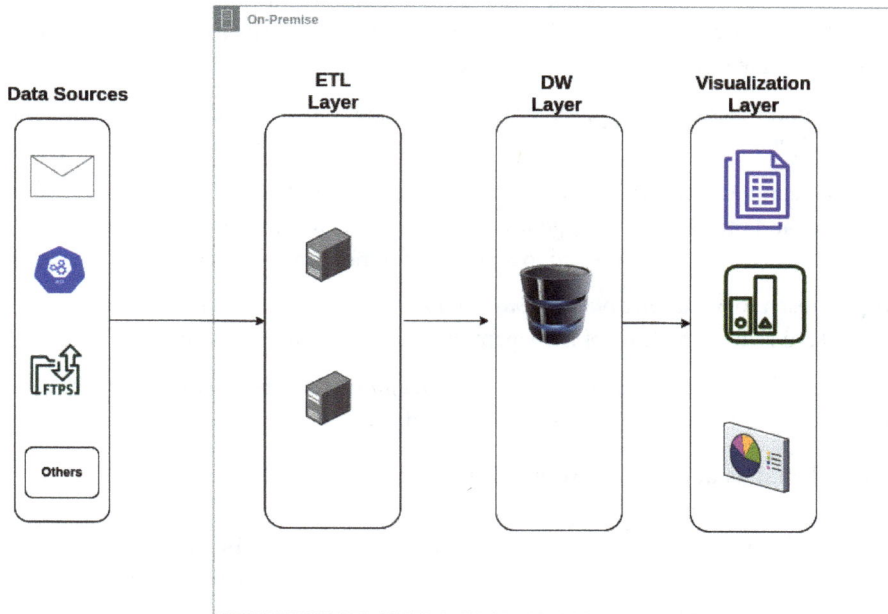

Figure 12.3 – PaaS case study initial solution diagram

In our on-premises setup, we use the following servers:

Server	Layer	Cores	RAM	OS Disk Size	Data Disk Size	Operating System
ETLlayer0	ETL layer	4	16	128 GB	128 GB	Linux
ETLlayer1	ETL layer	4	16	128 GB	128 GB	Linux
DWlayer0	Data warehouse layer	8	32	128 GB	2 TB	Linux + MySql

Table 12.15 – PaaS case study initial on-premises servers

Let's do a quick analysis of what we need to host this solution in the cloud. To do this, we have chosen the exact offerings in our cloud provider that match our VMs hosted on-premises.

Here are some considerations before going forward:

- We have used the AWS Pricing calculator (`https://calculator.aws/#/`) using retail prices in **July 2023 for Ireland region (eu-west-1)**. All cloud providers offer their own calculator so you can calculate the TCO of your solutions beforehand.

- For our VMs, we have chosen the *m5 virtual machine family*, which is used for general-purpose workloads and provides a great balance between cost and performance compared to older generations, while having the latest generation CPU chipsets.

- For our virtual machines' disks, we have also decided to use *gp3 disks*, which offer an IOPS baseline of 3,000.

- We will be using *MySQL as our database engine* as it is part of the original solution hosted on-premises and therefore no change will be required. As we already covered the BYOL model in the previous case study, we wanted to make this solution simpler from a licensing perspective.

- We have chosen to *modernize our database and move it to a PaaS service, in this case AWS RDS for MySQL*, to reduce the level of management needed on this resource.

- We will begin by using a *MultiAZ replication level for our database with one standby instance*, which is the one provided by default by our cloud provider, to ensure high availability.

With these considerations in mind, our service list in AWS would look like this:

Resources	Number	Service Selected	vCPU	RAM	Virtual Machine Size	EBS Disk Size (OS Disk)	EBS Disk Size (Data Disk)	OS	Price $/mo
ETLlayer	2	AWS EC2	4	16	m5.xlarge	GP3 (128 GB)	GP3 (128 GB)	Linux	$357.50
Dwlayer	1	AWS RDS	8	32	db.m5.2xlarge		GP3 (2,048 GB)	Linux + MySQL	$1,717.39
							Total		$2,074.89

Table 12.16 – PaaS case study initial services selected

Of course, this is the price of our would-be production environment. If we consider the same price for preproduction and development, we are at roughly $6,200/mo.

Using this as a starting point, we are going to try and optimize the solution by applying different initiatives, in the same manner as we did in the previous case study.

Initiatives covered

For this case study, we have decided to begin with an IaaS solution for our ETL tier, which we will evolve later to serverless. The idea behind this train of thought is to highlight serverless benefits as well.

Let us begin with some initiatives that are related to solution architecture such as disk rightsizing and limiting redundancy to production, and, once this part is clear and complete, we will proceed to other initiatives that let us generate additional savings such as data tiering.

Modernization to serverless architecture

As was already described in *Chapter 6*, one of the ways we have to modernize our applications while generating some savings is by leveraging **serverless** services.

In this case, we can consider our ETL layer a **stateless** service that is only used to process data and store it in our database. Stateless services usually are a great match for serverless offerings, so we will be moving the complete layer to AWS Glue jobs.

For our solution, we propose to use **AWS Glue 3.0** for our ETL layer instead of VMs. By doing this, we will avoid having idle VMs while our process is not running, and therefore we will have a much more optimal solution from a cost perspective. Another choice would have been to shut down these VMs while they're not in use, but having such automation in place takes work and effort, which we want to keep to a minimum.

AWS Glue is billed by **Data Processing Units** (**DPU**), which represent a VM with 4 vCPUs and 16 GB of RAM. The minimum number of DPUs we can use is two per execution. AWS Glue offers the possibility to execute Spark, Spark Streaming, Python, and Ray (currently in preview) jobs. For this example, we have chosen to move our processes to *Spark jobs*.

It is possible that, by using Spark for our processes instead of the scripts that were used on-premises, processing time could be reduced, but for the sake of simplicity we will consider *the same 60 minutes of execution for each of the 10 processes that we run after the change to serverless.*

With these considerations in mind, this is the final price of our ETL layer after moving the processes execution to AWS Glue using Spark jobs:

Resources	Number of DPUs	vCPU	RAM	ETL Engine	Execution time	Frequency	Jobs Number	Price $/mo
ETLlayer	2	8	32	Spark	60 minutes	Daily	10	$174.00

Table 12.17 – Serverless solution to ETL

Let's do a quick recap of the original and current total costs:

Environment	Original Cost before Optimization	Cost with Serverless Optimization
Development	$2,074.89	**$1,891.39**
Preproduction	$2,074.89	**$1,891.39**
Production	$2,074.89	**$1,891.39**
Total Cost	$6,224.67	$5,674.17
Total savings compared to original costs		**8.84%**

Table 12.18 – Total costs with Serverless optimization

Apart from the cost reduction, this solution improvement will reduce the administrative overhead needed for our solution, as we won't need to manage backups and system updates for example.

Let's now move forward to disk rightsizing.

Disk rightsizing

Once we have transformed our ETL layer to serverless, let's take a look again at our data warehouse layer. Do we really need to have 2 TB disks for storage in non-production environments? Having this big disk has a great impact on the overall costs of our solution.

We see it time and time again, especially in on-premises solutions where disk costs are not as evident. Databases grow without limits, and, over time, they become big and therefore difficult to operate and maintain. When databases such as this are moved to the cloud, we need to always consider whether or not all the data present is "hot" data that is used on a day-to-day basis, or if we could move some data to other storage services such as object storage or file storage, where we can make use of data temperature tiers.

For this case study, we won't go deeper into this data classification and database offload to other storage services, but we can limit the size of non-production environment data and reduce the disk size used in development and preproduction.

We will assume then that we have the following data retention policies:

- *2 years in production*
- *1 year in preproduction*
- *6 months in development*

With this consideration, assuming that the *amount of data generated every day is the same*, we end up with these disk capacities per environment:

- *Production: 2TB*

- *Preproduction: 1TB*

- *Development: 512 GB*

This results in the following costs for our DW layer, after the AWS EBS disk sizes are updated:

Environment	Original cost DW Layer	Disk Size (Data Disk)	Cost with Disk Rightsizing Optimization
Development	$1,717.39	GP3 (512 GB)	**$1,327.25**
Preproduction	$1,717.39	GP3 (1024 GB)	**$1,457.30**
Production	$1,717.39	GP3 (2048 GB)	$1,717.39
Total	$5,152.17		$4,501,94
Total savings compared to the original			12.65%

Table 12.19 – Total costs of DW layer before and after rightsizing

Let's do a quick recap on how our total cost evolution is going:

Environment	Original Cost before Optimization	Cost with Serverless Optimization	Cost with Disk Rightsizing Optimization
Development	$2,074.89	$1,891.39	**$1,501.25**
Preproduction	$2,074.89	$1,891.39	**$1,631.30**
Production	$2,074.89	$1,891.39	$1,891.39
Total Cost	$6,224.67	$5,674.17	$5,023.94
Total savings compared to original costs			19.28%

Table 12.20 – Total costs with disk rightsizing optimization

Let's now analyze how we can save costs by limiting redundancy and replication to secondary regions.

Limit replication to production

Replication is one of the key features we can use to ensure high availability. However, we must consider that using this feature has an important impact on TCO. Due to this fact, we should limit its use to environments that require high availability, such as production.

The preproduction environment concept varies from one company to the other. In some organizations, it is used as an exact copy of production with high-availability requirements as well (it is even used to test contingency scenarios), while in other organizations due to cost constraints these features are not used or limited to a minimum. For this example, we are going to *limit its use to production, assuming that this ETL process does not have any contingency plans that need to be tested.*

With this consideration in mind, let's assume that we *limit in our AWS RDS for MySQL the use of MultiAZ to production*, leaving preproduction and development environments with a *single Availability Zone*:

Environment	Original Cost DW Layer	MultiAZ	Cost with Rightsizing + MultiAZ Optimization
Development	$1,717.39	No	$710.34
Preproduction	$1,717.39	No	$775.37
Production	$1,717.39	Yes	$1,717.39
Total	$5,152.17		$3,203.10
Total Savings compared to the original			**37.83%**

Table 12.21 – Total costs with disk rightsizing and MultiAZ optimization

After this change, let's look again at the original and current total costs:

Environment	Original cost before optimization	Cost with Serverless optimization	Cost with Serverless + Disk rightsizing optimization	Cost with Serverless + Disk rightsizing + MultiAz optimization
Development	$2,074.89	$1,891.39	$1,501.25	**$884.34**
Preproduction	$2,074.89	$1,891.39	$1,631.30	**$949.37**
Production	$2,074.89	$1,891.39	$1,891.39	$1,891.39
Total Cost	$6,224.67	$5,326.17	$5,023.94	$3,725.1
Total savings compared to original costs				**40.15%**

Table 12.22 – Total costs after Serverless solution and disk rightsizing and MultiAz optimization

Once redundancy options have been limited to production, let's reconsider how we are storing our data in our case study. For these ETL scenarios, we can consider a data lake as an alternative to a traditional database.

For our use case, which is to have a foundation to perform data analysis and reporting, it may be more than sufficient to use AWS S3 storage instead of AWS RDS. Let's consider this change in the next section.

Move to AWS S3 storage

To move our data warehouse layer to S3 means to *move from a database to object storage*, in this case in the form of documents, which means essentially to *move from a relational database such as MySQL to a document store NoSQL database hosted in AWS S3*. In doing so, we also adopt a **data lake** approach instead of a traditional database as the basis for our data and analytics processes.

Let's ask ourselves a question. Could the format of the file (CSV, XML, or Parquet) affect the overall cost of the solution? Definitely yes, as some data formats, such as CSV files are not compressed, while Parquet files are. This fact results in much less storage capacity used and can even result in faster execution times as well.

Example – compressed versus non-compressed data formats

Let's use this image from Thomas Spicer's Medium article as an example (`https://blog.openbridge.com/how-to-be-a-hero-with-powerful-parquet-google-and-amazon-f2ae0f35ee04`) to illustrate the differences in how much storage 1 TB of data can take up:

Dataset	Size on Amazon S3	Query Run time	Data Scanned	Cost
Data stored as CSV files	1 TB	236 seconds	1.15 TB	$5.75
Data stored in Apache Parquet format*	130 GB	6.78 seconds	2.51 GB	$0.01
Savings / Speedup	87% less with Parquet	34x faster	99% less data scanned	99.7% savings

Figure 12.4 – 1 TB of data in CSV versus Parquet

In this case, we are not going to dig deeper into this fact, but it is important to consider this when doing these kinds of optimization exercises using document stores.

Let's begin by analyzing how much it costs to have *2 TB of data stored in AWS S3, using S3 Standard Hot*:

Resources	Size (TB)	Storage Class	Price $/mo
DWlayer	2	S3 Standard (Hot)	$50.18

Table 12.23 – S3 solution for DWlayer

This is amazing news, isn't it? However, capacity does not represent the whole picture when we consider AWS S3 costs. We need to also consider **indirect costs**, such as the number of read/write requests and data retrieval fees in some tiers.

Let's get the list prices and do some quick calculations to add indirect costs to our estimation by adding the following considerations to our 2 TB of storage capacity:

- *100 million of PUT, COPY, POST, or LIST*
- *100 million of GET, SELECT, and other requests*
- *10 TB of data returned*
- *10 TB of data scanned*

This, taking the price list from AWS S3, leaves us with these production prices for our DWlayer in production:

Cost driver	Environment	Tier	Cost
2048 GB capacity	Production	Standard Hot	$50.18
100 M PUT, COPY, POST, LIST operations	Production	Standard Hot	$540.00
100 M GET, SELECT and other request	Production	Standard Hot	$43.00
10 TB data returned by S3 select	Production	Standard Hot	$8.19
10 TB data scanned by S3 select	Production	Standard Hot	$23.04
		Total	$664.41

Table 12.24 – S3 solution for DWlayer

As we can observe in the preceding table, the final picture differs a great deal if we consider indirect costs. Make sure to incorporate these estimations, even if rough, into your cost calculations to prevent surprises later.

Please note that in this table, we are using *the color red to represent the hot tier* of data temperature.

Also, remember that we have reduced the data size in preproduction to 1 TB and development to 512 GB in previous steps. For simplicity in our calculations, *we have only modified the total capacity of these environments, leaving data processing and requests with the same totals.*

This leaves us with the following costs:

Environment	Original Cost before Optimization	Cost with Serverless Optimization	Cost with Serverless and Disk Rightsizing Optimization	Cost with Serverless and Disk Rightsizing and MultiAz Optimization	Cost with Serverless and AWS S3
Development	$2,074.89	$1,891.39	$1,501.25	$884.34	**$800.78**
Preproduction	$2,074.89	$1,891.39	$1,631.30	$949.37	**$813.72**
Production	$2,074.89	$1,891.39	$1,891.39	$1,891.39	**$838.41**
Total Cost	$6,224.67	$5,326.17	$5,023.94	$3,725.1	$2,452.91
Total savings compared to original costs					60.59%

Table 12.25 – Total costs with Serverless and changing DWlayer to S3 optimization

As we can see from the table, we have achieved major savings by applying this change in the storage platform for our solution. As the next step, we will illustrate how we can protect against data loss in our AWS S3 bucket by using the versioning feature.

Versioning

When we were using AWS RDS we chose MultiAZ to prevent unwanted data loss inside our bucket in the event of an issue in our primary Availability Zone.

Now that we have moved on to AWS S3 storage, we can make use of the **versioning** feature to store an additional version or copy of each object. If an object is deleted, we will still have the previous version available and, in the event of a file being modified by mistake, having the previous version can be a lifesaver in preventing data loss as well.

Using this feature will increase the cost slightly, but it is worth it in our opinion for our case study. Keep in mind that, if versioning is used, one of the most important policies to apply in buckets that have this feature is to *limit the number of versions per object* to prevent costs from growing exponentially if a lot of versions are generated.

Activating versioning adds additional costs that are justified as it adds additional protection for our files, which are the backbone of our reports and analysis processes after our move to AWS S3.

When using versioning, make sure to limit how many versions of an object will be stored, as costs can get out of hand easily if a limit is not in place, especially for environments where objects change frequently (remember that each small change in an object results in a new complete version of that object).

For our calculations, we have considered that all *objects in production have an additional version stored*, but we should limit this feature to the objects with high criticality and always based on how many changes to objects we expect for our data processing. The impact of storing an additional version of each object in production leaves us with a *production environment that is double the size*.

Activating versioning has the following impact on cost drivers for our DWlayer storage in production:

Cost driver	Environment	Tier	Cost
4096 GB capacity	Production	Standard Hot	$100.35
100 M PUT, COPY, POST, LIST operations	Production	Standard Hot	$540.00
100 M GET, SELECT and other request	Production	Standard Hot	$43.00
10 TB data returned by S3 select	Production	Standard Hot	$8.19
10 TB data scanned by S3 select	Production	Standard Hot	$23.04
		Total	$714.58

Table 12.26 – Total costs with disk rightsizing optimization

Let's do a quick recap of the original and the current total costs:

Environment	Original Cost before Optimization	Cost with Serverless Optimization	Cost with Serverless and Disk Rightsizing Optimization	Cost with Serverless and Disk Rightsizing and MultiAz Optimization	Cost with Serverless and AWS S3	Cost with Serverless and AWS S3 and Versioning
Development	$2,074.89	$1,891.39	$1,501.25	$884.34	$800.78	$800.78
Preproduction	$2,074.89	$1,891.39	$1,631.30	$949.37	$813.72	$813.72
Production	$2,074.89	$1,891.39	$1,891.39	$1,891.39	$838.41	**$888.58**
Total Cost	$6,224.67	$5,326.17	$5,023.94	$3,725.1	$2452.91	$2503.08
Total savings compared to original costs						59.78%

Table 12.27 – Total costs after Serverless and changing storage
to S3 and activating versioning optimization

As we can see, activating versioning did not result in a big impact on our total costs, and it is a great feature to use to prevent data loss.

Data tiering

As the final step in our optimization process, we are going to move different use cases of data to different storage temperature tiers offered in AWS S3 to further optimize our data warehouse layer costs. We should store files that are accessed frequently in Hot storage, while data that is rarely accessed or kept for regulatory requirements should be moved to Cold or Glacier storage.

Let's suppose we divide our data in each environment into the following data use cases:

- *50% of data stored is accessed regularly and requires low latency; this will be matched to the Standard Hot storage tier*

- *25% of data is accessed every trimester; this will be matched to the Standard-Infrequent access storage tier*

- *25% of data is stored only for legal requirements for 5 years but rarely accessed; this will be matched to the Glacier-Deep Archive storage tier*

This consideration is valid for production and preproduction, on which we will have cold data, but is not aligned with the use case of development. *For our development environment, we will only consider 512 GB of data stored and accessed regularly.*

The reason behind this decision is that for development purposes, we may not need the entire dataset we use in production. We can get by with using a subset of the original data, as the main goal of using this environment is to test out new features or fix existing bugs.

This leaves us with the following numbers on each environment:

- **Production:**

Cost driver	Environment	Tier	Cost
1024 GB capacity	Production	Standard Hot	$25.09
10 M PUT, COPY, POST, LIST operations	Production	Standard Hot	$540.00
10 M GET, SELECT and other request	Production	Standard Hot	$43.00
10 TB data returned by S3 select	Production	Standard Hot	$8.19
10 TB data scanned by S3 select	Production	Standard Hot	$23.04
512 GB of capacity	Production	Standard-IA	$6.91
1 M PUT, COPY, POST, LIST operations	Production	Standard-IA	$10.00
1 M GET, SELECT and other request	Production	Standard-IA	$1.00
10 K lifecycle requests	Production	Standard-IA	$0.10
0.5 TB data retrievals	Production	Standard-IA	$5.12
0.5 TB data returned by S3 select	Production	Standard-IA	$5.12
0.5 TB data scanned by S3 select	Production	Standard-IA	$1.15
512 GB of capacity	Production	Glacier-Deep Archive	$0.93
100 K PUT, COPY, POST, LIST operations	Production	Glacier-Deep Archive	$6.00
100 K lifecycle requests	Production	Glacier-Deep Archive	$6.00
10 K restore requests (bulk)	Production	Glacier-Deep Archive	$0.30
0.5 TB data retrievals (bulk)	Production	Glacier-Deep Archive	$2.56

Table 12.28 – Total costs with disk rightsizing optimization

- **Preproduction:**

Cost driver	Environment	Tier	Cost
512 GB capacity	Preproduction	Standard Hot	$12.54
5 M PUT, COPY, POST, LIST operations	Preproduction	Standard Hot	$270.00
5 M GET, SELECT and other request	Preproduction	Standard Hot	$21.50
10 TB data returned by S3 select	Preproduction	Standard Hot	$8.19
10 TB data scanned by S3 select	Preproduction	Standard Hot	$23.04
256 GB of capacity	Preproduction	Standard-IA	$3.46
500 K PUT, COPY, POST, LIST operations	Preproduction	Standard-IA	$5.00
500 K GET, SELECT and other request	Preproduction	Standard-IA	$0.50
5 K lifecycle requests	Preproduction	Standard-IA	$0.05
0.25 TB data retrievals	Preproduction	Standard-IA	$2.56
0.25 TB data returned by S3 select	Preproduction	Standard-IA	$2.56
0.25 TB data scanned by S3 select	Preproduction	Standard-IA	$2.56
256 GB of capacity	Preproduction	Glacier-Deep Archive	$0.47
50 K PUT, COPY, POST, LIST operations	Preproduction	Glacier-Deep Archive	$3.00
50 K lifecycle requests	Preproduction	Glacier-Deep Archive	$3.00
5 K restore requests (bulk)	Preproduction	Glacier-Deep Archive	$0.15
0.25 TB data retrievals (bulk)	Preproduction	Glacier-Deep Archive	$1.28

Table 12.29 – Total costs with disk rightsizing optimization

- **Development:**

Cost driver	Environment	Tier	Cost
512 GB capacity	Development	Standard Hot	$12.54
5 M PUT, COPY, POST, LIST operations	Development	Standard Hot	$270.00
5 M GET, SELECT and other request	Development	Standard Hot	$21.50
10 TB data returned by S3 select	Development	Standard Hot	$8.19
10 TB data scanned by S3 select	Development	Standard Hot	$23.04

Table 12.30 – Total costs with disk rightsizing optimization

Please note that, on all these tables, we are using the following color code:

- *Red represents the hot tier* of data temperature
- *Yellow represents the mild tier* of infrequent access
- *Blue represents the colder tiers* of data temperature

Let's analyze how much this change has impacted our overall costs:

Environment	Original Cost before Optimization	Cost with Serverless Optimization	Cost with Serverless and Disk Rightsizing Optimization	Cost with Serverless and Disk Rightsizing and MultiAz Optimization	Cost with Serverless and AWS S3	Cost with Serverless and AWS S3 and Versioning	Cost with Serverless and AWS S3 and Versioning and Tiering
Development	$2,074.89	$1,891.39	$1,501.25	$884.34	$800.78	$800.78	$509.28
Preproduction	$2,074.89	$1,891.39	$1,631.30	$949.37	$813.72	$813.72	$533.86
Production	$2,074.89	$1,891.39	$1,891.39	$1,891.39	$838.41	$888.58	$858.51
Total Cost	$6,224.67	$5,326.17	$5,023.94	$3,725.1	$2452.91	$2503.08	$1,379.65
Total savings compared to original costs							77.83%

Table 12.31 – Total costs after data tiering

Doing this exercise not only resulted in greater savings but also in better data governance, as we are aligning the storage used for our data with the use case of each type of data we store.

Summary of initiatives and final results

After all the changes that were made to our infrastructure, the following diagram shows the final result:

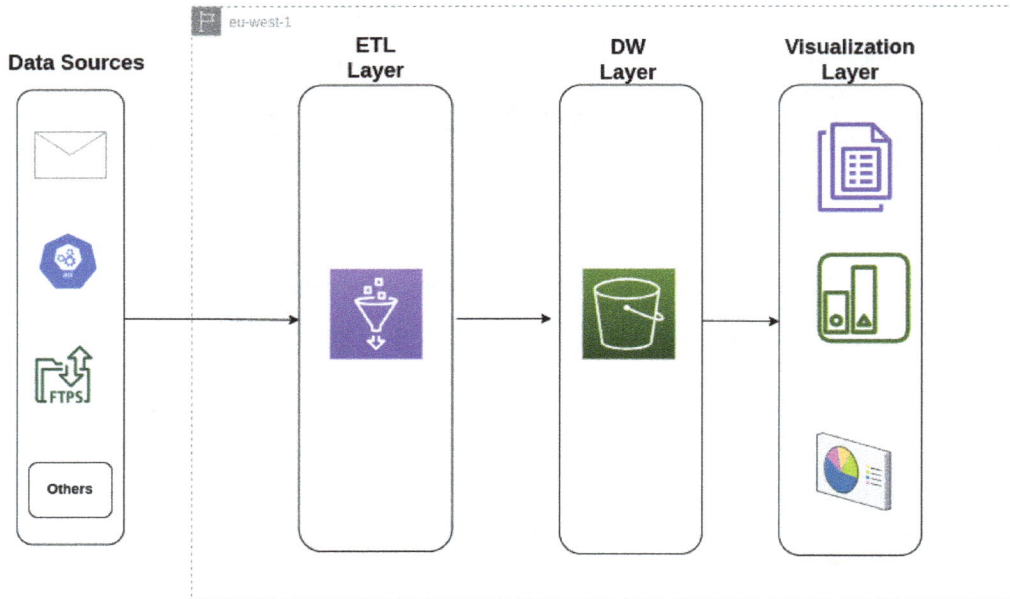

Figure 12.5 – Final storage solution diagram

We managed to transform our solution into a much simpler one that uses both serverless and object storage technologies, and that should be much easier to manage as well.

From the cost perspective, our TCO went down by *77.83%* during the process. Please remember that savings are not the whole picture; we should also consider the following after the changes:

- We replaced VMs with serverless processes that we can pay for as we use them. By doing so, we also reduced the administrative overhead as we don't need to take care of backup or virtual machine permissions, management, patching, and so on anymore.

- We converted a database hosted in IaaS to a PaaS AWS RDS service, which also simplifies database management and provides high availability and backups, among other features, out of the box.

- We also proposed to change from a traditional relational database to a NoSQL document store based in AWS S3, which offers more flexibility in the long run for our data warehouse solution.

- We proposed the use of data temperature tiers in AWS S3, to align the use case of our different types of data to the storage on which they are stored. Apart from the cost perspective, having data distributed across different tiers helps us manage their different life cycles and simplifies data management and governance.

Keep in mind, as in the first case study, that the order in which we have applied the initiatives is key here as some initiatives depend on others. Here are some closing thoughts and considerations:

- The technology and products we choose for our applications will have a direct impact on our costs and project budgets.

- Replication and redundancy should be limited. We should ensure that these features are only used where organizational policies require us to do so. If no policy is in place regarding replication and redundancy as part of cloud governance, we should work with technical teams on defining one.

- We should analyze all the initiatives and the cost drivers of each service before implementing any changes in solution architectures. There are often hidden or indirect costs that are not easily detected at first glance in many cloud services.

- Moving to serverless is a good choice for some workloads due to the way we are billed for these services (only when we use them).

- Aligning the data life cycle with storage tiers is an important initiative to optimize storage costs in every organization. It can take a lot of work to catalog data but is a worthwhile exercise for sure.

Apart from solution-specific conclusions, we want to highlight some general recommendations to consider when doing these cost-optimization exercises:

- Don't try to do everything at once. It is always better to do this iteratively. Apply the simplest initiative and then move forward to other options, if they are available. This will almost always yield great results.

- We know that tables and Excel are an inconvenience. As technology lovers, we struggle with that. But keep in mind that we need to calculate the potential savings of each initiative beforehand in order for us to have a key driver to justify changing the solution. We recommend using tables like the ones we used to illustrate the optimization process of these case studies to compare costs, while carefully making sure that every number matches what is expected. By doing this and by being thorough, you may be able to discover hidden cost discrepancies that can be invaluable for your organization.

- Before deciding on one final architecture, consider all the options and calculate the impact that choosing one architecture or another has on costs. You need to have the complete picture, not only of technical features and constraints, before making a final decision.

Summary

We hope that these two case studies have illustrated the train of thought that we, as FinOps practitioners, should follow when working on analyzing technical solutions. While the specifics are important as well, we wanted this chapter to also exemplify the process of optimization.

With these considerations in mind, let's move to the final chapter of this book, in which we will summarize the knowledge that we have presented throughout the book, and in which we will look at future challenges FinOps practitioners will need to face. We will also arrive at some conclusions, such as the dos and don'ts of cost optimization, as well as interesting insights.

13
Wrapping up and Looking ahead

And, just like that, we have reached the end of this book. We hope all this content has proven useful for you, as we have tried to put every ounce of knowledge we have in this domain into it, leaving nothing behind.

In this chapter, apart from summarizing what we have learned and getting to some conclusions, we will analyze the synergies between FinOps and other rising fields, such as sustainability, machine learning (ML), and artificial intelligence (AI). To close this chapter, we will have a self-assessment or knowledge check section, where we will include some questions that you should be able to answer, now that the content of each chapter has been processed and understood so that you can evaluate how much you have learned throughout this book.

In this chapter, we will cover the following topics:

- FinOps summary and future challenges: how to keep up
- Cloud sustainability and FinOps
- Machine learning, artificial intelligence, and FinOps
- Self-assessment/knowledge check

FinOps summary and future challenges – how to keep up

Throughout this book, using a practical approach and all three major clouds (Azure, GCP, and AWS), we have covered everything that we believe is needed to build a strong FinOps practice from scratch.

First, we explained what FinOps is and why it is important for organizations, as it seeks to solve some challenges that are inherent to cloud adoption, especially around cloud financial management and making the most out of the resources we have.

We also explained how FinOps can fit and synergize with multiple methodologies that are used widely in organizations across the world, such as **Well Architected Framework**, **Infrastructure-as-Code**, and **Agile project management**.

Inform (Chapters 3, 4, and 5)

After these introductory chapters, we have delved into the first pillar of FinOps, which is the **Inform** pillar. In this domain, we have gotten to understand why **naming conventions** and **tagging strategies** are important for FinOps practices, as well as how to design and create our own, which will serve as a strong foundation going forward.

We have also learned how to estimate the TCO for our solutions and how to use the different tools available from each cloud provider to make these analyses. In addition to this, we have presented the idea of **potential savings**, which can serve as an amazing driver to fuel FinOps initiatives by projecting possible savings to come from future initiatives.

To close the work on the Inform pillar, we also explained what a **dashboard** is and what a **report** is, the differences between them, and how they can help us increase the visibility of our FinOps work results for key stakeholders and interested parties. To top it off, we even covered some **financial basics** to help support the work that we did around costs. We also covered the concept of **unit economics**, which can create amazing value by correlating technical and business KPIs together.

Optimize (Chapters 6, 7, 8, and 9)

These chapters were fully dedicated to increasing the visibility of cloud costs in general, as well as of cost optimization awareness. Here, we jumped to the more technical part of our book, which was dedicated to the **Optimize** pillar.

We started with a chapter that was dedicated to **Infrastructure-as-a-Service** (**IaaS**) optimization, where we analyzed multiple strategies and initiatives for cost optimization, such as **rightsizing**, the use of **Reserved Instances** and **Saving Plans**, and other initiatives, such as **virtual machine family standardization** and making use of **Spot Instances**.

With the IaaS part covered, we went through multiple initiatives for PaaS services, such as **Managed Kubernetes cost optimization**, the use of **serverless offerings**, and other forms of optimization that should be also considered, such as **licensing optimization** and **data transfer costs**.

Furthermore, we also reviewed two more specific domains where there is a lot of room for optimization in our experience, which are databases and storage services. As regards databases, we analyzed different ideas and concepts that are essential to understanding how to optimize databases and the key differences between **structured** and **unstructured databases**, as well as specific ways to optimize those two database paradigms, such as **database grouping**, **licensing optimization**, and the use of **serverless** services when applicable. As regards storage services, we covered the different types of storage – **block**, **object**, and **file storage** – and how they work, as well as optimization initiatives that can make a difference to storage services costs, such as **disk rightsizing**, **data temperature tiering**, and **logs and backup storage optimization**.

Operate (Chapters 10 and 11)

After the more technical chapters, we delved into the last FinOps pillar, **Operate**, which aims to make FinOps practices part of the DNA of organizations and not just one-time exercises.

To begin the work on this pillar, we dedicated a full chapter to **FinOps KPIs**. From what a KPI is and the process we need to go through to define our own, to the main types of KPIs and the value they provide, we covered everything there is to know from a theoretical perspective. In addition to this, we also introduced the concept of **OKR methodology** and how we can use this methodology to help us define clear goals for our KPIs. Once the concepts were introduced, we provided clear **examples of FinOps KPIs of each type** that we can use to boost FinOps practices and bring new insights to light.

In addition to this, we also took a deep dive into enterprise governance, analyzing what a **target operating model** is and how to adapt this idea to create our own **FinOps operating model**. As part of the FinOps operating model, we have explored how to use **functions**, **capabilities**, and **roles** to establish proper governance and define responsibility lines and processes to support FinOps work.

Case studies

We didn't want to end this book without providing some practical examples in the form of **case studies**, in which we analyzed two different solutions, based on past projects we have worked on, from a cost optimization perspective. For both solutions, we described the solution architecture and its components in a detailed way before proposing some of the initiatives covered in the chapters dedicated to the Optimize phase. We went through all these initiatives, analyzing how their application impacts the overall cost of the solution, in a step-by-step process. Through these examples, we went from non-optimized solutions to solutions that were more cost-effective and whose costs could be much more efficiently controlled.

With this summary of what we covered, let's move on to the challenges we feel that, as FinOps practitioners, we will be facing in the future.

FinOps future challenges

Of all the challenges that we have faced in our FinOps experience, we want to highlight two of them, as we feel they will still be present going forward for the future generations of FinOps practitioners. We refer to the following:

- The first one we want to highlight is the lack of resources or, in other words, the lack of enough budget dedicated to FinOps work. Without committing resources and budget to FinOps resources in the form of people, projects, and tools, FinOps practices won't be able to truly take off and make an impact in organizations. Cloud professionals are already scarce and hard to hire, and most probably are also fully booked with migrations, application modernization, and other similar lines of work. If FinOps is just another side project for these teams that keeps being deprioritized for more important things, we will never be able to do the work that is needed to transform the entire organizational culture.

- The second challenge we want to cover is the **ever-changing nature of the cloud**. New services and features are coming out every day, and it is harder and harder for cloud professionals to keep up with this pace of technological evolution. It requires constant training and adaptation to new ways of building and implementing solutions. In the cost optimization world, FinOps practitioners will need to keep up with the times, analyzing and detecting new initiatives to optimize and control the cloud costs of cloud services that we can only imagine right now. The good news, from our point of view, is that the concepts and ideas of FinOps will still stand, and this challenge will only affect how we implement these principles in future environments. As an example, we are almost sure that everything we covered in *Chapters 6, 7, 8*, and *9* will most probably be outdated in the short term, and that is okay. In our opinion, it's just how technology works nowadays.

With these challenges in mind, let's move on to the first section dedicated to FinOps' synergy with other emergent fields that are on the rise, such as cloud sustainability.

Cloud sustainability and FinOps

Sustainability, from an environmental perspective, refers to the ability to fulfill our needs as the human race without compromising the future of coming generations.

As regards sustainability, climate change is perceived as the major threat, but there are also other threats, such as biodiversity loss, drought and resource depletion, pollution, and deforestation.

Sustainability is far from a new concept. For millennia, it has been one of the main concerns of humanity, as we have tried to be in tune with the environment and its natural limits, cycles, and changes. But how can this idea apply to companies and businesses?

Even though the concept of sustainability was first formulated at a United Nations conference in 1972, it wasn't until 1987 that it really took shape, when the Brundtland report titled *Our Common Future* was published, which clarified the goals of sustainable development.

We can evaluate sustainability in three dimensions:

- **Environmental**: This dimension revolves around preserving and protecting the natural environment over time, seeking to minimize our impact on it as much as possible
- **Social**: This consists of improving our society by focusing on the well-being of people, which can be done by promoting human rights, equity, and access to education, among things
- **Business**: This focuses on the sustainability of economic and business activities, by balancing economic growth with resource efficiency and financial well-being

Even though these three dimensions can be developed in parallel, they are all contained within each other. Sustainability policies should try to make an impact on each dimension. We can summarize both these ideas in the following diagram:

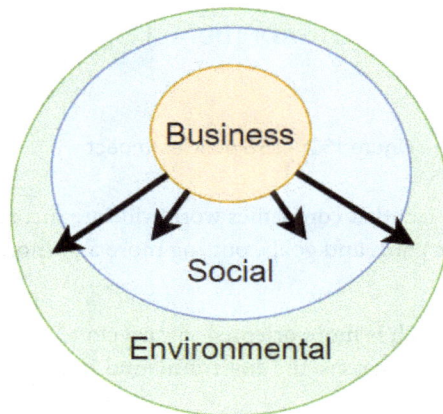

Figure 13.1 – Sustainability policies impact

For example, for an organization to make an impact on the social dimension, they can focus on eliminating discrimination, prejudice, and social exclusion. With the same idea, they can also work on having an impact on the environmental dimension by reducing carbon dioxide emissions that result from their operations.

Companies worldwide are also adopting **Environmental, Social, and Governance** (ESG) policies, which seek to improve companies' performance while being accountable and transparent to investors and customers; they have been on the rise for a few years now.

The idea behind ESG policies is to improve the reputation of organizations by having them be transparent about global concerns such as helping the environment, supporting diversity and well-being, and ensuring ethical business decisions.

In summary, ESG seeks to have an impact on the same dimensions as sustainability, with the difference being the return that the company gets from it – in this case, more clients or customers. We can represent the idea of ESG with the following figure:

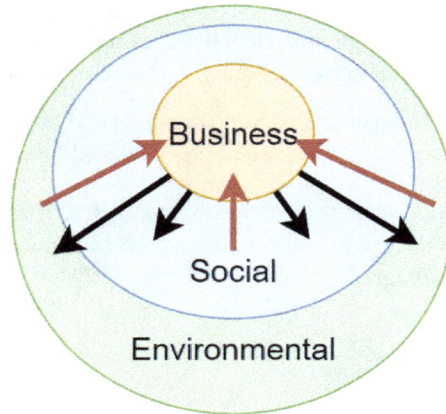

Figure 13.2 – ESG policies impact

Regardless of the reasons, it is a fact that companies worldwide are incorporating sustainability and ESG policies into their strategic plans and goals, putting more and more effort and resources into these fields.

For the purpose of this book, which is more oriented toward cloud financial optimization, we will solely focus on one of the three dimensions, the **environmental impact**.

Let's begin by understanding how companies apply environmental sustainability policies, which we will do in the next section.

How environmental sustainability policies work

In the **Paris Agreement** in 2015, world leaders (195 parties, consisting of 194 states and the European Union) at the **United Nations Climate Change Conference (COP21)** reached an agreement that included the following goals:

- Substantially reduce greenhouse gas emissions to limit global temperature increase, specifically, 50% reduction by 2030

- Review countries' commitments every five years

- Provide additional financing for developing countries to mitigate climate change and preserve the environment

Since this groundbreaking agreement, major companies have supported these initiatives and created partnerships with governments to mobilize investment, technology, and innovation to make it happen. As a result, the **We Mean Business** coalition (https://www.wemeanbusinesscoalition.org/) was created.

From then on, there have been a lot of companies working on reducing greenhouse gas emissions and being more environmentally friendly. Besides companies that are fully dedicated to sustainability, other big organizations worldwide have created sustainability divisions that also contribute to what was agreed, such as Netflix, Microsoft, and Amazon. Most of the efforts have been put into reducing carbon emissions, as carbon dioxide is the greenhouse gas that accounts for most of the warming associated with human activity.

In general, carbon emissions are reduced by executing different projects all around the world, such as reforestation, regenerative agriculture, and renewable energy projects, that reduce current emissions (**avoidance or reduction**) or capture carbon dioxide before it gets into the atmosphere (**removal or sequestration**).

These projects generate carbon emissions reductions that are represented in the carbon market in the form of **carbon credits** or **offset credits**, which companies can purchase for a price. Carbon credits can also be generated from within companies, by reducing their emissions through specific projects or initiatives to reduce emissions structurally by changing the infrastructure or hardware that creates such emissions.

This market works as shown in the following figure:

Carbon emissions

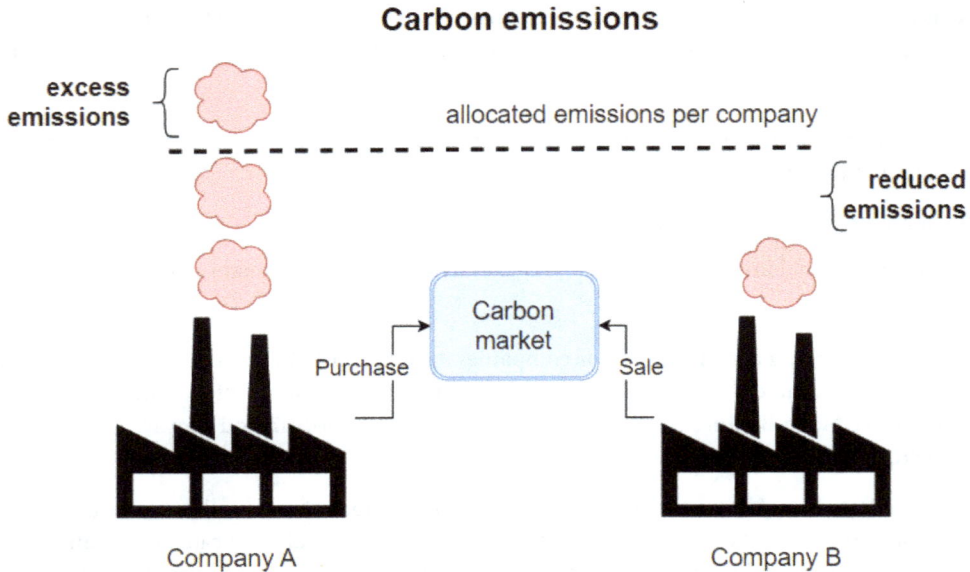

Figure 13.3 – Carbon market

As we can see from the figure, **Company A** has exceeded its allocated carbon emissions and will be in need of purchase credits to compensate for these excess emissions. **Company B**, on the other hand, has reduced its emissions way below the baseline, which allows the company to sell these emissions reductions on the carbon market.

There are two main markets of carbon:

- **Mandatory**: This applies to governments and companies that are legally mandated to offset their emissions. Each ton of carbon dioxide that is not emitted to the atmosphere equals one **Certified Emissions Reduction (CER)** unit, which needs to be verified and certified by third parties.

- **Voluntary**: This applies to any organization that wants to participate in carbon emissions for different reasons: certification, reputation, or environmental or social benefits. Voluntary emissions are measured in **Verified Emissions Reduction (VER)** units, which, like the CER unit, equal one ton of carbon dioxide not emitted to the atmosphere. Many companies have clear sustainability goals, such as becoming carbon neutral sometime in the foreseeable future, and this voluntary market is one of the means to track and reach these goals.

Companies acquiring carbon credits are accountable for their quality, as they must come from verifiable and validated sources. Otherwise, they can risk their reputation by being accused of **greenwashing**, which means publicly supporting sustainability from a marketing perspective while their actions speak otherwise. For example, buying cheap carbon credits from non-reputable companies may not result in any tangible climate benefits and can come across as greenwashing.

The rules of how to determine the amount of additional carbon emissions that each company has allocated are defined by protocols/checks, which in turn are defined by the carbon registries. The rules have evolved over time, becoming more restrictive to circumvent loopholes/shortcomings that scientists have found, which have resulted in over-crediting (i.e., selling more credits that were actually sequestered).

Still, project developers can be sneaky and find ways to incur over-crediting due to the economic incentive that selling a higher number of credits represents. A series of companies have come about in order to **bring transparency to the carbon market**, such as **Pachama**, **Renoster**, and **Sylvera**. These companies rate the carbon credits of other companies by running tests based on state-of-the-art scientific standards that leverage ML and remote sensing. Ultimately, this provides transparency to clients interested in purchasing high-quality carbon credits, in a similar way to how it is done on the stock market, for example.

As an example of sustainability policies and how a company can make an impact on carbon emissions, we want to showcase **Salesforce**, which is currently one of the world leaders in this field. You can find all the environmental and sustainability initiatives (that include a clear climate action plan and even some tooling they have developed to this effect that other companies can use) that are being currently carried out here: `https://www.salesforce.com/company/sustainability/`.

With these basics in mind about sustainability, let's now describe how it can affect the public cloud in general.

Public cloud and sustainability – GreenOps

At this point, you may be wondering: *how is all this related to the public cloud and FinOps?* FinOps and these sustainability practices go hand in hand, as they are about understanding the impact that our cloud operations have, on both cost and the environment.

All the major cloud providers are taking steps to make their data centers more environmentally friendly. We have many examples of this area of work, such as the presence of sustainability as the sixth pillar in the **AWS Well-Architected Framework**, or the study that Microsoft carried out with WSP on how using Microsoft Azure and M365 contributes to reducing carbon emissions compared to on premises-hosted infrastructure (you can download the fully study here: `https://www.microsoft.com/en-us/download/details.aspx?id=56950`).

All these ideas of working in the cloud in a sustainable manner have been encapsulated, and, as a result, a new methodology has been born based on these concepts. It is often called GreenOps, green cloud, or cloud sustainability, which is how the *FinOps Foundation* refers to this idea.

GreenOps encompasses everything that companies can do to reduce the environmental footprint of their cloud use, in the form of specialized initiatives focused on waste reduction and carbon emissions reduction, for example.

GreenOps responsibilities are not exclusive to companies using the cloud (indirect emissions), but also on the cloud provider's side (direct emissions):

- **Cloud providers** should ensure that the cloud is sustainable, in the form of carbon emissions reduction and the optimization of facilities and data centers.

- **Organizations that make use of the public cloud** should, on the other hand, ensure that their workloads are efficient and optimal and make use of the least amount of resources necessary, which will contribute to reducing carbon emissions. These emissions are also the hardest to track and control.

We want to demonstrate this idea with a quote from AWS: *AWS is responsible for the sustainability of the cloud, while AWS customers are responsible for sustainability in the cloud.* This is part of the **AWS shared responsibility model**, and you can find more information about it at the following link: `https://aws.amazon.com/blogs/aws/sustainability-pillar-well-architected-framework/`.

FinOps and GreenOps are strongly related to each other. When we work in cost optimization initiatives, such as powering off virtual machines during off-hours, or rightsizing resources to maximize usage, we are effectively also reducing the carbon emissions that our cloud resources generate. This is a win-win situation for organizations, as they both benefit from the cost optimization side and the improvement to their sustainability.

GreenOps also can benefit from FinOps in other ways: we can, for example, use FinOps practices on dashboards, reports, and the design and definition of KPIs, including carbon footprint metrics and sustainability KPIs in the picture and fostering both FinOps and GreenOps at the same time by creating common work areas and assets between these two interconnected methodologies.

FinOps and GreenOps also seek to increase organizational awareness and visibility of both cloud costs and carbon emissions, respectively. By increasing awareness, we will also increase accountability and give everyone a common goal.

To sum it up, GreenOps is about the following:

- Reducing cloud carbon emissions by eliminating waste and optimizing the use of cloud resources

- Improving sustainability reporting (in the form of sustainability KPIs, dashboards and reports, and other related assets)

- Creating organizational awareness of carbon footprint

Regarding the point related to the reporting and tracking of carbon emissions, there are a number of tools available from cloud providers and other companies to help on this sustainability journey. Some examples are as follows:

- **Azure:** Microsoft Emissions Impact dashboard (`https://learn.microsoft.com/en-us/power-bi/connect-data/service-connect-to-emissions-impact-dashboard`)

- **Azure:** Sustainability assessment (`https://learn.microsoft.com/en-us/assessments/a24b1079-29a4-4d22-b678-376e84884f76/`)

- **AWS:** Customer Carbon Footprint tool (`https://aws.amazon.com/es/aws-cost-management/aws-customer-carbon-footprint-tool/`)

- **GCP:** Carbon footprint dashboard (`https://cloud.google.com/carbon-footprint?hl=en`)

- **Commitments.cloud's sustainability calculator for AWS, Azure, and GCP** (`https://commitments.cloud/sustainability-calculator`)

With these tools, we close this section on sustainability, FinOps, and GreenOps. We will now delve into another interesting and emergent domain, which is machine learning and artificial intelligence.

Machine learning, artificial intelligence, and FinOps

There is no denying that both artificial intelligence and machine learning are on everyone's lips. With the rise of tools and digital assistants such as OpenAI's ChatGPT, Google Bard, Microsoft Copilot, Apple's Siri, and Amazon's Alexa, it is closer than ever to our everyday lives. Let's begin by getting a general understanding of what these terms refer to.

Artificial Intelligence (**AI**) refers to improving the ability of computers and machines to perform human-like tasks in real-world environments and their ability to mimic human thought and cognition, through the use of deep math and logic. From a scientific perspective, we can consider AI as a cross-sectional study field that has as a goal to produce automatic programs that can develop human-like capabilities in single or multiple tasks.

Machine Learning (**ML**), on the other hand, refers to a subfield of AI focused on technologies and algorithms that are able to improve themselves through constant learning, with the ultimate goal of providing AI systems with the ability to acquire their own knowledge by extracting patterns from raw data.

These two terms are often referred to interchangeably, though this is not exactly correct. ML is actually an application or subfield of AI.

A key thing we should consider is that for models to be able to process information, we must transform our data or information into a format that the machine or model can read. We refer to this concept as **data representation**. As an example of data representation, the color of a single pixel in an image can be represented in **RGB format**, which represents how much red, green, and blue the color has using three digits ranging from 0 to 255 (for example, yellow can be represented as 100 red, 100 green, and 0 blue).

Coming back to AI and ML, we can use the categories presented in the book *Deep Learning*, by Ian Goodfellow, Yoshua Bengio, and Aaron Courville, which illustrate the relationship between AI and ML in the following manner:

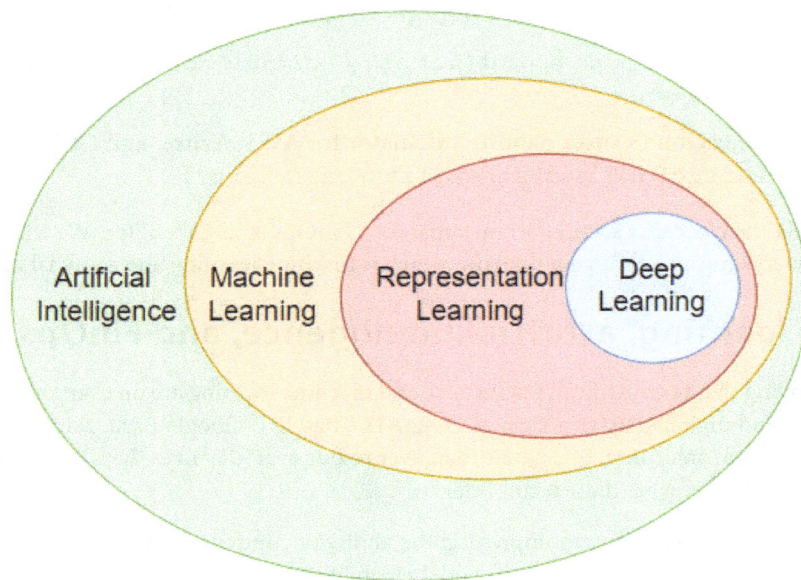

Figure 13.4 – AI, ML, and its subcategories

From this figure, we can also introduce two additional paradigms that are contained within ML:

- **Representation learning**: In this paradigm, multiple techniques are used to improve the performance of AI systems by not limiting map inputs to outputs, but by also creating an optimal data representation of these inputs during the learning process.

- **Deep learning**: In this paradigm, complex representations are solved by using simpler representations that the algorithms can work with. Deep learning tries to emulate the human brain's way of thinking by using multi-layered neural networks that are able to learn on their own.

We can illustrate representation learning with the following diagram:

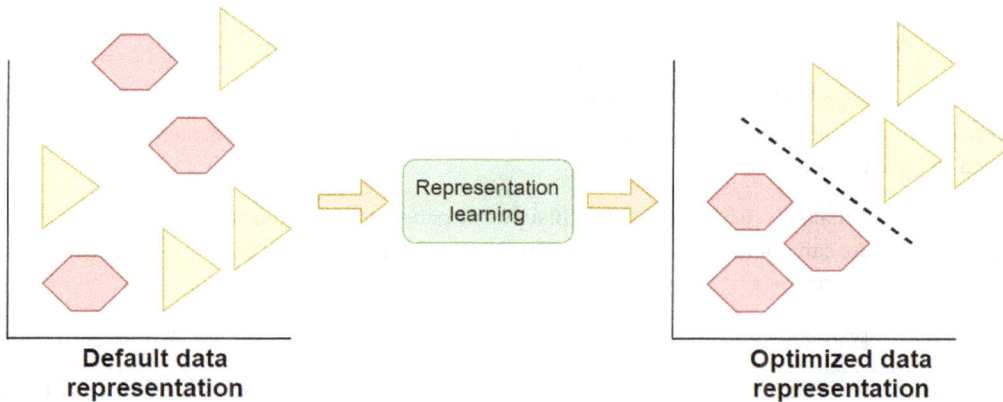

Figure 13.5 – Representation learning example

With these basic ML and AI concepts and categories in mind, let's proceed to their relationship with FinOps practices and cloud cost optimization. For the purposes of this book, we are going to solely focus on ML, as it is the AI field most related to FinOps.

ML has become increasingly important: as the volumes of data that humans manage and create continue to grow, with devices such as cell phones and computers generating enormous amounts of data, we are increasingly incapable of interpreting or even processing such amounts of information. In modern enterprises, having this ability is also often the edge over the competition.

ML has many applications:

- Text-to-speech or speech-to-text conversion
- Self-driving cars
- Diagnosing patients' illnesses based on their medical history and other patients' data and past diagnoses
- Segmenting people with similar interests based on location, acquisitive power, interests, and past purchases for ad-targeting and marketing purposes

As a small side note, given that we have just described GreenOps and cloud sustainability, there are many voices claiming that ML model training is having a big impact worldwide on carbon emissions. Training models often involves using highly powerful processors that, during the training phase of the models (in which huge datasets are ingested and processed), result in huge amounts of power consumption and carbon emissions. Just some food for thought.

In the next section, we will provide a brief description of how ML works.

How ML works

ML works by using pre-made algorithms (created and built by humans) that we call **models**. The main difference between traditional algorithms and ML models is that ML models learn from data and do not need to have rules defined explicitly.

Models are able to make decisions and detect insights or patterns with datasets never seen before by the model. Datasets can contain many different forms of data, from pure numerical data to images, videos, and voice recordings.

To begin the ML process, we need to train our models using datasets that contain two types of data:

- **Labeled data**: Data that has already been marked with labels to be used as a reference for our model. For example, if our model is trying to determine which vehicles are motorcycles in traffic camera images, we can label our data with *motorcycle/not-motorcycle*.

- **Unlabeled data**: Data that comes without labels.

There are different types of ML:

- **Supervised learning**: The training is supervised by data scientists, who provide the inputs and outputs and the key variables to be used for the model to assess for correlation. Models are trained using labeled datasets in a guided and continuous way. Taking the previous motorcycle example, we could use as key variables the *number of wheels* or *vehicle length*.

- **Unsupervised learning**: In this non-supervised training, the computing algorithm or process looks for unusual patterns or anomalies in unlabeled datasets.

- **Semi-supervised learning**: The training consists of both labeled and unlabeled data, which is a hybrid of supervised and unsupervised learning.

- **Reinforcement learning**: The training uses trial and error as part of the learning process. After each algorithm decision, we use rewards to steer the algorithm in the right direction. The algorithm then aims to maximize the rewards it gets from its decisions.

Depending on the analysis we can perform with our models, they can be classified as follows:

- **Predictive**: The models are focused on forecasting or foreseeing patterns and trends that will happen in the future

- **Descriptive**: The models are focused on explaining what has happened and getting insights from it

- **Prescriptive**: The models are focused on analyzing all possible outcomes and explaining what should be done

- **Diagnostic**: The models are focused on explaining why things are happening by extracting key insights and reasons that have caused the current situation

We could go on and on about this topic, but we won't get any deeper as it goes far beyond the purpose of this book. With this brief introduction in mind, let's look at how ML and FinOps are related to each other.

FinOps applications

In this section, we are going to analyze ML and FinOps synergies and observe how they are related to each other. Once the synergies are clear, we will also delve into some possible applications of ML in the FinOps field.

Let's begin with the basics: FinOps is a methodology for cost optimization, and it is heavily supported by data in multiple formats, such as resource usage and billing information. Let's now go back to the introduction to ML earlier in this chapter, where we stated how important it is, mainly because we are unable to manually process the huge volumes of data that we generate as humans.

When we work in the cloud, especially in big companies, we generate huge amounts of data just by using the cloud, mainly in the following forms:

- Activity logs

- Resource usage logs

- Billing information

- Specific configurations

This data can be increasingly difficult to process and control, making it harder for us to oversee and govern these really big environments without external help. If, for example, we are not able to detect a service that has doubled its spend overnight until one or two months after the fact, that small configuration issue can have a huge impact on costs.

FinOps is also about finding insights behind billing and usage data and making key information for organizations visible, to help them control their spend and make more optimal use of cloud resources.

So, what other field makes heavy use of data, is able to bring hidden insights to light, and can help us detect patterns among big volumes of information? ML, of course!

We should also consider that, to simplify the process of creation and training of ML models, there are ML services available in all three major public clouds that we can use, such as **Azure Machine Learning**, **AWS Sagemaker**, and **GCP Vertex AI/AutoML**.

Due to these reasons, ML can be a great ally of FinOps practices. Let's analyze the possible synergies and specific areas where ML can help us go the extra mile.

Optimization recommendation engines

In this book, we have covered tools, such as **Azure Advisor** and **AWS Trusted Advisor**, that review our cloud resources, look for inefficiencies, and recommend possible mitigations to those inefficiencies.

These tools are essentially recommendation engines, and most of them use ML models or similar algorithms to perform these recommendations in real time. They act as a black box that uses as inputs our resource usage and billing information, producing as outputs possible recommendations on how to optimize these resources.

With these references in mind, we could create our own prescriptive models that take as inputs our resource usage and costs and produce as output the best configuration to either minimize the costs, have a balance between cost and performance, or fully focus on maximizing performance.

We can summarize this idea with the following diagram:

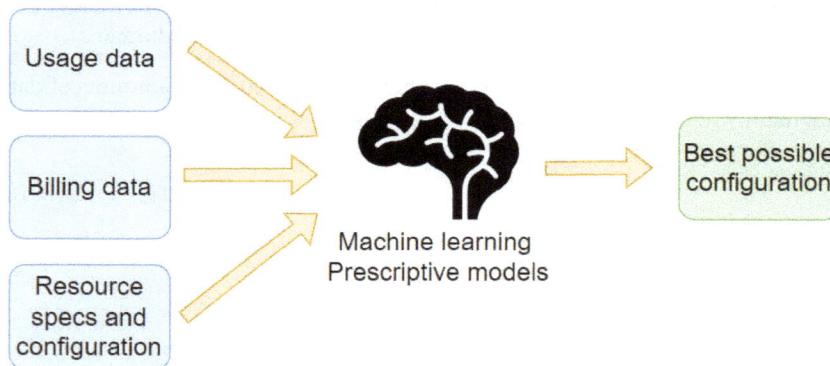

Figure 13.6 – Optimization recommendation engines

We could use these ML models to support our FinOps practitioners' decisions, adding new perspectives and insights to the decision-making that, as humans, we can sometimes miss.

With this idea in mind, let's now move to another application – a cost-anomaly detector.

Cloud-cost-anomaly detection

In the world of cloud financial management, an unexpected spike in cloud costs, which we often refer to as a **cost anomaly**, can have a big impact on organizations.

Let's imagine that, due to a misconfiguration in some cloud resources, the cost of a specific solution doubles or triples. In this situation, it is essential to detect and analyze this cost drift as soon as possible, as every minute that passes without us knowing could be costing our organization a lot of money. On the other hand, it might also be possible that this increase in cost is justified, and it comes because of changes in the environment or major project milestones such as a new release or a new application (we covered this in *Chapter 5*). For this use case, setting up budget limits is just not enough, and alerts when budgets are surpassed are not going to fit our needs.

Due to this fact, it is essential to build, as part of our FinOps work, some sort of cost monitoring to detect cost anomalies and act on them as soon as possible. If we do this, we could have an alert be triggered whenever sudden or unforecasted spikes in cost appear, sending us an email or registering a ticket in our ITSM tool for the cloud operations or FinOps team to review. We can see an example of cost anomalies in the following figure:

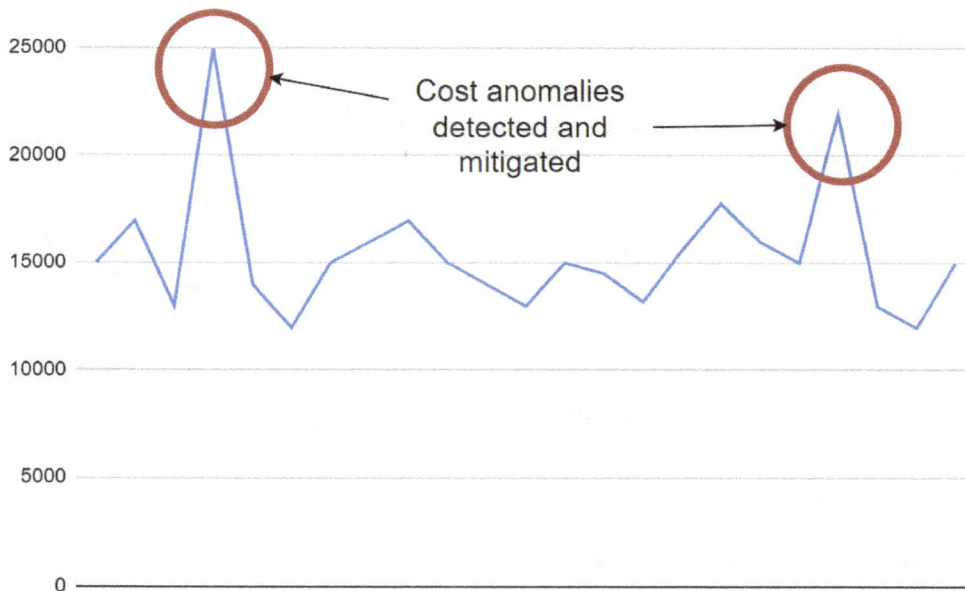

Figure 13.7 – Cloud-cost-anomaly detection

This monitoring can be reactive and deterministic, based on thresholds and specific rules, as is usually the case, and act only after the cost increase. On the other hand, when deterministic models fall short, we could alternatively use ML to look for unfamiliar patterns or cost drifts before they actually happen, saving us a lot of money in the process.

Of course, this idea can also be used in other non-FinOps fields, such as preventing security breaches or monitoring the health of our systems, for example.

For this use case, we could use either **supervised or unsupervised predictive models** to detect outliers or cost anomalies in our billing data.

Cloud providers are already integrating these detection algorithms into their cost management products:

- **Azure Cost Management Anomaly detection**: `https://learn.microsoft.com/en-us/azure/cost-management-billing/understand/analyze-unexpected-charges#create-an-anomaly-alert`

- **AWS Cost Anomaly detection**: `https://docs.aws.amazon.com/cost-management/latest/userguide/manage-ad.html`

- **AWS CloudWatch metrics Anomaly Detection** (for any metric): `https://docs.aws.amazon.com/AmazonCloudWatch/latest/monitoring/CloudWatch_Anomaly_Detection.html`

Outside of the cost management world, there are specific APIs and ML services to detect anomalies in a dataset:

- **Azure Cognitive Services Anomaly Detector API**: `https://learn.microsoft.com/en-us/rest/api/anomalydetector/`

- **AWS OpenSearch Anomaly detector**: `https://docs.aws.amazon.com/opensearch-service/latest/developerguide/ad.html`

- **GCP BigQuery ML**: `https://cloud.google.com/blog/products/data-analytics/bigquery-ml-unsupervised-anomaly-detection`

In addition to these tools, most FinOps commercial tools, such as **Aria Cost Powered by CloudHealth** and **Apptio Cloudability**, already offer these features as they are essential tools that must be part of every FinOps practitioner toolset.

In the next section, we will learn how ML can also help us in forecasting our cloud costs.

Estimations and forecasting

Cloud cost forecasting is a key component of FinOps practices. With the rise of cloud computing, cost forecasting has become much more difficult (with the shift to OPEX costs instead of CAPEX costs, which we covered in *Chapter 1*) than it was before.

We need to understand, for every new project, how much the solution is going to cost before it is approved. Without this key information, it is going to be much harder to get approval from C-level management and stakeholders, as it will be perceived that we are asking for a blank check.

In addition to this, we also need to prevent, as much as possible, the cost drifts that might occur when specific project milestones are reached or planned changes are implemented in our cloud environments.

We can do this forecasting **manually**, using the calculator tools offered by cloud providers that we covered in *Chapter 4*. We must consider, though, that doing these forecasts manually is going to take a lot of time and effort from our FinOps practitioners that could be used elsewhere. As we also covered in *Chapter 4*, **indirect costs** should not be dismissed during this process, which can take even additional work.

So, how could we do this automatically? The solution is to use ML models.

This use case is a textbook example for a **predictive model**, using which we could process the following data:

- Activity logs
- Resource usage logs
- Billing information
- Pricing APIs

Once this data is processed, we could use ML to detect hidden patterns and insights in our datasets, such as seasonal trends of usage or inflation in the cloud cost, based on key variables such as price, usage, and other relevant metrics. With this information, our model will be able to predict, in a much more precise manner.

We can summarize this idea with the following figure:

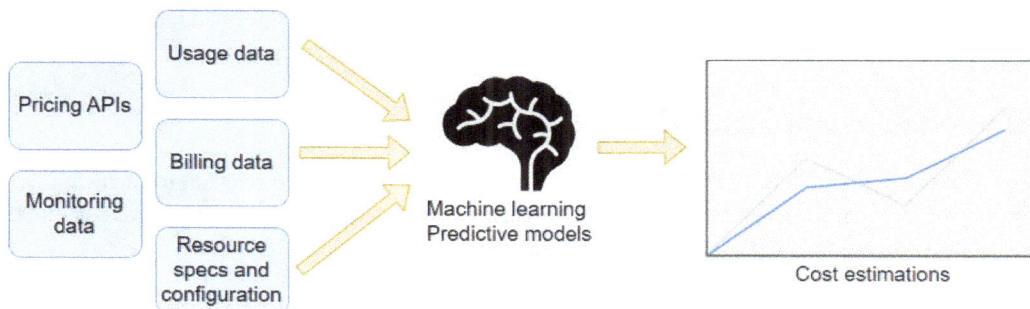

Figure 13.8 – Forecasting through ML models

Cloud providers' cost analysis tools already provide some forecasting options:

- **Azure Cost Management forecasting**: `https://learn.microsoft.com/en-us/azure/cost-management-billing/costs/cost-analysis-common-uses#view-forecast-costs`

- **AWS Cost Explorer forecasting**: `https://docs.aws.amazon.com/cost-management/latest/userguide/ce-forecast.html`

- **GCP Cloud Billing cost forecast**: `https://cloud.google.com/blog/products/gcp/predict-your-future-costs-with-google-cloud-billing-cost-forecast`

However, the cloud providers don't disclose how these models are built and the parameters that are used as inputs. Due to this fact, if we want to have our own approach based on the criteria that we choose, we may be forced to build our own ML models for this task.

These ML models are also offered out of the box by some cloud providers:

- **AWS Forecast**: `https://docs.aws.amazon.com/forecast/latest/dg/what-is-forecast.html`
- **GCP AutoML forecasting**: `https://cloud.google.com/vertex-ai/docs/tabular-data/forecasting/overview`
- **Azure Machine Learning AutoML forecasting**: `https://learn.microsoft.com/en-us/azure/machine-learning/concept-automl-forecasting-methods?view=azureml-api-2`

With this last ML use case, we have concluded this interesting section, in which we have opened some new paths for FinOps practitioners to delve into the AI domain with interesting use cases and ideas that can be developed further.

In the next section, we provide a knowledge check for you in the form of questions, which is a great way to review the key ideas presented in each chapter and check whether or not you have the level of understanding that we expect our readers to have reached after reading this book.

Self-assessment/knowledge check

To conclude this great adventure of a book, we have included some questions relevant to each chapter in this section, for you to reflect on what you have learned about each of the key topics and concepts that have been covered throughout the different chapters of this book.

Keep in mind that there will be no self-assessment questions for *Chapter 12, Case Studies for Mastering Cost Optimization*, as it is solely dedicated to examples.

Chapter 1

- Can you briefly explain what FinOps is?
- Why is FinOps necessary?
- Can you describe the key differences between traditional on-premises computing and cloud computing?
- What do the terms CAPEX and OPEX mean?

- What are the differences, from a cost perspective, between cloud and on-premises?
- What is the FinOps Foundation?
- What are the areas of work that the FinOps pillars Inform, Optimize, and Operate cover?

Chapter 2

- What is the Well-Architected Framework and what does it try to achieve?
- What are the pillars that the Well-Architected Framework is based on?
- What is the Crawl, Walk, Run methodology?
- What synergies can we find between Agile methodologies and FinOps?
- What is Infrastructure as Code and what are its benefits?
- Can you name a few examples of how Infrastructure as Code can boost FinOps practices?
- Why is change management important for IT organizations and how is it related to FinOps practices?
- Can you name a few different approaches of FinOps depending on the situation of organizations in regard to the cloud?
- Can you name a few base tools included in each cloud provider for cost analysis and cost optimization?
- Can you name a few commercial tools that can help organizations in their FinOps journey?

Chapter 3

- Why are naming conventions important in cloud computing?
- Can you name a few examples of information that a naming convention can provide from the name of a resource?
- What are the differences between naming conventions and tagging strategies?
- In what way are naming conventions different in Amazon Web Services?
- How can naming conventions and tagging strategies be enforced?
- Can you name a few benefits of resource tagging in the public cloud?
- What are the main purposes of enforcing tagging strategies?
- Can you provide an example of possible tag keys and their use?
- How can a name generator help organizations?
- What are compound tags in Azure?
- How does tagging inheritance work and how can it be set up in each cloud provider?

- What is cost allocation and how is it related to tagging strategies?

- How can we avoid repeating code in Terraform when using the same tags in multiple resources?

- What are the differences between AWS-generated tags and user-defined tags?

Chapter 4

- What is TCO?

- What are the benefits of preparing a TCO before cloud budgets are decided?

- What are the costs that disappear from the TCO when moving to the cloud?

- What is vendor lock-in?

- What are the 6 Rs of cloud migration? Can you provide an example of each one?

- Can you name a few factors that should be considered when deciding which one of the 6 Rs is more suitable for a specific workload?

- What are the differences between direct and indirect costs in the cloud?

- What level of cloud understanding do we need when using cloud providers' calculator tools?

- What is a REST API and how is it related to cloud resource pricing?

- What methodology can we use when estimating the potential savings that can be obtained from a FinOps initiative?

- Which two factors should be considered when analyzing FinOps initiatives prioritization?

- What are the steps required for automated cost estimation?

Chapter 5

- How are billing data and cost drivers important for FinOps practices?

- Could you name a few fields that are commonly used in cloud invoices from all cloud providers?

- What are the differences between a dashboard and a report and what are the benefits that both of them can bring?

- What is row-level security in a report?

- How can we leverage dashboards for what-if scenarios?

- What is a cost evolution dashboard or report and what kind of information does it show?

- How can we measure both savings and cloud waste?

- In which two ways can we obtain savings?

- Apart from pay-as-you-go, what other purchase model can we use in the cloud?

- To what financial KPIs do we refer by the acronyms YTD, MoM, and YoY?

- What are the benefits of using unit economics and how do they provide value to a business?

- Can you provide a few examples of unit economics?

- What FinOps dashboards and reports are offered out of the box in Azure and AWS?

Chapter 6

- What does the concept of the quick win represent in FinOps practices?

- What are the main differences between the IaaS, PaaS, and serverless service models?

- Can you provide a few examples of IaaS, PaaS, and serverless services?

- What is the difference between stateful and stateless workloads?

- What is an orphaned resource? Can you provide some examples of orphaned resources in your cloud provider of choice?

- How does an orphan resource impact our cloud bills?

- What are the benefits of using newer-generation virtual machines?

- What should we analyze before upgrading a virtual machine to a newer generation?

- What are the benefits of using ARM and AMD processors and what should we consider before using each one?

- What is rightsizing? What tools can we use to get information on this cost optimization initiative?

- What three decisions can we take after a rightsizing analysis?

- What are the benefits of virtual machine family standardization initiatives? What should we consider before making any changes to our virtual machines?

- How does the region where cloud resources are hosted affect their cost?

- What are virtual machine power scheduling initiatives?

- How much can we save if we shut down our virtual machines during weekends and out-of-office hours?

- How can ephemeral environments boost cost optimization?

- What is virtual machine scaling, and which two flavors of scaling can we use in cloud resources?

- What are the advantages and disadvantages of each scaling flavor?

- How can burstable virtual machines boost cost optimization and in which situations can we benefit from their use?

- Can you provide an example of a burstable virtual machine in each cloud provider?

- What are the benefits of using Reserved Instances on the cloud and what are the disadvantages of using this purchase model?

- What is the Reserved Instance break-even point and what does it represent?

- Can you name one advantage and one disadvantage of broad-scoped Reserved Instances and specific-scoped Reserved Instances?

- What is the difference between Reserved Instances and Saving Plans?

- Which two KPIs should we consider when working when Reserved Instances and what are their desired values?

- Which optimization initiatives should we implement before considering purchasing Reserved Instances?

- In which environments should we prioritize Reserved Instances use?

- What are Spot virtual machines, and in which cases is their use recommended?

Chapter 7

- What are the differences between provisioned compute and serverless? Can you name a few examples of each in Azure, GCP, and AWS?

- Can you provide a few examples of architectures that can benefit from serverless resources?

- Can you provide some examples of Managed Kubernetes services in Azure, AWS, and GCP?

- What are the differences between a regular installation of Kubernetes and Managed Kubernetes services offered in the cloud?

- Can you name a few optimization initiatives for Managed Kubernetes services in Azure, AWS, and GCP?

- What are the differences between ingress and egress data transfer? Which of them is free and why?

- Can you name a few recommendations to reduce data transfer charges in each cloud provider?

- How do bring-your-own license models work in cloud resources?

- What is Azure Hybrid Benefit?

- What services allow for BYOL licensing in AWS and GCP?

- How can tools such as GCP License Tracker and AWS License Manager help in licensing management for cloud resources?

- What are the key differences between an Enterprise Agreement and a CSP contract in Azure? What is the minimum number of users needed to consider an Enterprise Agreement?

- What are the best practices for organizing AWS management resources such as organizations, accounts, and billing accounts?

- What are the best practices for organizing GCP management resources such as organizations, folders, and projects?

Chapter 8

- What are the differences between structured and unstructured data? Can you provide an example of each data type?

- What are the main categories of non-relational databases?

- Can you highlight the differences between a relational and a non-relational database?

- What are the database management system and the database engine in a relational database? Can you provide some examples of DMBSs?

- What are the benefits of PaaS database services over IaaS?

- Can you name a few recommendations for database use in IaaS services?

- What type of storage should we prioritize for IaaS database backups from a cost perspective?

- What kind of disk in AWS is most suitable for backup storage purposes?

- In what way can a cluster protect our databases from failures?

- What are the differences between a shared-disk and a shared-nothing cluster? Which one of these architectures results in better pricing and what are the implications?

- What are the features in each cloud provider that support disk-sharing capabilities across multiple virtual machines?

- What is database shrinking?

- Can you briefly describe how database grouping can help optimize the cost of our databases?

- What are the main metrics we should use for rightsizing exercises in PaaS resources?

- What is an Azure SQL Database elastic pool and how does it work?

- In what way is vertical scaling limited?

- What are the benefits of using horizontal scaling in databases?

- What are the benefits, from a cost perspective, of auto-scaling databases?

- Can you provide some examples of relational and non-relational PaaS and serverless databases in Azure, AWS, and GCP?

- What are the database services in each cloud provider that support using Reserved Instances or Reserved Capacity?
- What Azure feature can we use to reduce the cost of development subscriptions if we already have Visual Studio licenses for our developers?

Chapter 9

- What are the main storage types that are offered in the public cloud and what are their strengths and weaknesses? Can you provide an example of each cloud storage type for Azure, AWS, and GCP?
- Can you briefly explain how each storage type works?
- Can you describe the main cost drivers for each storage type?
- Can you name a few cost optimization initiatives for block, file, and object storage cloud services?
- How can the concept of thin provisioning optimize how we allocate disk space in the cloud?
- Is there a way to reduce the size of an existing disk in Azure, AWS, and GCP?
- What is a snapshot and why is it different from a backup?
- In what scenarios should we use storage redundancy and what is its impact on costs?
- What are the metrics IOPS, latency, and throughput?
- What are data temperature tiers and how do these tiers impact our storage costs?
- How can versioning and soft delete affect our object storage costs?
- What is the difference between an incremental snapshot and a full snapshot?
- What are the advantages and disadvantages of using ephemeral disks?
- What are the two angles we can use to rightsize disks?
- What are the benefits of offloading storage to file or object storage?
- Which cloud provider is the only one currently offering reserved capacity for block storage and file storage?
- How do lifecycle policies work in object storage?
- In what way can observability systems impact our overall cost? In what ways can we reduce their cost?
- What is the difference between RTO and RPO?
- What does the concept of selective disk backup describe and how can it help cost optimization?

Chapter 10

- What is a KPI?
- Do you know what metrics, measures, and KPIs are and how they are related to each other?
- What are the main cloud services in each cloud provider that we can use to check metrics for the services we use in the cloud?
- What are the phases we should go through to define and create a KPI?
- What are the main types of KPIs? Can you provide an example of each one?
- What is OKR methodology?
- What does the idea of SMART goal definition methodology describe?
- Can you provide examples of FinOps KPIs for each KPI category?

Chapter 11

- What does a target operating model describe?
- What specific areas should we work on when defining a FinOps operating model?
- What are the different organizational models we can implement for FinOps practices and what are the advantages and disadvantages of each one?
- What is a cloud center of excellence and how is it related to FinOps practices?
- Which phases should we follow to define a FinOps rollout and execution plan?
- What are functions, capabilities, and processes and how are they related to each other? Can you provide an example of each one?
- What is a RACI matrix and how can it help us define roles and responsibilities?
- What tools can we use to enforce naming conventions and tagging strategies in an organization?
- What is the best way to track a FinOps initiative's progress? Why should we keep historical data in mind?
- Is there a way to combine the progress of multiple initiatives into one KPI? If so, how can we do it?
- How can unit economics and potential savings improve our FinOps dashboards and reports?

Chapter 13

- What are the main future challenges we identified for FinOps practices?
- Why is sustainability important for modern-day enterprises?
- What is the main difference between sustainability and ESG policies?

- What does the term GreenOps describe and how is it related to FinOps?

- What are the differences between ML and AI?

- Can you name a few applications where ML can synergize with FinOps practices?

- What are the main data sources that can be used for ML models in FinOps practices?

Summary

With the self-assessment done, we have reached the end of this book. First and foremost, we want to thank you for having gone through all the content we prepared for you. We hope that everything we've covered here has been useful for you and has enriched your knowledge of cloud financial management and cost optimization.

As possible next steps, we recommend getting in touch with other FinOps practitioners and sharing common experiences and challenges. In our view, this methodology revolves around collaboration; as such, we feel that teaching by example is a good choice here. One of the best ways to get acquainted with other practitioners is to join the **FinOps Foundation**, where there is a Slack workspace that you will be invited to with a myriad of channels filled with constant discussion. Additionally, all manner of events worldwide covering FinOps experiences and practices are organized by the foundation, which represents a great opportunity for networking and getting in touch with other FinOps professionals.

Getting certified in FinOps through the foundation is also a good idea, as you will both gain professional recognition for having the certification and learn a great deal in the process. We also recommend Azure, AWS, and GCP certifications as, even though they are no substitute for actual experience in each public cloud, they cover a lot of content that is essential for cloud engineers and practitioners in general.

As we've talked about throughout the book, the cloud is ever-changing, so we also encourage you to stay on top of cloud news and technologies, so you will be able to adapt the concepts covered here to newer products, tools, and cloud services.

Please feel free to reach out to us on LinkedIn for any questions or feedback about the book – we would be happy to hear from you!

- Alfonso San Miguel Sánchez (`https://www.linkedin.com/in/alfonso-san-miguel-s%C3%A1nchez-42822635/`)

- Danny Obando García (`https://www.linkedin.com/in/danny-obando/`)

Index

F

Other Books You May Enjoy

If you enjoyed this book, you may be interested in these other books by Packt:

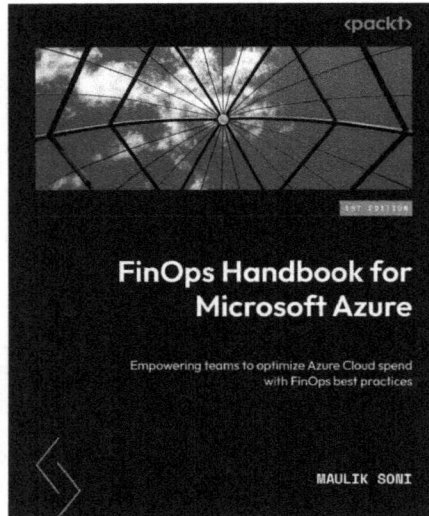

FinOps Handbook for Microsoft Azure

Peter Chung

ISBN: 9781801810166

- Get the grip of all the activities of FinOps phases for Microsoft Azure
- Understand architectural patterns for interruptible workload on Spot VMs
- Optimize savings with Reservations, Savings Plans, Spot VMs
- Analyze waste with customizable pre-built workbooks
- Write an effective financial business case for savings
- Apply your learning to three real-world case studies
- Forecast cloud spend, set budgets, and track accurately

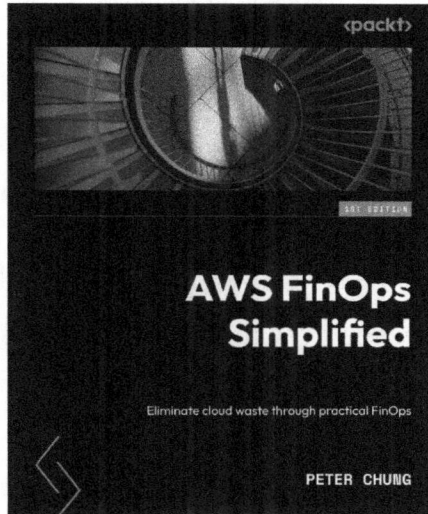

AWS FinOps Simplified

Peter Chung

ISBN: 9781803247236

- Use AWS services to monitor and govern your cost, usage, and spend
- Implement automation to streamline cost optimization operations
- Design the best architecture that fits your workload and optimizes on data transfer
- Optimize costs by maximizing efficiency with elasticity strategies
- Implement cost optimization levers to save on compute and storage costs
- Bring value to your organization by identifying strategies to create and govern cost metrics

Packt is searching for authors like you

If you're interested in becoming an author for Packt, please visit authors.packtpub.com and apply today. We have worked with thousands of developers and tech professionals, just like you, to help them share their insight with the global tech community. You can make a general application, apply for a specific hot topic that we are recruiting an author for, or submit your own idea.

Share Your Thoughts

Now you've finished *Efficient Cloud FinOps*, we'd love to hear your thoughts! Scan the QR code below to go straight to the Amazon review page for this book and share your feedback or leave a review on the site that you purchased it from.

https://packt.link/r/1805122576

Your review is important to us and the tech community and will help us make sure we're delivering excellent quality content.

Download a free PDF copy of this book

Thanks for purchasing this book!

Do you like to read on the go but are unable to carry your print books everywhere?

Is your eBook purchase not compatible with the device of your choice?

Don't worry, now with every Packt book you get a DRM-free PDF version of that book at no cost.

Read anywhere, any place, on any device. Search, copy, and paste code from your favorite technical books directly into your application.

The perks don't stop there, you can get exclusive access to discounts, newsletters, and great free content in your inbox daily

Follow these simple steps to get the benefits:

1. Scan the QR code or visit the link below

https://packt.link/free-ebook/978-1-80512-257-9

2. Submit your proof of purchase
3. That's it! We'll send your free PDF and other benefits to your email directly

www.ingramcontent.com/pod-product-compliance
Lightning Source LLC
Chambersburg PA
CBHW072009230326
41598CB00082B/6894